Sustainability and Reliability of Networks

The text emphasizes the methodologies and strategies for finding maintainability, sustainability, dependability, and reliability. It discusses how Industry 5.0 affects networks and gives ideas about diversification and an interdisciplinary approach in the field of network reliability.

Key Features:

- Covers different methods for enhancing the network's reliability, maintainability, and performance based on applications of the universal generating function, L_z-transform, and Monte Carlo simulation methodology.
- Discusses the concepts of network reliability in Industry 5.0, in which we can enhance the maintainability, sustainability, reliability, and dependability of the different complex networks and systems.
- Provides a comparative framework for the network reliability used in real life for future work.
- Explains how Industry 5.0 affects networks and gives ideas about diversification and an interdisciplinary approach in the field of network reliability.
- Highlights future research and problems in network reliability and focuses on how emerging networks are used in the agriculture field, the food supply chain network, and multistage interconnection networks.

The text is primarily written for graduate students and academic researchers in industrial engineering, systems engineering, production engineering, reliability engineering, mechanical engineering, and mathematics.

Advanced Research in Reliability and System Assurance Engineering

Series Editor: Mangey Ram, Professor, Graphic Era University, Uttarakhand, India

Social Networks
Modeling and Analysis
Niyati Aggrawal and Adarsh Anand

Operations Research
Methods, Techniques, and Advancements
Edited by Amit Kumar and Mangey Ram

Statistical Modeling of Reliability Structures and Industrial Processes
Edited by Ioannis S. Triantafyllou and Mangey Ram

Industrial Reliability and Safety Engineering
Applications and Practices
Edited by Dilbagh Panchal, Mangey Ram, Prasenjit Chatterjee, and Anish Kumar Sachdeva

Reliability and Maintenance Modeling with Optimization
Advances and Applications
Edited by Mitsutaka Kimura, Satoshi Mizutani, Mitsuhiro Imaizumi, and Kodo Ito

Intelligent Prognostics for Engineering Systems with Machine Learning Techniques
Edited by Gunjan Soni, Om Prakash Yadav, Gaurav Kumar Badhotiya and Mangey Ram

Statistical Methods and Applications in Systems Assurance and Quality
Edited by Sotiris Bersimis, Polychronis Economou, and Athanasios Rakitzis

Designing Faultless Mechanical Products Based on Advanced Reliability Analysis
Seongwoo Woo

Operations Research: Recent Advances and Applications
Edited by Shreekant Varshney, Jyotiranjan Nayak, Mayank Gupta, Chandra Shekhar and Balakrushna Tripathy

Sustainability and Reliability of Networks
Technology and Applications
Edited by Akshay Kumar, Mangey Ram, Soni Bisht and Yigit Kazancoglu

For more information about this series, please visit: https://www.routledge.com/Advanced-Research-in-Reliability-and-System-Assurance-Engineering/book-series/CRCARRSAE

Sustainability and Reliability of Networks

Technology and Applications

Edited by

Akshay Kumar, Mangey Ram, Soni Bisht, and Yigit Kazancoglu

CRC Press
Taylor & Francis Group
Boca Raton London New York

CRC Press is an imprint of the
Taylor & Francis Group, an **informa** business

Designed cover image: ©Shutterstock Images

First edition published 2027
by CRC Press
2385 NW Executive Center Drive, Suite 320, Boca Raton FL 33431

and by CRC Press
4 Park Square, Milton Park, Abingdon, Oxon, OX14 4RN

CRC Press is an imprint of Taylor & Francis Group, LLC

ISBN: 9781041333838 (hbk)
ISBN: 9781041333845 (pbk)
ISBN: 9781003787525 (ebk)

DOI: 10.1201/9781003787525

Typeset in Sabon
by codeMantra

Contents

Preface viii
About the Editors ix
List of Contributors xii

1 Recent advances in software reliability growth models 1
DAVID D. HANAGAL

2 Risk-informed planning for maritime network reliability under operational uncertainty 25
OLEKSIY MELNYK, SERGIY VOLIANSKY, OLEG ONISHCHENKO, VALERIY SHEVCHENKO, HENNADY SHCHENIAVSKY, AND INNA SHAPOVALOVA

3 Plyboard manufacturing plant with perfect switchover device: Reliability and availability analysis 39
SEEMA SAINI, ARVIND KUMAR LAL, SUNIL KUMAR, RAJENDRA KUMAR SHARMA, AND JAI SINGH

4 Bayesian availability analysis of repairable *k*-out-of-*n* systems using Weighted Exponential-Lindley model 57
SUNITA SHARMA

5 Reliability assessment of cloud security system using Markov model 73
NUPUR GOYAL, NIKHIL NAWANI, MANGEY RAM, AKANSHA GUPTA, AND NEERAJ SENGAR

6 Multistate network reliability – an overview 94
PARTHA CHAKRABARTI AND NEERAJ KUMAR GOYAL

7 Maximum reliability travelling salesman tour in a stochastic network 115

TRUST TAWANDA, ADARSH ANAND, AND SANTOSH KUMAR

8 Improvement of the agro-food supply chain by asset management 127

FATEMEH AFSHARNIA

9 Multi-objective risk-based optimization of single-story steel frames: Balancing lifecycle costs and reliability for sustainable industrial buildings 137

BAC AN HOANG

10 Evaluating k-out-of-$(p + w + c)$ system under sequential standby policy via L_z-transform technique 158

MANPREET KAUR, SONI BISHT, SURENDRA VIKRAM SINGH PADIYAR, AND VAIBHAV BISHT

11 Reliability and availability analysis of 4×4 Shuffle Exchange Network using L_z-transform 191

VAIBHAV BISHT, S. VISHAL VIGNESH, G. HEMALATHA, PRIYANSHU ARYA, AND AYUSH SINGH

12 Mathematical modeling and reliability attributes of feeding system in sugarcane industry 215

GEETANJALI RAJPUT, SONI BISHT, MANJINDER KAUR, AND SURENDRA VIKRAM SINGH PADIYAR

13 AI-driven UAV solutions for sustainable and reliable infrastructure monitoring in Industry 5.0 240

IREM BAYRAKTAR AND MURAT BAKIRCI

14 Sustainable availability and MTTF estimation of machine repair systems under fuzzy conditions 261

AMIT KUMAR

15 Reliability analysis of an off-grid fire safety system 287

MOHD ANAS, MANGEY RAM, AND SHRISTI KHAROLA

16 Availability and reliability analysis of an interchangeable three-unit series-parallel system 300

VENKATA SESHACHALA SARMA YADAVALLI, BALAKRISHNAN THILAKA, SANKARALINGAM HEMANTH KUMAR, AND MICHAEL AYOMOH

Index 337

Preface

In today's world, the reliability of network is one of the most important pillars upon which all industries' networks, communications networks, telecommunication networks, cloud computing networks, etc. depend. Network reliability is basically a subset of system reliability, and the network is reliable if it provides connectivity without any failures.

Moreover, this includes an extensive idea about software reliability used in cloud networks, review of software reliability growth models, and use of different methodologies to find the reliability of binary and multistate networks. Artificial intelligence infrastructure and monitoring networks are used in Industry 5.0 to find the reliability from different probabilistic and statistical approaches.

Recent advancements in the mathematical network reliability modelling are developed and given in detail across this book. Various probabilistic and statistical approaches are implemented to investigate the maintainability, dependability, and sustainability of the networks used in real life.

About the Editors

Akshay Kumar received his B.Sc. and M.Sc. degrees in Science from Chaudhary Charan Singh University, Meerut, India, in 2010 and 2012, respectively, and his Ph.D. degree major in Mathematics and minor in Computer Science from G. B. Pant University of Agriculture and Technology, Pantnagar, India, in 2017. He is currently working as an Assistant Professor in the Department of Mathematics, Graphic Era Hill University, Dehradun, India. He has more than 8.6 years of teaching experience. He is a regular reviewer for more than 60 international journals. He has published 100 plus research papers in many national and international journals of repute and presented his works at national and international conferences. Also, he has published seven books (edited) with international publishers. His fields of research are reliability theory, fuzzy reliability, signature reliability, and applied mathematics. He has recently been awarded "Outstanding Researcher Award" in 2019–2020 at Graphic Era Hill University, Dehradun, India, and in the 15th and 16th Uttarakhand State Science and Technology Congress 2020–2022 for his significant contribution in academics and research at Graphic Era Deemed to be University, Dehradun, India.

Mangey Ram received his Ph.D. degree major in Mathematics and minor in Computer Science from G. B. Pant University of Agriculture and Technology, Pantnagar, India, in 2008. He has been a Faculty Member for around 16 years and has taught several core courses in Pure and Applied Mathematics at undergraduate, postgraduate, and doctorate levels. He is currently the *Research Professor* at Graphic Era (Deemed to be University), Dehradun, India. Before joining Graphic Era, he was a Deputy Manager (Probationary Officer) with Syndicate Bank for a short period. He is Editor-in-Chief of the *International Journal of Mathematical, Engineering and Management Sciences; Journal of Reliability and Statistical Studies*; *Journal of Graphic Era University*; Series Editor of six book series; and the Guest Editor and Associate Editor with various journals. He has published 400 plus publications (journal articles/books/book chapters/conference articles) in many other national and international journals and conferences. Also, he has published around 100 books (authored/edited) with international publishers. His fields of research are reliability theory and applied mathematics. Dr. Ram is a Senior Member of IEEE, Senior Life Member of the Operational Research Society of India, the Society for Reliability Engineering, Quality and Operations Management in India, and the Indian Society of Industrial and Applied Mathematics. He has been a member of the organizing committee of a number of international and national conferences, seminars, and workshops. He was awarded the "Young Scientist Award" by the Uttarakhand State Council for Science and Technology, Dehradun, in 2009; the "Best Faculty Award" in 2011; the "Research Excellence Award" in 2015; and the "Outstanding Researcher Award" in 2018 for his significant contribution to academics and research at Graphic Era Deemed to be University, Dehradun, India. Also, he has received the "Excellence in Research of the Year-2021 Award" by the Honourable Chief Minister of Uttarakhand State, India, and the "Emerging Mathematician of Uttarakhand" state award by the Director, Uttarakhand Higher Education. Recently, he received the "Distinguished Service Award-2023" for the subject and nation development by Vijñāna Parishad of India and "Excellence in Research of the Year-2025 Award" by Uttarakhand State Council for Science and Technology.

Soni Bisht is an Assistant Professor in the Department of Mathematics at Eternal University, Baru Sahib, Himachal Pradesh, India. She received her B.Sc. and M.Sc. degrees from H.N.B. Garhwal Central University, Srinagar, in 2012 and 2014, respectively, and she earned her Ph.D. degree in 2019 from G. B. Pant University of Agriculture and Technology, Pantnagar, India, with a major in Mathematics and a minor in Computer Science. She has guided several postgraduate students and is currently supervising postgraduate and doctoral research scholars. Her teaching interests include Functional Analysis, Vector Calculus, Topology, Analysis, Statistical Methods, Quantitative Methods, Operations Research, and Research Methodology at the postgraduate level. She has published more than 25 research papers in reputable national and international journals, including *Reliability Engineering and System Safety*, *OPSEARCH*, *MDPI*, *IEEE Xplore*, *Emerald*, the *International Journal of Mathematical, Engineering, and Management Sciences*, and the *Journal of Reliability and Statistical Studies*. She has also presented her research work at numerous national and international conferences. She serves as a regular reviewer for leading journals and publishers. In recognition of her research contributions, she was awarded the Young Scientist Award by the Uttarakhand State Council for Science and Technology, Dehradun, in 2018. Her research interests include Fuzzy Reliability Theory, System Reliability Theory, Operations Research, Applied Mathematics, Network Reliability Theory, and Parallel and Distributed Systems.

Yigit Kazancoglu received his B.S. degree from Eastern Mediterranean University, Department of Industrial Engineering. He has graduated from Coventry University with an MBA and Izmir University of Economics with an MBA. He received his Ph.D. from Ege University (Production and Operations Management). He is a Full Professor at Yasar University, Logistics Management Department, and Head of the department.

His research areas include Operations Management and Supply Chain Management. He is working on sustainability, circular economy, green supply chain management, and digitalization. He has published more than 100 articles in the SCI & SCI-E & SSCI & ESCI and SCOPUS indexed journals. He has papers on related topics that were published in *International Journal of Production Research*, *Production Planning and Control*, *Business Strategy and the Environment*, *Journal of Cleaner Production*, *International Journal of Logistics Management*, *Expert Systems with Applications*, *Annals of Operations Research*, *Transportation Research Part – E*, *Technological Forecasting and Social Change*, *Sustainable Cities and Society*, *Sustainable Production and Consumption*, *Resource Conservation and Recycling*, *Environmental Science and Pollution Research*, *Sustainable Development*, *Journal Material Cycles and Waste Management*, *Computers and Operations Research*, *Journal of Global Information Management*, *Science of Total Environment*, *Operations Management Research*, *Journal of Enterprise Information Management*, *International Journal of Logistics Research and Applications*, *Industrial Data and Management System*, and *International Journal of Quality and Reliability Management.*

He is associate editor of *International Journal of Mathematical, Engineering and Management Sciences* (IJMEMS) (Scopus, ESCI), special issue guest editor for *Journal of Enterprise Information Management* (JEIM) "Data driven design and management of sustainability for emerging modes of supply chain", *International Journal of Productivity and Performance Management* (IJPPM) "Circular Economy and Sustainable Business Performance Management in the era of Digitalization", *Journal of Business Research* "AI and Data Analytics for Omnichannel Health Care Business", *Sustainability* "Smart and Sustainable Food Supply Chain Management", *International Journal of Global Business and Competitiveness* "The competitiveness of SMEs in the era of circular and digital economy", and *Computers & Industrial Engineering* "Carbon Neutrality through Industry 4.0 based Smart Manufacturing".

He has completed eight projects funded by the EU under Erasmus+, Transversal, EU, Grundtvig, and Newton Fund Research Environment Links Project titled "Developing capacity and research network on circular and Industry 4.0 driven sustainable solutions for reducing food waste in supply chains in Turkey" and USERC – Indian Government Funded Research Project – "Food Waste Management in Circular Economy".

Contributors

Fatemeh Afsharnia
Department of Biosystems Engineering
Ferdowsi University of Mashhad
Mashhad, Iran

Adarsh Anand
Department of Operational Research
University of Delhi
Delhi, India

Mohd Anas
Department of Mathematics
Graphic Era Deemed to be University
Dehradun, India

Priyanshu Arya
Department of Mathematics
Graphic Era Hill University
Dehradun, India

Michael Ayomoh
Department of Industrial & Systems Engineering
University of Pretoria
Pretoria, South Africa

Murat Bakirci
Unmanned/Intelligent Systems Lab
Faculty of Aeronautics and Astronautics
Tarsus University
Mersin, Turkey

Irem Bayraktar
Department of Aerospace Engineering
Turkish Air Force Academy
National Defence University
Istanbul, Turkey

Soni Bisht
Department of Mathematics
Eternal University
Baru Sahib, Himachal Pradesh, India

Vaibhav Bisht
Department of Mathematics
Jaypee University of Engineering and Technology
Guna, Madhya Pradesh, India

Partha Chakrabarti
Subir Chowdhury School of Quality and Reliability
IIT Kharagpur

Geetanjali Rajput
Department of Mathematics
Eternal University
Baru Sahib, Himachal Pradesh, India

Neeraj Kumar Goyal
Subir Chowdhury School of Quality and Reliability
IIT Kharagpur

Nupur Goyal
Department of Mathematics
Graphic Era Deemed to be University
Dehradun, India

Akansha Gupta
Department of Computer Science and Engineering
Graphic Era Deemed to be University
Dehradun, India

David D. Hanagal
Department of Statistics
Savitribai Phule Pune University
Pune, India

G. Hemalatha
Department of Mathematics
College of Engineering and Technology
SRM Institute of Science and Technology
Kattankulathur, Tamil Nadu, India

Sankaralingam Hemanth Kumar
Department of Mathematics
Sri Sairam Engineering College
West Tambaram, Tamil Nadu, India

Bac An Hoang
Civil Engineering Department
University of Architecture Ho Chi Minh City
Ho Chi Minh City, Vietnam

Manjinder Kaur
Department of Economics
Eternal University
Baru Sahib, Himachal Pradesh, India

Manpreet Kaur
Department of Mathematics
Eternal University
Baru Sahib, Himachal Pradesh, India

Shristi Kharola
Department of Mathematics
Graphic Era Deemed to be University
Dehradun, India

Amit Kumar
Department of Mathematics
Chandigarh University
Mohali, Punjab, India

Arvind Kumar Lal
Department of Mathematics
T. I. E. T.
Patiala, India

Santosh Kumar
Department of Mathematical and Geospatial Sciences
RMIT University
Melbourne, Australia

Sunil Kumar
Department of Mathematics
Graphic Era Hill University
India

Oleksiy Melnyk
Odesa National Maritime University
Odesa, Ukraine

Nikhil Nawani
Department of Computer Science and Engineering
Graphic Era Deemed to be University
Dehradun, India

Oleg Onishchenko
National University
Odesa Maritime Academy
Odesa, Ukraine

Mangey Ram
Department of Mathematics
Computer Science and Engineering
Graphic Era Deemed to be University
Dehradun, India

Seema Saini
Department of Mathematics
Computer Science and Engineering
Graphic Era Deemed to be University
Dehradun, India

Neeraj Sengar
Department of Mechanical Engineering
IIT Delhi, Delhi
Graphic Era Deemed to be University
Dehradun, India

Inna Shapovalova
Odesa State University of Internal Affairs
Odesa, Ukraine

Rajendra Kumar Sharma
Department of Computer Science and Engineering
T. I. E. T.
Patiala, India

Sunita Sharma
Manipal Institute of Technology
Manipal Academy of Higher Education
Manipal, Karnataka, India

Hennady Shcheniavsky
Odesa National Maritime University
Odesa, Ukraine

Valeriy Shevchenko
National University
Odesa Maritime Academy
Odesa, Ukraine

Ayush Singh
Department of Mathematics
Lovely Professional University
Phagwara, Punjab, India

Jai Singh
Department of Mathematics (Retired)
N.I.T.
Kurukshetra, India

Surendra Vikram Singh Padiyar
Department of Mathematics
H.N.B.G.P.G. College
Khatima, Uttarakhand, India

Trust Tawanda
Department of Statistics and Operations Research
National University of Science and Technology
Bulawayo, Zimbabwe

Balakrishnan Thilaka
Department of Applied Mathematics
Sri Venkateswara College of Engineering
Sriperumbudur, Tamil Nadu, India

Vishal Vignesh S
Department of Mathematics
College of Engineering and Technology
SRM Institute of Science and Technology
Kattankulathur, Tamil Nadu, India

Sergiy Voliansky
Navigation and Maritime Safety Department
Odesa National Maritime University
Odesa, Ukraine

Venkata Seshachala Sarma Yadavalli
Department of Industrial & Systems Engineering
University of Pretoria
Pretoria, South Africa

Chapter 1

Recent advances in software reliability growth models

David D. Hanagal

1.1 INTRODUCTION

With the advancement of technology, systems in safety-critical fields usually contain both hardware components and software components. These types of systems can be found in nuclear plants, aircrafts, and financial institutes. It is one of the attributes of software quality, a multidimensional property including customer satisfaction factors such as functionality, usability, performance, capability, serviceability, maintainability, installation, and documentation. Software reliability (SR), however, is generally accepted as the key factor in software quality since it quantifies software failures, which can make a powerful system inoperative. As a result, reliability is an essential ingredient in customer satisfaction.

1.2 NON-HOMOGENEOUS POISSON PROCESS SOFTWARE RELIABILITY GROWTH MODELS

In this section we shall briefly describe some non-homogeneous Poisson process (NHPP) SR models. The models described here are only a small subset of those which appear in the literature.

The NHPP models provide another analytical framework for describing the software failure phenomenon during testing. The NHPP model that represents the number of failures experienced up to time t is an NHPP $\{N(t), t \geq 0\}$. The main issue in the NHPP model is to determine an appropriate mean value function (MVF) to denote the expected number of failures experienced up to a certain time. Let $N(t)$ denote the cumulative number of faults detected by time t, and $m(t)$ denote its expectation. Then $m(t) = E[N(t)]$. Under such models the reliability of the software for a mission of duration t is simply $P\ (N(t) = 0)$.

DOI: 10.1201/9781003787525-1

The NHPP model is based on the following assumptions:

1. The failure process has an independent increment, i.e., the number of failures during the time interval $(t, t+s)$ depends on the current time t and the length of the time interval s, and does not depend on the past history of the process.
2. The failure rate of the process is given by:

$$P\{\text{exactly one failure in } (t, t+\Delta t)\} = P\{N(t+\Delta t) - N(t) = 1\}$$
$$= \lambda(t)\Delta t + o(\Delta t)$$

   where $\lambda(t)$ is the intensity function.
3. During a small interval Δt, the probability of exactly one failure is negligible, that is, $P\{\text{two or more failures in } (t, t+\Delta t)\} = o(\Delta t)$
4. The initial condition is $N(0) = 0$.

On the basis of these assumptions, the probability of exactly n failures occurring during the time interval $(0, t)$ for the NHPP is given by:

$$P\{N(t) = n\} = \frac{[m(t)]^n}{n!} e^{-m(t)} \quad n = 0,1,2,\ldots\infty$$

The Goel and Okumoto (1979a) model, referred to hereafter as GO, is an NHPP variant of the JM model, i.e., failure detection described as an NHPP with an exponentially decaying rate function. The GO model assumes that the cumulative number of failures detected by time t is an NHPP and based on few assumptions. The following differential equation is given by

$$\frac{\partial m(t)}{\partial t} = b[a - m(t)] \quad a > 0, b > 0 \quad t \geq 0, \tag{1.1}$$

where a is the expected total number of faults that exist in the software before testing and b is the failure detection rate. Solving the above differential equation with respect to $m(t)$ under the initial condition $m(0) = 0$, we get

$$m(t) = a\left(1 - e^{-bt}\right) \tag{1.2}$$

Further, considering the counting process $\{N(t), t \geq 0\}$ representing the number of faults detected up to testing time t, we get:

$$P\{N(t)=n\}=\frac{[m(t)]^n}{n!}e^{-m(t)} \quad n=0,1,2,\ldots\infty \tag{1.3}$$

where $m(t)$ is given by:

$$m(t)=\int_0^t \lambda(x)dx \tag{1.4}$$

where $\lambda(t)=\frac{\partial m(t)}{\partial t}=abe^{-bt}$

This represents the number of faults detected per unit testing time.

This model cannot be applied to situations where new faults might be introduced in the process of debugging. Some NHPP model can incorporate the situation where new faults may be added during repairs; these models are the infinite failure models. It means that $m(t) \to \infty$ as $t \to \infty$.

The Duane (1964) model referred to hereafter as DU originally devised for hardware reliability model is an infinite failure model. This model is an NHPP with the expected number of failures

$$m(t)=at^b \tag{1.5}$$

and the intensity function

$$\lambda(t)=abt^{b-1} \tag{1.6}$$

This function is increasing for $b > 1$, decreasing for $b < 1$, and constant for $b = 1$. The DU model could be stochastically represented as Weibull process, allowing statistical procedures to be used in the application of this model in reliability growth. In particular, this model is the counting process of the record values from a Weibull distribution.

In these NHPP models, usually parameter a represents the mean number of software failures that will eventually be detected and parameter b represents the probability that a failure is detected in a constant period.

Musa and Okumoto (1985) proposed another model for infinite failures. This NHPP also called the logarithm Poisson execution time model implies that the expected number of failures is a logarithmic function of time. This model represents a situation where the failures in an early stage of program execution reduce failure more than those repaired in a later stage. The MVF is

$$m(t)=a\ln(1+bt), \quad t>0 \tag{1.7}$$

And the intensity function is derived as

$$\lambda(t) = \frac{ab}{1+bt} \tag{1.8}$$

Musa et al. (1987) stated that the logarithmic Poisson execution time model combined simplicity with high predictive validity.

Observing that the best distribution to be applied varies in different projects, Schneidewind (1972) proposed an empirical SR approach. He suggested different reliability functions and distributions and presented a choice of the exponential, normal, gamma, or Weibull distributions be applied to different projects. In a later paper, Schneidewind (1975) treated the process of faults detected per time interval as an NHPP having an exponential MVF. Either least squares or MLE could be applied for the determination of the parameters of this process.

Musa and Okumoto (1985) presented an execution time SR model, which is called as the basic execution time model. Musa postulated execution time as being superior to calendar time for modeling because it characterizes the stress that is placed on the software due to failures. In this model, consideration of debugging personnel was not necessary since Musa observed that when analysis was performed in terms of execution time, the fault correction rate was proportional to the fault detection or hazard rate. These concepts were tested on four software projects with very good results. Hecht (1981) verified the superiority of looking at SR as a function of execution rather than calendar time. Under this new formulation, the variability observed in the distribution functions of different projects had not occurred, thus the execution time modeling approach became the standard (Musa et al., 1987).

Obha (1984) proposed the hyper-exponential model to describe the fault detection process (FDP) in module structured software, assuming that the software consists of different modules. Each module has its own unique characteristics and therefore the faults uncovered in a particular module have their own peculiarities. Consequently, the fault removal rate for each module is not the same. Obha (1984) recommended that a separate modeling for each module can be established and the total fault removal phenomenon is the linear sum of the fault removal process of all modules.

A similar extension of the exponential growth model has been suggested by Yamada and Osaki (1985) by dividing the software into k modules. The failure intensity of faults within different modules was assumed to be different while the failure intensity of faults within the same module is assumed to be same and they also assumed that the expected number of faults detected for each module is exponential. The estimation of model parameters is carried using MLE method.

In the mid-1980s many NHPP models with S-shaped SR growth curve models were presented, which means that the curve crosses the exponential curve from below and the crossing occurs once and only once. The detection rate of faults, where the error detection rate changes with time, becomes the greatest at a certain time after testing begins, after which it decreases exponentially. In other words, some faults are covered by other faults at the beginning phase and therefore these faults are actually removed, the covered faults remain undetected. Obha and Yamada (1984) proposed an inflection S-shaped SR model based on the assumptions of error detection rate differing among faults by taking fault detection rate and inflection factor and also assuming that each time a software failure occurs, the software error which caused it is immediately removed and no new faults are introduced. The model parameters are estimated using MLE method.

Yamada et al. (1984) presented an NHPP delayed S-shaped model by considering that the software error detection process described by an S-shaped curve can be characterized as a learning process in which test team members become familiar with the test environment, testing tools, or project requirement, and their test skills gradually improve. The MVF of this model is derived by stating few assumptions and MLEs are also obtained for model parameters.

In the 1090s error generation (or imperfect debugging) activity is introduced into NHPP SR model by Yamada et al. (1992) by arguing that the assumption of perfect debugging is considered to be unrealistic in actual software testing and operation environments. Therefore, it is necessary to develop a software reliability growth model (SRGM) with imperfect debugging and they proposed two SR assessment models with imperfect debugging by assuming that new faults are sometimes introduced when the faults originally latent in a software system are corrected and removed during the testing phase. It is assumed that the fault detection rate is proportional to the sum of the number of faults remaining originally in the system and faults introduced by imperfect debugging. These two models were described by an NHPP. Several quantitative measures for reliability assessment were derived and the MLEs of unknown model parameters were presented. Finally, numerical examples of SRGM analysis based on these two models were showed.

Pham (1993) developed SRGM with imperfect debugging and multiple failure types in software development. Pham and Zhang (1997) have developed a SRGM considering error introduction rate as exponential and error detection rate function as of nondecreasing function of time. They derived MVF and also obtained MLEs of model parameters.

Pham et al. (1999) discussed about linear fault introduction function, and fault detection of nondecreasing function of time with S-shape is considered. They derived MVF of SRGM. Later Pham (2000) proposed inflection S-shaped SRGM with error detection rate function, and model parameters are derived using method of least squares. He also derived MVF of SRGM.

Zhang et al. (2003) addressed the problem of fault removal efficiency and fault introduction rate in NHPP model by stating fault removal efficiency is a percentage of bugs eliminated by reviews, inspections, tests, and its important and convenient metric in software development practice. Incorporating this metric into SRGM analysis will not only improve the SRGM assessment but also change quality from an amorphous term to a tangible factor and they were able to derive MVF of SRGM.

Generally two research directions for SRGM are usually considered: parameter estimation and unification of SRGM. In fact, a unified modeling framework comprising some typical reliability growth patterns should be developed for robust SR assessment. There are some but only few model unification schemes in the literature. Langberg and Singpurwalla (1985) first showed that several SRGM can be comprehensively described by adopting a Bayesian point of view. Miller (1986) and Thompson (1988) extended Langberg and Singpurwalla's idea and developed the generalized order statistic (GOS) model. They claimed that the model selection problem is reduced to a simple selection problem of a fault detection time distribution within the framework of GOS and record values especially for NHPP models. The MVF in these models can be characterized by a theoretical probability distribution function of the fault detection time.

Gokhale et al. (1996) and Gokhale and Trivedi (1999) introduced the concept of testing coverage and proposed a modeling structure similar to GOS theory. Chen and Singpurwalla (1997) proved that all SRGM, as well as the NHPP models developed in the past literature, can be unified by self-exciting point processes. Apart from the probabilistic approach, Haung et al. (2003) explained the deterministic behavior of NHPP models, namely the MVF, by introducing several kinds of mean operation. Pham and Normann (1997) solved a generalized differential equation by which the MVF in the NHPP model is governed by an NHPP with a generalized mean value parameter. Inoue and Yamada (2002) applied infinite server queuing theory to the basic assumption of delayed S-shaped SRGM, i.e., fault detection phenomenon, consisting of successive failure detection and isolation processes considering time distribution of fault isolation process and obtained several NHPP models describing the FDP as a two-stage process. Dohi et al. (2004) proposed a unification method for NHPP models describing test input and program path searching times stochastically by an infinite server queuing model. Hwang and Pham (2009) developed a generalized NHPP SRGM considering quasi-renewal time-delay fault removal.

Kapur et al. (2011) recently proposed two general frameworks for deriving several SRGMs based on an NHPP in the presence of imperfect debugging and error generation. The proposed models are initially formulated for the case when there is no differentiation between failure observation and fault removal testing processes and then extended for the case when there is a clear differentiation between failure observation and fault removal testing processes and finally numerical illustrations shown for performance of estimation procedure on real data sets.

Hanagal and Bhalerao (2016a, 2016b, 2017, 2019) proposed NHPP SRGMs based on inverse Weibull, generalized inverse Weibull, extended inverse Weibull, and generalized extended inverse Weibull distributions. They analyzed these NHPP SRGM using three different data sets.

1.3 DELAYED S-SHAPED SRGMS

The delayed S-shaped models were developed based on the assumption that faults detected in the testing phase are removed immediately with no debugging time delay and no new faults introduced into the software. In other words, it is assumed that whenever an attempt is made to remove a fault, it is removed with certainty and this is referred to as perfect debugging. But the debugging activity is not always perfect because of a number of factors like the tester's skill and expertise. The testing team, however, may not be able to remove a fault with certainty when a software fault is observed and the original fault may remain, leading to a phenomenon known as imperfect debugging. Another possibility is that while correcting a software error additional errors may be generated and these errors may get into the software. Such models may be referred to as error generation models. In the case of error generation, the total fault content increases as testing progresses because new faults are introduced into the system while removing the original faults.

Goel (1985) has introduced the concept of imperfect debugging for the first time in literature. Obha and Chou (1989) extended this work by introducing error generation into the GO model and named it as imperfect debugging model. Kapur and Garg (1990) introduced imperfect debugging in GO model in that they considered fault detection rate per remaining faults being reduced due to imperfect debugging. Thus the number of failure observed and detected by an infinite time is more than the initial fault content. Yamada et al. (1992) extended the GO model by assuming that fault introduction rate is linear and exponential time dependent. Further, several authors have used this approach to model error generation activity in the SR model. Pham and Zhang (1997) have developed an SR model considering error introduction rate as exponential and error detection rate as nondecreasing type. Pham et al. (1999) have discussed linear fault introduction function and fault detection is of nondecreasing time-dependent type. The above discussed models do not always reflect real testing environments. It is of interest to construct a SRGM by considering introduction of error generation during testing and central processing unit (CPU) execution time. Hanagal and Bhalerao (2019) proposed NHPP delayed S-shaped SRGMs and allowing both the error content function and the error detection rate to be time dependent.

1.3.1 NHPP delayed S-shaped SRGM

In the NHPP S-shaped model, the SR growth curve is an S-shaped curve. The detection rate of faults, where the error detection rate changes with time, becomes the greatest at a certain time after testing begins, after which it decreases exponentially. In other words, some faults are covered by other faults at the beginning of the testing phase, before these faults are actually removed, the covered faults remain undetected. Yamada et al. (1984) and Yamada (2014) also determined that the software testing process usually involves a learning process where testers become familiar with the software products, environments, and software specifications. Several S-shaped models (Yamada et al. 1984; Pham and Zhang, 1997) such as delayed S-shaped, inflection S-shaped, etc., exist in the literature. Kareer et al. (1990) proposed an S-shaped SRGM with two types of errors. The errors have been classified depending upon their survey. Gupta et al. (2011) obtained SR estimation using delayed S-shaped model under imperfect debugging and time lag. Ahmad et al. (2011) developed an inflection SRGM considering log-logistic testing effort and imperfect debugging. Kaur and Sharma (2015) predict the time between failure and accuracy by using CASRE tool for SRGMs.

We now discuss a stochastic model for a software error detection process based on NHPP in which the growth curve of the number of detected software errors for the observed failure data is S-shaped, called delayed S-shaped NHPP model (Yamada et al. 1984). The software error detection procedure described by an S-shaped curve can be characterized as a learning process in which the test team members become familiar with the test environment, testing tools or project requirements, i.e., their test skills gradually improve. The delayed S-shaped model is based on the following assumptions:

1. All the faults in the program are mutually independent from the failure detection point of view.
2. The probability of failure detection at any time is proportional to the current number of faults in a software.
3. The proportionality of failure detection is constant.
4. The initial error content of the software is a random variable.
5. A software system is subject to failures at random times caused by errors present in the system.
6. The time between $(i-1)$th and ith failures depends on the time to the $(i-1)$th failure.
7. Each time a failure occurs, the error, which caused it, is immediately removed and no other errors are introduced.

   On the basis of these assumptions we have the following differential equation:

$$\frac{\partial m(t)}{\partial t} = b(t)\left[a - m(t)\right]. \quad a > 0, b(t) > 0 \quad t \geq 0, \tag{1.9}$$

a = expected total number of faults that exists in the software before testing.

$b(t)$ = failure detection rate function, which also represents the average failure rate of a fault. In delayed S-shaped model, $b(t) = \frac{b^2 t}{1+bt}$ where b is the error detection rate per error in the steady state.

Solving the differential equation given in (1.9) with respect to $m(t)$ under the initial condition $m(0) = 0$, Yamada et al. (1984) obtained the MVF given by:

$$m(t) = a\left[1-(1+bt)e^{-bt}\right] \tag{1.10}$$

which shows an S-shaped curve. This model is called delayed S-shaped NHPP model. Hanagal and Bhalerao proposed three models in this chapter which are compared with delayed S-shaped NHPP model by Yamada et al. (1984) and Yamada (2014). Further, the counting process $N(t)$, $t \geq 0$ representing the cumulative number of software faults detected up to testing time t is a stochastic process. Basic assumption about this process leads to the commonly accepted conclusion that, for any fixed $t \geq 0$, $N(t)$ is a Poisson-distribution with time-dependent Poisson parameter $m(t)$, the MVF.

$$P\{N(t)=n\} = \frac{[m(t)]^n}{n!} e^{-m(t)} \quad n = 0,1,2,\ldots\infty \tag{1.11}$$

where $m(t)$ is given by:

$$m(t) = \int_0^t \lambda(x)dx \tag{1.12}$$

The MVF represents the expected number of software errors that have accumulated up to time t, or estimated cumulative faults up to time t.

$\lambda(t)$ is called the intensity function of the NHPP given by:

$$\lambda(t) = ab^2te^{-bt} \tag{1.13}$$

This represents the number of faults represented per unit testing time. In general the second assumption (1.2), i.e., the detected faults are removed with certainty, is not always true (see Yamada et al., 1992, Pham and Zhang, 1997, Pham, 1999, Yamada, 2014). If additional introduced faults affect the fraction of the fault content, we have to modify this assumption by introducing the concept of error generation while debugging (or testing) process in action. That is, total fault content (or introduction) rate function, $a(t)$, representing the sum of expected number of initial software faults and introduced faults by time t. Replacing a by $a(t)$ in equation (1.9), we have differential equation (see Pham and Zhang, 1997).

$$\frac{\partial m(t)}{\partial t} = b(t)\left[a(t) - m(t)\right]. \quad a(t) > 0, b(t) > 0 \quad t \geq 0 \tag{1.14}$$

We get the generalized MVF by solving the above differential equation (Pham and Zhang, 1997) as follows:

$$m(t) = e^{-B(t)}\left[m_0 + \int_0^t a(\tau)b(\tau)e^{B(\tau)}\,d\tau\right] \tag{1.15}$$

where $B(t) = \int_{t_0}^{t} b(\tau)d\tau$, with initial condition $m(t_0) = m_0$ and t_0 is the time to begin the debugging process. Yamada et al. (1992) proposed NHPP SRGM with two types of $a(t) = \alpha \exp(\beta t)$ and $a(t) = \alpha(1 - \gamma t)$ with $b(t) = b$ which are not delayed S-shaped SRGMs because $b(t)$ is a constant. Pham (2007) obtained NHPP delayed S-shaped SRGM with $a(t) = \alpha(1 + \gamma t)^2$ and $b(t) = \frac{\gamma^2 t}{(1 + \gamma t)}$; in this case both $a(t)$ and $b(t)$ have common parameter γ. In our proposed models we consider the $a(t)$ and $b(t)$ have independent parameters instead of dependent parameters.

Many existing SRGMs (Pham, 2006, Pham and Zhang, 2003, Yamada et al., 1984, Yamada and Osaki, 1985) can be considered as a special case of the above generalized model given in equation (2.15). An increasing function $a(t)$ implies an increasing total number of faults (note that this includes those already detected and removed and those inserted during the debugging process) and reflects imperfect debugging. An increasing $b(t)$ implies an increasing fault detection rate, which could be either attributed to a learning curve phenomenon, or to software process fluctuations, or a combination of both.

Different $a(t)$ and $b(t)$ functions also reflect different assumptions of the software testing processes. In the simplest model, the function $a(t)$ and $b(t)$ are both constants. A constant $a(t) = a$ stands for the assumption that no new errors are introduced during the debugging process. A constant $b(t) = b$ implies that the proportional factor relating the error detection rate $\lambda(t)$ to the total number of remaining errors is constant. In a general model, the functions $a(t)$ and $b(t)$ are both functions of time. Hanagal and Bhalerao (2018) proposed NHPP SRGMs based on three different fault content rate functions and analyzed these three models using three different data sets (Hanagal and Bhalerao, 2021).

1.4 FDP AND FAULT CORRECTION PROCESS

The accuracy and functionality of software is very important in order to ensure high reliability, during which faults in software manifest by causing failures and can be detected and removed by debuggers. It is almost not possible to make the software bug-free even though disciplined and scientific development practices are followed. SRGMs are very useful for making critical decisions, for example, testing resource allocation and the determination of software release time.

Testing consumes a huge amount of resources, such as CPU hours and manpower, which are usually not constantly allocated during testing phase. The testing-effort function (TEF) is a function which describes how testing resources are distributed, and it has been incorporated into SR studies by Stikkel (2006), Kapur et al. (2008, 2009), and Yamada et al. (1992). Weibull TEF curve can fit well to most data and is often used in SR modeling (Huang, 2005). Parr (1980), Demarko (1982), and Huang and Kuo (2002) used logistic TEF which seems to be fairly accurate in describing the consumption of testing effort.

It will take time to correct detected fault, and the time required to correct a detected fault is called debugging lag or delay. Schneidewind (1975) proposed fault correction process (FCP), which was modeled by a separate process following FDP with a constant time lag. Based on this framework, Xie et al. (2007) and Wu et al. (2008) proposed several paired FDP and FCP models through incorporating other variants of debugging delay. Later, Hwang and Pham (2009) developed a generalized NHPP model considering quasi-renewal time-delay fault removal. Jia et al. (2010) proposed a Markovian SR model considering FCP. However, the influence of testing effort on debugging lag is not considered in these papers. Intuitively, the time needed to correct a detected fault, or the debugging lag, tends to be shorter if more testing effort is allocated during the period between detection and correction of the fault. Thus it is more reasonable to incorporate TEF into the modeling framework on both FDP and FCP. Moreover, debugging process is usually far from perfect, and actually many faults encountered by customers are those introduced during debugging. It is essential to incorporate imperfect debugging into FDP and FCP models. Peng et al. (2018) discussed in detail FDP and FCP with applications.

1.4.1 The modeling process

The expected total number of faults at time t is denoted by the fault content rate function $a(t)$, which is the sum of the number of initial faults in the software $a(= a(0))$ and the number of faults introduced during time interval $[0, t)$. We use $w(t)$ to denote the time-dependent testing-effort rate and $W(t)$ to denote the cumulative testing effort consumed till time t.

1.4.2 FDP model

MVF md(t) is used to depict the expected number of faults detected till time t and $\lambda_d(t)=\frac{dm_d(t)}{dt}$ is used to denote the fault intensity function. The number of faults detected during time interval $[t, t+\Delta t)$ by current testing-effort expenditure is usually assumed to be proportional to the number of remaining faults at time t . Hence we have

$$\lambda_d(t)=\frac{dm_d(t)}{dt}=b(t)w(t)\left[a(t)-m_d(t)\right] \tag{1.16}$$

where $b(t)$ is the current fault detection rate per unit of testing effort at time t and $w(t)$ is the current testing-effort expenditure at time t. Substituting the marginal condition $m_d(0) = 0$ into (1.16) gives

$$m_d(t)=a(t)-ae^{-\int_0^t b(x)w(x)dx}-\left(e^{-\int_0^t b(x)w(x)dx}\right)\times\int_0^t a'(x)e^{\int_0^t b(y)w(y)dy}dx \tag{1.17}$$

where $a'(x)=\frac{da(x)}{dx}$. Various $m_d(t)$ can be derived based on different assumptions of $a(t)$, $b(t)$, and $w(t)\lambda_d(t)$ can be obtained by substituting (1.17) into the right-hand side of (1.16) as

$$\begin{aligned}\lambda_d(t)&=\frac{dm_d(t)}{dt}\\&=ab(t)w(t)e^{-\int_0^t b(x)w(x)dx}\left(1+\int_0^t\frac{a'(x)}{a}e^{\int_0^t b(y)w(y)dy}dx\right)\end{aligned} \tag{1.18}$$

1.4.3 FCP model

MVF $m_c(t)$ is used to denote the expected number of faults corrected till time t and $\lambda_c(t) = dm_c(t)/dt$ is used to denote the fault correction intensity function. Since a corrected fault must first be detected, FCP can be modeled as a separate process following FDP with a debugging delay. For convenience, the testing effort consumed during the period from detection of a fault to the final correction of the fault is termed as correction effort of the fault. Generally correcting different faults requires different amounts of testing resources; hence correction effort can be modeled as a random variable with probability density function (PDF) and the cumulative distribution function (CDF) denoted as $f(x)$ and $F(x)$. Thus it can be obtained that

$$m_c(t) = \int_0^t \lambda_d(y) F\big(W(t) - W(y)\big) dy \tag{1.19}$$

where $F(W(t) - W(y))$ is the probability that the fault detected at y is corrected before t. Different $m_c(t)$ can be derived based on $m_d(t)$ and different $f(x)$. We have

$$\begin{aligned}\lambda_c(t) &= \int_0^t \lambda_d(y) f\big(W(t) - W(y)\big) w(t) dy \\ &= \int_0^{W^*(t)} \lambda_d\big(W^{-1}(W(t)-x)\big) f(x) w(t) d\big(W^{-1}(W(t)-x)\big) \\ &= \int_0^{W^*(t)} \lambda_d\big(W^{-1}(W(t)-x)\big) f(x) w(t) \frac{dx}{w\big(W^{-1}(W(t)-x)\big)}\end{aligned} \tag{1.20}$$

1.4.4 Some specific models

Fault detection rate $b(t)$ is usually assumed to be constant, and it is denoted as b (Lin and Huang, 2008). From (1.17) we have

$$m_d(t) = a(t) - ae^{-bW^*(t)} - e^{-bW^*(t)} \int_0^t a'(x) e^{bw^*(x)} dx \tag{1.21}$$

The total number of faults $a(t)$ was usually assumed to be an exponential or linear function of time. Yamada et al. (1992) proposed two FDP models, with consideration of imperfect debugging, by assuming that the expected total number of faults increases exponentially and linearly with the testing time, respectively. An S-shaped concave FDP model was proposed in Pham et al. (1999) assuming that the total number of faults is a linear function of the testing time. The following are the various TEF-dependent FDP and FCP models derived based on different assumptions on $a(t)$ and $f(x)$.

1.4.4.1 Exponential total testing effort

We assume that the total number of faults increases exponentially with the total testing effort consumed and taking exponential distribution for correction effort variable as given below.

$$a(t) = ae^{\alpha W^*(t)}, \quad \alpha \geq 0 \tag{1.22}$$

$$f(x) = ce^{-cx} \tag{1.23}$$

The MVF of fault detection model based on the above exponential correction effort and $a(t)$ is given by

$$m_d(t) = \frac{ab}{b+\alpha}\left(e^{\alpha W*(t)} - e^{-bW*(t)}\right) \tag{1.24}$$

and the MVF of fault correction model based on exponential correction effort is given by

$$m_c(t) = \begin{cases} \dfrac{ab}{(b+\alpha)^2}\left(be^{\alpha W*(t)} + \alpha e^{-bW*(t)}\right) \\ -\dfrac{ab}{(b+\alpha)}\left(1 + bW*(t)\right)e^{-bW*(t)}, & c = b \\ \dfrac{a}{1+\alpha/b}\left(\dfrac{ce^{\alpha W*(t)} + \alpha e^{-cW*(t)}}{c+\alpha} \right. \\ \left. + \dfrac{ce^{-bW*(t)} - be^{-cW*(t)}}{b-c}\right) & , \ c \neq b \end{cases} \tag{1.25}$$

Actually (1.24) can be obtained by combining (1.21) and (1.22). Equation (1.25) can be obtained by substituting (1.24) into (1.18) and (1.19). When $W*(t) = t$, equation (1.24) is the same as the FDP model obtained by Yamada et al. (1992) for the case when the total number of faults is an exponential function of testing time. When $\alpha = 0$ and $W*(t) = t$, (1.24) and (1.25) are the same as the paired model obtained by Wu et al. (2008) for the case of exponential debugging delay.

We assume that the total number of faults increases exponentially with the total testing effort consumed as given in equation (1.22) and taking gamma distribution for correction effort variable as given below.

$$f(x) = \frac{\mu e^{-\mu x}(\mu x)^{c-1}}{\Gamma c} \tag{1.26}$$

The MVF of fault detection model is similar to (1.24) and the MVF of gamma fault correction model based on gamma correction effort is given by

$$m_c(t)=\begin{cases}\dfrac{ae^{\alpha W^*(t)}\Gamma\left(c,o,(b+\alpha)W^*(t)\right)}{(1+\alpha/b)^{c+1}\Gamma(c)} \\ -\dfrac{a\Gamma\left(c,0,bW^*(t)\right)}{(1+\alpha/b)\Gamma(c)}+\dfrac{a\Gamma\left(c+1,0,bW^*(t)\right)}{(1+\alpha/b)c\Gamma(c)}, & \mu=b \\ \dfrac{ae^{\alpha W^*(t)}\Gamma\left(c,o,(\mu+\alpha)W^*(t)\right)}{(1+\alpha/b)(1+\alpha/\mu)^c\Gamma(c)} \\ \dfrac{ae^{bW^*(t)}\Gamma\left(c,o,(\mu-b)W^*(t)\right)}{(1+\alpha/b)(1-b/\mu)^c\Gamma(c)}, & \mu\neq b\end{cases} \tag{1.27}$$

where $\Gamma(a,b,c)=\int_b^c e^{-y}y^{a-1}$ is a generalized incomplete gamma function. When $\alpha = 0$ and $W*(t) = t$, equations (1.26) and (1.27) are the same as the paired model obtained by Wu et al. (2008) for the case of gamma debugging delay.

1.4.4.2 Linear total testing effort

We assume that the total number of faults increases linearly with the total testing effort consumed and taking exponential distribution for correction effort variable as given below:

$$a(t)=a+sW^*(t), \quad s\geq 0 \tag{1.28}$$

$$f(x)=ce^{-cx} \tag{1.29}$$

The MVF of fault detection model based on the above exponential correction effort and

$a(t)$ is given by

$$m_d(t)=(a-s/b)(1-e^{-bW^*(t)})+sW^*(t) \tag{1.30}$$

and the MVF of fault correction model based on exponential correction effort is given by

$$m_c(t)=\begin{cases}(a-2s/b)\left(1-e^{-bW^*(t)}\right)+sW^*(t)+sW^*(t)\left(1-e^{-bW^*(t)}\right), & c=b\\ (a-s/b)\left(1+\dfrac{be^{-cW^*(t)}-ce^{-bW^*(t)}}{c-b}\right) & \\ +sW^*(t)-s/c\left(1-e^{-cW^*(t)}\right), & c\neq b\end{cases} \tag{1.31}$$

The MVF of fault detection model is similar to equation (1.30) and the MVF of gamma fault correction model based on gamma correction effort is given by

$$m_c(t)=\begin{cases}\dfrac{sw(t)}{b\Gamma(c)}\left[bW*(t)\Gamma\left(c,0,bW^*(t)\right)-\Gamma\left(c+1,0,bW^*(t)\right)\right] & \\ +\dfrac{(a-s/b)}{c\Gamma(c)}\Gamma\left(c,0,bW^*(t)\right), & \mu=b\\ \dfrac{s}{\mu\Gamma(c)}\left[\mu W^*(t)\Gamma\left(c,0,\mu W^*(t)\right)-\Gamma\left(c+1,0,\mu W^*(t)\right)\right] & \\ +\dfrac{(a-s/b)}{\Gamma(c)}\Gamma\left(c,0,\mu W^*(t)\right) & \\ -\dfrac{(a-s/b)e^{-bW^*(t)}}{\Gamma(c)(1-b/\mu)^c}\Gamma\left(c,0,(\mu-b)W^*(t)\right), & \mu\neq b\end{cases} \tag{1.32}$$

When $s = 0$ and $W*(t) = t$, (1.31) and (1.32) are the same as the paired model obtained by Wu et al. (2008) for the case of gamma debugging delay.

1.4.4.3 Some TEFs

Some TEFs which are used more often in the literature are constant, exponential, Rayleigh, Weibull, and logistic curves. Weibull curve includes exponential curve and Rayleigh curve. The details are given below.

The constant TEF is $w(t) = 1$ and the cumulative TEF is $W(t) = t$, $w > 0$. The Weibull TEF is very flexible, and can fit well to most of the data. The cumulative TEF $W(t)$ is given by

$$W(t)=N\left[1-e^{-\alpha t^{\beta}}\right]$$

where N is the expected total amount of testing effort that is required by software testing, α and β are the scale and shape parameters respectively. The corresponding TEF function is

$$w(t) = N\alpha t^{\beta-1}e^{-\alpha t^{\beta}}$$

When $\beta = 1$, it reduces to exponential TEF and when $\beta = 2$, it reduces to Rayleigh TEF. Rayleigh testing-effort rate first increases to its peak and then decreases to zero as time increases. Logistic curve is similar to Rayleigh curve, except during the initial period. The logistic cumulative TEF, $W(t)$, is given by

$$W(t) = \frac{N}{1+\alpha e^{-\beta t}}$$

where α is a constant parameter and β is the consumption rate of testing-effort expenditure. The corresponding TEF is given by

$$w(t) = \frac{N\alpha\beta}{\left[e^{\beta t/2} + \alpha e^{-\beta t/2}\right]^2}$$

$w(t)$ reaches its maximum value at $t = \ln \alpha/\beta$.

Peng et al. (2018) analyzed the data set given by Musa et al. (1987) which contains both fault detection data and fault correction data. It also contains the information on testing effort, which is characterized by computer time (CPU hours) consumed in each week.

1.4.5 Software release policies

Many researchers studied software release policies (SRPs) based on reliability and cost criteria. The calculation of the optimal release time is one of the most crucial decisions for software modeling and has been studied by Boland and Chuiv (2007), Inoue and Yamada (2007), and Li et al. (2010). As the reliability and the cost requirements are of great concern, they are often used to determine the time to stop the testing and release the software (Lin and Huang, 2008).

1.4.5.1 SRP based on reliability criterion

SR is defined as the probability that no failure occurs during time interval $(T, T + \Delta T]$ given that the software is released at time T. Assuming that software normally does not change in operational phase, the reliability function is

$$R(\Delta T / T) = e^{-\lambda_d(T)\Delta T}$$

If R_1 is the reliability target and T_{LC} is the length of the software life cycle, the time when the reliability of the software reaches R_1 can be obtained as

$$T_1 = \inf\{\lambda_d(T) \le \ln(1/R_1)/\Delta T : T \in [0, T_{LC}]\}$$

1.4.5.2 SRP based on cost criterion

Now we discuss the optimal release time based on the total cost during the software testing phase and the operational phase. With the incorporation of FCP $m_c(t)$, the cost model can be expressed as

$$C(T) = c_1 m(T) + c_2 [m_d(T_L C) - m_c(T)] + c_3 W^*(T)$$

where c_1 is the cost of fixing a fault during the testing phase, c_2 is the cost of fixing a fault during the operational phase ($c_2 > c_1 > 0$), and c_3 is the unit cost for testing effort consumed during testing. By minimizing the cost model with respect to T, the optimal release time T_c can be obtained. Differentiating both sides of $C(T)$, we have

$$C'(T) = c_3 w(T) - (c_2 - c_1)\lambda_c(T)$$

Furthermore, we have $C'(0) = c_3 w(0) > 0$. Let $z_1 \le z_2 \le \ldots \le z_n$ be all the solutions to $\lambda_c(T)/w(T) = c_3/(c_2 - c_1)$ during $(0, T_{LC})$. If $n = 2k$, $(k \ge 0)$, T_c can be determined as $T_c = \mathrm{argmin}_{T=0,z}2,\ldots,z2k\ C(T)$. Otherwise $n = 2k + 1$ and T_c can be determined as $T_c = \mathrm{argmin}_{T=0,z}2,\ldots,z2k,TLC\ C(T)$.

1.4.5.3 SRP based on mixed criterion

When both reliability requirements and the total cost are considered, our goal is to determine the optimal release time $T*$ which minimizes the total cost without compromising the reliability requirements. Thus the problem can be formulated as

$$\text{Minimize } C(T) = c_1 m_c(T) + c_2 [m_d(T_{LC}) - m_c(T)] + c_3 W^*(T)$$

$$\text{subject to } R(\Delta T|T) = e^{-\lambda_d(T)\Delta T} \ge R_1$$

The time axis (T_1, T_{LC}) can be divided into four types of intervals such that both $R(\Delta T | T)$ and $C(T)$ increase on type 1 intervals; both $R(\Delta T | T)$ and $C(T)$ decrease on type 2 intervals; $R(\Delta T | T)$ increases, while $C(T)$ decreases on type 3 intervals; and $R(\Delta T | T)$ decreases, while $C(T)$ increases on type 4 intervals. The candidates for $T*$ comprise of the minimum T in each type 1 interval that satisfies $R(\Delta T | T) \ge R_1$, the maximum T in each type 2 interval that satisfies $R(\Delta T | T) \ge R_1$, the end points of type 3 intervals which satisfy $R(\Delta T \mid T) \geq R_1$, and the initial points of type 4 intervals which satisfy $R(\Delta T \mid T) \geq R_1$. $T*$ equals to the candidate which incurs the lowest cost.

1.5 SOME MORE GENERALIZATION ON SRGM

The SR model based on Markov renewal processes that covers interdependence among successive software runs was introduced by Goseva-Popstojanova and Trivedi (1999). Sahinoglu (1992) presented the Compound-Poisson SR model and considered multiple failures that occur simultaneously. A new SRGM proposed by Singh et al. (2007) assumes the presence of two types of faults in the software: leading and dependent. Leading faults can be removed when a failure is observed, but dependent faults are masked by leading faults and can be removed only after the corresponding leading faults are removed.

One important problem is to minimize the total cost subject to a constraint on SR. For example, a series of theoretical findings regarding when a software version should be released and when fault dependency and debugging time lag are considered have been provided by Huang and Lin (2006). In addition, the optimal release timing in an SRGM with Burr-type X TEFs has been analyzed by Ahmad et al. (2009). Another stream adopts the multi-attribute utility theory (MAUT) to decide the release time. For instance, Singh et al. (2012) built an SRGM based on the Yamada two-stage model and focused on two different attributes to investigate the release time. Pachauri et al. (2014) studied an SRGM incorporating a generalized modified Weibull TEF in an imperfect debugging environment. They examined the optimal release policy using both genetic algorithm (GA) and MAUT. Li and Yi (2016) propose a new model that incorporates a power-law TEF and three generations of interdependent errors.

Specifically, the detection process for the first-generation errors is independent, and the detection rate is proportional to the instantaneous testing-effort expenditure and the mean of remaining first-generation errors in the system. By contrast, the detection process of the second-generation errors depends on the first generation and the detection rate is proportional to not only the instantaneous testing-effort expenditure and the mean of remaining second-generation errors but also the ratio of removed first-generation errors. Similarly, the detection process of the third-generation errors relies on the second-generation errors and the detection rate is proportional to the instantaneous testing-effort expenditure, the mean of remaining third-generation errors, and the ratio of removed second-generation errors.

Singh et al. (2007) assumed the detection/correction rate as a power-law function of testing time. That is, the instantaneous testing-effort expenditure at time t, $w(t)$, takes the following form:

$$w(t) = bt^k$$

where b and k are parameters to be estimated by real-world data sets. The cumulative amount of testing-effort expenditures is by time t, that is,

$$W(t) = \frac{bt^{k+1}}{k+1}$$

It is essential to consider fault interdependency when modeling SR growth processes. Singh et al. (2007) propose three generations of interdependent errors. That is, the detection of the second-generation errors relies on the first generation, and the detection of the third-generation errors depends on the second generation. Moreover, the second-generation errors are detectable only if the first-generation errors have been removed, and the third-generation errors are detectable only if the second-generation errors have been removed.

The MVFs for all three generations' errors satisfy the following differential equations:

$$\frac{dm_1(t)}{dt} = w(t)\left[a_1 - m_1(t)\right]$$

$$\frac{dm_2(t)}{dt} = w(t)\left[a_2 - m_2(t)\right]\frac{m_1(t)}{a_1}$$

$$\frac{dm_3(t)}{dt} = w(t)\left[a_3 - m_3(t)\right]\frac{m_2(t)}{a_2}$$

The corresponding three MVFs for all three generations are

$$m_1(t) = a_1\left(1 - e^{-bt^{k+1}} / (k+1)\right)$$

$$m_2(t) = a_2\left(1 - e^{-bt^{k+1}} / (k+1) - e^{-bt^{k+1}/(k+1)} + 1\right)$$

$$m_3(t) = a_3\left(1 - e^{-bt^{k+1}} / (k+1) + e^{1-bt^{k+1}/(k+1)} - 1\right)$$

where a_i is the initial content of ith generation errors in the system and $m_i(t)$ represents the expected number of ith generation errors detected in time $(0, t]$ with initial conditions $m_i(0) = 0$ and $m(\infty) = a$ for $i = 1, 2, 3$. The overall MVF $m(t)$ is the mean value of all three generations of errors. Let $a_i = ap_i$, $i = 1, 2, 3$ and $p_1 + p_2 + p_3 = 1$, we have

$$m(t) = m_1(t) + m_2(t) + m_3(t)$$

When $k = 0$, $p_1 = 1$, $p_2 = p_3 = 0$, the above model reduces to the Goel and Okumoto (1979b) model.

BIBLIOGRAPHY

Ahmad, N., Khan, M. G. M. and Rafi, L. S. (2011). Analysis of an inflection S-shaped software reliability model considering log-logistic testing-effort and imperfect debugging. *International Journal of Computer Science and Network Security*, 11(11), 161–171.

Ahmad, N., Khan, M. G. M., Quadri, S. M. K. and Kumar, M. (2009). Modelling and analysis of software reliability with Burr type X testing-effort and release-time determination. *Journal of Modelling in Management*, 4(1), 28–54.

Boland, P. J. and Chuiv, N. N. (2007). Optimal times for software release when repair is imperfect. *Statistics and Probability Letters*, 77, 1176–1184.

Chen, Y. and Singpurwalla, N. D. (1997). Unification of software reliability models by self-exciting point processes. *Advances Application Probability*, 29, 337–352.

Demarko, T. (1982). *Controlling Software Projects: Management, Measurement and Estimation*. Englewood Cliffs, Prentice-Hall.

Dohi, T., Osaki, S. and Trivedi, K. S. (2004). An infinite server queuing approach for describing software reliability growth unified modelling and estimation framework. *Proceedings of the 11th Asia-Pacific Software Engineering Conference (APSEC'04)*, 110–119.

Duane, J. T. (1964). Learning curves approach to reliability monitoring. *IEEE Transactions on Aerospace*, 2, 563–566.

Goel, A. L. (1985). Software reliability models: Limitations and applicability. *IEEE Transactions Software Engineering*, 2(12), 1411–1423.

Goel, A. L. and Okumoto, K. (1979a). A Markovian model for reliability and other performances measures of software system. *Proceedings of COMPCON* (IEEE Computer Society Press, Los Angeles), 769–774.

Goel, A. L. and Okumoto, K. (1979b). Time dependent error-detection rate model for software reliability and other performance measures. *IEEE Transactions on Reliability*, 28(3), 206–211.

Gokhale, S. and Trivedi, K. S. (1999). Time/structure based software reliability model. *Annals of Software Engineering*, 8, 85–121.

Gokhale, S., Marinos, P. N. and Trivedi, K. S. (1996). Important milestones in software reliability modeling. *SEKE '96: The 8th International Conference on Software Engineering and Knowledge Engineering, Proceedings*, 345–352.

Goseva-Popstojanova, K. and Trivedi, K. (1999). Failure correlation in software reliability models. *Proceedings of the 10th International Symposium on Software Reliability Engineering (ISSRE 1999)*, 232–241.

Gupta, A., Choudhary, D. and Sexena, S. (2011). Software reliability estimation using Yamada delayed S-shaped model under imperfect debugging and time lag. *International Journal of Computer Applications*, 23(7), 49–52.

Hanagal, D. D. and Bhalerao, N. N. (2016a). Analysis of NHPP software reliability growth model. *International Journal of Statistics and Reliability Engineering*, 3(2), 53–67.

Hanagal, D. D. and Bhalerao, N. N. (2016b). Modeling and statistical inference on generalized inverse Weibull software reliability model. *Journal of Indian Society of Probability and Statistics*, 17(2), 145–160.

Hanagal, D. D. and Bhalerao, N. N. (2017). Modeling on extended inverse Weibull software reliability growth model. *Indian Association for Product, Quality and Reliability Transactions*, 42(2), 75–95.

Hanagal, D. D. and Bhalerao, N. N. (2018). Analysis of delayed S-shaped software reliability growth model with time dependent fault content rate function. *Journal of Data Science*, 17(3), 573–589.

Hanagal, D. D. and Bhalerao, N. N. (2019). Modeling on generalized extended inverse Weibull software reliability growth model. *Journal of Data Science*, 16(4), 857–878.

Hanagal, D. D. and Bhalerao, N. N. (2021). *Software Reliability Growth Models*. Springer Nature, Singapore.

Haung, C. Y., Lyu, M. R. and Kuo, S. Y. (2003). A unified scheme of some non-homogeneous Poisson process models for software reliability estimation. *IEEE Transactions on Software Engineering*, 29(3), 261–269.

Hecht, H. (1981). Allocation of resources for software reliability. *Proceedings COMP-CON Fall*, Washington, DC, 74–82.

Huang, C. Y. (2005). Performance analysis of software reliability growth models with testing-effort and change-point. *Journal of Systems and Software*, 76(2), 181–194.

Huang, C. Y. and Kuo, S. Y. (2002). Analysis and assessment of incorporating logistic testing effort function into software reliability modeling. *IEEE Transactions on Reliability*, 51(3), 261–270.

Huang, C. Y. and Lin, C. T. (2006). Software reliability analysis by considering fault dependency and debugging time lag. *IEEE Transactions on Reliability*, 55(3), 436–450.

Hwang, S. and Pham, H. (2009). Quasi-Renewal time-delay fault removal consideration in software reliability modelling. *IEEE Transaction on Systems, Man and Cybernetics - Part A*, 39(1), 200–209.

Inoue, S. and Yamada, S. (2002). An infinite server queuing model considering time distribution of fault isolation process for software reliability assessment. *IEICE Technical Report Reliability*, 102(454), 119–124.

Inoue, S. and Yamada, S. (2007). Generalized discrete software reliability modeling with effect of program size. *IEEE Transactions on Systems, Man, and Cybernetics: Part A – Systems and Humans*, 37(2), 170–179.

Jia, L. X., Yang, B., Guo, S. C. and Park, D. H. (2010). Software reliability modeling considering fault correction process. *IEICE Transactions on Information and Systems*, E93D(1), 185–188.

Kapur, P. K. and Garg, R. B. (1990). Optimal software release policies for software reliability growth model under imperfect debugging. *Recherche Opertionnelle/ Operations Research (RAIRO)*, 24, 295–305.

Kapur, P. K., Goswami, D. N., Bardhan, A. and Singh, O. (2008). Flexible software reliability growth model with testing effort dependent learning process. *Applied Mathematical Modelling*, 32, 1298–1307.

Kapur, P. K., Pham, H., Anand, S. and Yadav, K. (2011). A unified approach for developing software reliability growth models in the presence of imperfect debugging and error generation. *IEEE Transactions on Reliability*, 60(1), 331–340.

Kapur, P. K., Shatnawi, O., Aggarwal, A. and Kumar, R. (2009). Unified framework for developing testing effort dependent software reliability growth models. *WSEAS Transactions on Systems*, 8(4), 521–531.

Kareer, N., Kapur, P. K. and Grover, P. S. (1990). An S-shaped software reliability growth model with two types of errors. *Microelectronics and Reliability*, 30(6), 1085–1090.

Kaur, D. and Sharma, M. (2015). Software reliability models: Time between failures and accuracy estimation. *International Journal of Computer Science and Information Technologies*, 6(4), 3370–3373.

Langberg, N. and Singpurwalla, N. D. (1985). Unification of some software reliability models via the Bayesian approach. *SIAM Journal on Scientific and Statistical Computing*, 6, 781–790.

Li, F. and Yi, Z. L. (2016). A new software reliability growth model: Multigeneration faults and a power-law testing-effort function. *Mathematical Problems in Engineering*, 2016, 1–13.

Li, X., Xie, M. and Ng, S. H. (2010). Sensitivity analysis of release time of software reliability models incorporating testing effort with multiple change-points. *Applied Mathematical Modelling*, 34(11), 3560–3570.

Miller, D. R. (1986). Exponential order statistic models of software reliability growth. *IEEE Transaction on Software Engineering*, 12(1), 12–24.

Musa, J. D. and Okumoto, K. (1985). Applications of basic and logarithmic Poisson execution model in software reliability measure. The challenges of advanced computing technology to system designs methods. *NATO Advanced Study Institute*, 34(1), 68–100.

Musa, J. D., Lannino, A. and Okumoto, K. (1987). *Software Reliability: Measurement, Prediction and Application*. McGraw-Hill, New York.

Obha, M. (1984). Software reliability analysis models. *IBM Journal of Research Development*, 28(4), 428–443.

Obha, M. and Chou, X. (1989). Does imperfect debugging affect software reliability growth models? *Proceeding 4th International Conference on Software Engineering*, Pittsburg, 237–244.

Obha, M. and Yamada, S. (1984). S-shaped software reliability growth models. In *Proceedings of 4th International Conference Reliability and Maintainability* (pp. 430–436).

Pachauri, B., Kumar, A. and Dhar, J. (2014). Software reliability growth modeling with dynamic faults and release time optimization using GA and MAUT. *Applied Mathematics and Computation*, 242, 500–509.

Parr, F. N. (1980). An alternative to the Rayleigh curve for software development effort. *IEEE Transactions on Software Engineering*, 6(3), 291–296.

Peng, R., Li, Y. F. and Liu, Y. (2018). *Software Fault Detection and Correction: Modeling and Applications*. Springer, Singapore.

Pham, H. (1993). Software reliability assessment: Imperfect debugging and multiple failure types in software development. EGRAND-RAAM-10737, Idaho National Engineering Laboratory.

Pham, H. (1999). Reliability analysis for dynamic configurations of systems with three failure modes. *Reliability Engineering and System Safety*, 63, 13–23.

Pham, H. (2000). *Software Reliability*. Springer, New York.

Pham, H. (2006). *System Software Reliability*. Springer, New York.

Pham, H. (2007). An imperfect debugging fault-detection dependent parameter software. *International Journal of Automation and Computing*, 4(4), 325–328.

Pham, H. and Normann, L. (1997). A generalized NHPP software reliability growth model. *Proceedings 3rd ISSAT (International Society of Science and Applied Technologies) International Conference on Reliability and Quality in Design*, August, ISSAT Press, Anaheim.

Pham, H. and Zhang, X. (1997). An NHPP software reliability model and its comparison. *International Journal of Reliability. Quality and Safety Engineering*, 4, 269–282.

Pham, H. and Zhang, X. (2003). NHPP software reliability and cost models with testing coverage. *European Journal of Operational Research*, 145(2), 443–454.

Pham, H., Nordmann, L. and Zhang, X. (1999). A general imperfect software debugging model with S-shaped fault detection rate. *IEEE Transactions on Reliability*, 48(2), 169–175.

Sahinoglu, M. (1992). Compound-Poisson software reliability model. *IEEE Transactions on Software Engineering*, 18(7), 624–630.

Schneidewind, N. F. (1972). An approach to software reliability prediction and quality control. *Fall Joint Computer Conference, AIFPS Conference Proceedings, 41*, AFIPS Press, Montvale, NJ, 837–847.

Schneidewind, N. F. (1975). Analysis of error processes in computer software. *Proceedings of International Conference on Reliable Software*, IEEE Computer Society, 337–346.

Singh, O., Kapur, P. K. and Anand, A. (2012). A multi-attribute approach for release time and reliability trend analysis of a software. *International Journal of Systems Assurance Engineering and Management*, 3(3), 246–254.

Singh, V. B., Yadav, K., Kapur, R. and Yadavalli, V. S. S. (2007). Considering the fault dependency concept with debugging time lag in software reliability growth modeling using a power function of testing time. *International Journal of Automation and Computing*, 4(4), 359–368.

Singpurwalla, N. D. and Wilson, S. W. (1999). *Statistical Methods in Software Engineering: Reliability and Risk*. Springer-Verlag, New-York.

Stikkel, G. (2006). Dynamic model for the system testing process. *Information and Software Technology*, 48, 578–585.

Thompson, Jr., W. A. (1988). *Point Process Models with Application to Safety and Reliability*. Chapman and Hall, New York.

Wu, Y. P., Hu, Q. P., Xie, M. and Ng, S. H. (2008). Modeling and analysis of software fault detection and correction process by considering time dependency. *IEEE Transactions on Reliability*, 56(4), 629–642.

Xie, M., Hu, Q. P., Wu, Y. P. and Ng, S. H. (2007). A study of the modeling and analysis of software fault-detection and fault-correction processes. *Quality and Reliability Engineering International*, 23, 459–470.

Yamada, S. (2014). *Software Reliability Modeling Fundamentals and Applications*. Springer, New York.

Yamada, S. and Osaki, S. (1985). Reliability growth modelling: Models and applications. *IEEE Transactions on Software Engineering*, 11(12),1431–1437.

Yamada, S., Obha, M. and Osaki, S. (1984). S-shaped software reliability growth models and their applications. *IEEE Transactions on Reliability*, 33, 289–292.

Yamada, S., Tokuno, K. and Osaki, S. (1992). Imperfect debugging models with fault introduction rate for software reliability assessment. *International Journal of Systems Sciences*, 23(12), 2241–2252.

Zhang, X., Teng, H. and Pham, H. (2003). Considering fault removal efficiency in software reliability assessment. *IEEE Transactions on System, Management, and Cybernetics-Part A*, 33(1), 114–120.

Chapter 2

Risk-informed planning for maritime network reliability under operational uncertainty

Oleksiy Melnyk, Sergiy Voliansky, Oleg Onishchenko, Valeriy Shevchenko, Hennady Shcheniavsky, and Inna Shapovalova

2.1 INTRODUCTION

Modern maritime logistics is characterized by the increasing complexity of operations and the rapid growth of global trade, which creates new challenges for ensuring the safety and reliability of port-ship interaction. Increasing vessel size, limited maneuvering space in port waters, and increased traffic density reduce the effectiveness of traditional, deterministic models for planning mooring operations that do not take into account risk dynamics and real-world operating conditions.

The analysis of unexpected situations, such as a sharp change in weather conditions, technical failures of towing equipment, delays in communication between the crew and port services, indicates a high level of operational uncertainty. This requires the implementation of adaptive solutions based on the principles of sustainability and risk-based approach.

High-profile incidents in recent years, such as the collision of the container ship DALI with a bridge pillar in the port of Baltimore (2024) and the accident of the Milano Bridge in the port of Busan (2020), have demonstrated systemic vulnerabilities in existing mooring procedures. The causes of the accidents go beyond technical errors and point to a lack of situational awareness, fragmented application of risk assessment methods, and insufficient integration of data into the decision-making process.

Research in the field of search and rescue operations outlines the main challenges for planning and demonstrates the potential of AI-oriented algorithms to reduce response times and improve the accuracy of incident localization [1]. These ideas correlate with the work on autonomous navigation, where adaptive multi-source quantitative risk assessment allows to take into account the dynamics of the environment and reduce the likelihood of collisions [2]. At the macro level of industry policy, the focus is shifting to environmental goals: research on achieving greenhouse gas emission reductions emphasizes the risks of not meeting international standards [3], while a review of ML applications in maritime risk management [4] outlines methodological gaps and prospects for explanatory AI. Expanding

DOI: 10.1201/9781003787525-2

the analytical framework through game theory and network approaches [5] allows for the integration of reliability issues in complex transportation systems.

In the area of sensor networks, the use of LoRaWAN for decentralized monitoring, which increases the stability of data transmission channels in marine areas, is attracting attention [6]. This approach is enhanced by the coordinated route planning of unmanned vehicles (UAVs and USVs), which can provide real-time monitoring despite the uncertainty in travel time [7]. Security studies demonstrate the effectiveness of probabilistic networks in modeling the role of barriers [8], assessing the risks of transporting dangerous goods, including electric vehicles [9], and assessing the sustainability of green shipping policies [10]. At the same time, deep learning models provide forecasting of offshore winds, which are critical for safe navigation [11].

An important area is the integration of cognitive maps with causal analysis methods and expert opinions, which allows to reproduce complex human factors [12]. Algorithms based on transformer networks show potential for predicting "near-miss" events and explaining patterns [13]. This correlates with the use of fuzzy logic and Bayesian networks in the study of reliability of inert gas operations [14] and general overviews of BN applications in the maritime industry [15]. The safe trajectory of autonomous vessels is also built through risk-based planning [16] and multi-scale scenario analysis [17]. In parallel, FRAM combined with AIS data shows how system interactions lead to incidents in port operations [18].

Real-time monitoring of containers is realized through hybrid IoT networks [19], while the resilience of LNG chains is analyzed by methods ranging from static Bow-tie to dynamic epidemiological modeling [20]. Bayesian networks are increasingly being used to evaluate trade routes under uncertainty [21], optimize the placement of SAR equipment [22], plan AUV courses [23], and enhance process security on a regional scale [24].

Against this background, it is worth noting the contribution of Ukrainian research. Papers [25–27] form the basis of a probabilistic approach to hydrometeorological risks and navigation safety assessment. The development of a methodology for expert assessment of the quality of risk analysis of operations [27] is directly integrated into modern DSS systems. The analysis of the efficiency of the Trumpet fleet [28] and the impact of hull geometry on maneuverability [29] add a practical engineering component, while the study of ship systems' efficiency [30] emphasizes the reduction of energy consumption.

Special mention should be made of the work on improving the fuel efficiency of engines through additives [31] and ensuring the safe operation of diesel plants [32]. These are examples of applied technical solutions that are consistent with the general paradigm of sustainable development. Although the works on mathematical analysis [33] or optimization of port equipment [34] are more of a fundamental nature, they demonstrate the connection between accurate models and applied logistics problems. Finally, new methods based on Markov processes and fuzzy sets [35–37] bring reliability assessment to a level capable of taking into account uncertainty and multicriteria. Studies in the field of law and policy [38,39] introduce into the discussion the dimension of "gray areas" and cluster management, which is important for the institutional context of risk management.

Hydrodynamic aspects are revealed in the modeling of the impact of cold-water pipes and mooring systems on OTEC-type platforms [40], which emphasizes the importance of physical models and numerical simulations in the overall safety loop: application of neural networks for predicting ship maneuverability [41], use of 3D virtual environments for training and risk modeling [42], structural methods for visualization and diagnostics of materials [43], assessment of the impact of shock loads on alloys through dynamic modeling [44], analysis of the behavior of hull structures during collisions [45], and improvement of the efficiency of restorable hydrodynamic systems [46]. Together, these approaches form a comprehensive framework for risk-informed planning and improving the reliability of maritime transportation networks under conditions of operational uncertainty.

2.2 PROBLEM STATEMENT

The increasing complexity of maritime transportation, combined with new environmental requirements and the transition to autonomous solutions, necessitates a systematic review of current approaches to risk management.

Despite the rapid development of digital technologies in maritime logistics, most modern mooring operations' planning systems remain deterministic, poorly adapted to uncertain conditions and dynamic risks. Decisions on routing, tug allocation, or approach times are often made on the basis of static norms, without proper consideration of variable factors such as weather conditions, navigation environment, vessel condition, or port congestion.

In this case, even minor failures in one link of the chain, such as a tugboat delay or a change in wind load, can lead to cascading effects and serious accidents. Risks associated with mooring are often not accurately predictable, so planning must take into account scenario variability and support a risk-based approach.

Existing models usually either do not take into account the interrelationships between factors (e.g., between reduced visibility and tug delay) or do not have mechanisms for prompt reassessment of risk in the face of changing environmental parameters.

This creates a gap between regulatory planning and the actual conditions of mooring operations. Thus, there is a need for a formalized model that will allow to quantify the level of risk in real time, take into account uncertainty scenarios, and reasonably adapt planning to changed conditions based on predefined safety criteria.

2.3 MATERIALS AND METHODS

In this study, a risk-informed methodology was developed to support the process of planning mooring operations under conditions of operational uncertainty. The main idea of the approach is to build a formalized model that allows not only to quantify the level of risk in real time but also to adapt decision-making in accordance with changes in the environment and operational parameters. The proposed methodology combines elements of dynamic monitoring, scenario analysis, and the concept of risk management based on critical thresholds.

2.4 MODEL VARIABLES AND PARAMETERS

The model is based on three integrated indicators that are crucial for assessing risk during mooring:

$S(t)$ – integral risk index at time t, which is a function of the cumulative impact of partial risks;
$\delta(t)$ – deviation that may be associated with deformations of structural elements, including control systems, fasteners, or towing equipment;
$\tau r(t)$ – resonance time during which the object is in the danger zone of frequency coincidence of external and internal vibrations;
Scrit – critical risk value, the excess of which indicates the need to immediately change the operation plan.

2.5 FORMALIZATION OF INTEGRAL RISK

The calculation of the integral risk is based on the weighted sum of partial risks, each of which reflects a specific component of the impact – technical, navigational, human, or hydrometeorological. The mathematical expression is as follows:

$$S(t) = \sum_{i=1}^{n} w_i \cdot r_i(t), \tag{2.1}$$

where w_i are weighting factors set by experts or based on historical data, and $r_i(t)$ are partial risks modeled as a function of time and environmental variables. The value of the weights can be adjusted according to the priority of each of the factors in a particular port or during a special type of shipping.

2.6 ALGORITHM OF THE MODEL FUNCTIONING

The model functioning involves the following sequence of steps:

1. Initialization of input data – collection of data on the current state of the system: vessel characteristics, weather conditions, type of operation, load level, etc.
2. Estimation of partial risks $r_i(t)$ – calculation based on empirical functions, trends of previous incidents, or using expert tables.
3. Aggregation of results – combining partial risks into a single indicator $S(t)$, taking into account the weighting factors.
4. Comparison with threshold levels – determining the degree of permissibility of the operation:
   if $S(t) \leq S$add, then the operation is allowed;
   if $S(t) > S$crit, an alternative strategy is activated or the operation is suspended.

If the safe interval of the resonance time $\tau r(t)$ is exceeded, the system initiates a recalculation of the risk taking into account new external data.

2.7 VISUALIZATION AND VALIDATION OF RESULTS

For the practical implementation of the model, we used the Python environment with the appropriate visualization libraries (Matplotlib, Numpy). A series of graphs were built (Figures 2.1–2.3), which demonstrate

- a decrease in the risk index in the process of the system's transition to a stable state;
- change in deviation depending on the time of operation;
- time intervals during which the system is in the resonance zone.

Additionally, a criterion comparison with the classical risk matrix was implemented to justify the risk categories on the "probability × consequences" scale. The model was validated on the basis of statistical data on accidents in the ports of Northern Europe, as well as through an expert survey of marine pilots.

2.8 RESULTS

Within the framework of numerical modeling, a simulation analysis of the dynamics of key risk parameters that determine the reliability of the maritime network under conditions of uncertainty is carried out. The main

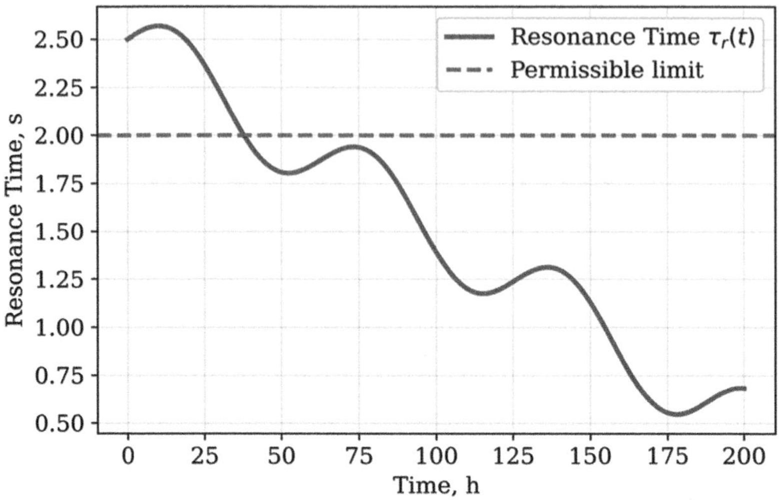

Figure 2.1 Risk index dynamics over time.

attention is paid to three components: the integral risk index $S(t)$, the transverse deviation parameter $\delta(t)$, and the time indicator of the system's stay in the resonance zone $\tau r(t)$. The models are built using exponential and harmonic functions to take into account the influence of environmental fluctuations (hydrometeorological conditions, equipment disturbances, etc.).

Figure 2.1 shows the change in the integral risk for three scenarios: baseline (Scenario 1), moderately destabilized (Scenario 2), and highly unstable (Scenario 3). In all cases, there is a decrease in the value of $S(t)$ over time, which indicates the natural stabilization of the system after the initial impact of external factors. However, for Scenario 3, the risk level remains in the zone above the critical value of 0.3 for a long time, which requires prompt intervention. Scenario 1 demonstrates the fastest decline in the risk index, which indicates high efficiency of management measures or stable operating conditions. However, it also has significant fluctuations, which may indicate sensitivity to short-term external factors. Scenario 2, on the other hand, is smoother: the risk is reduced gradually and the level of fluctuations is lower, making it more predictable. Scenario 3 is characterized by the slowest decline in the risk index, and even after a long time, the system is closer to the critical threshold. This indicates a weaker effectiveness of the measures taken or unfavorable operating conditions. Thus, curve analysis allows assessing the system's resilience to risks and identifying scenarios that require enhanced monitoring (Table 2.1).

Table 2.1 Comparison of scenarios by risk reduction efficiency

Scenario	*Characteristics*	*Strengths*	*Weaknesses*
1	Rapid decrease with high fluctuations	High efficiency of risk management	Sensitivity to short-term external factors
2	Moderate decrease, more stable dynamics	Predictability, lower fluctuations	Slower transition to the safe zone
3	Slow decrease, proximity to critical threshold	Long-term stability above 0.3	Risk of remaining in the danger zone

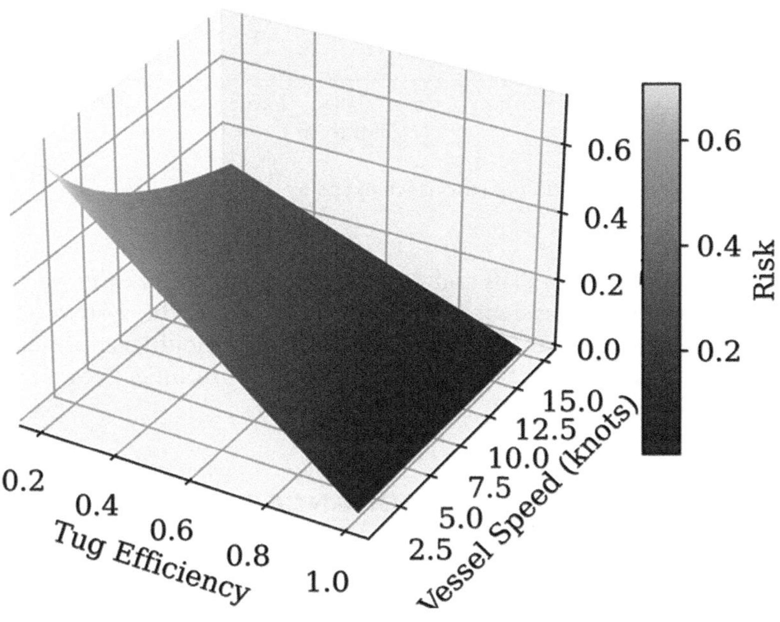

Figure 2.2 Transverse displacement dynamics.

2.9 DEVIATION $\delta(t)$

Figure 2.2 demonstrates the change in the deviation $\delta(t)$ over time, which is conventionally interpreted as the transverse displacement of a ship's hull or other structural element under the influence of cyclic external loads. The applied model based on a sinusoidal signal adequately reflects the behavior during changes in operating conditions. It is noticeable that the amplitude is approaching a critical level (0.3 mm), which may indicate a decrease in geometric stability and a potential loss of technical resource efficiency.

The graph shows the loading history curve of the transversal displacement of the ship's hull or any structural element under cyclic loads. Displacement

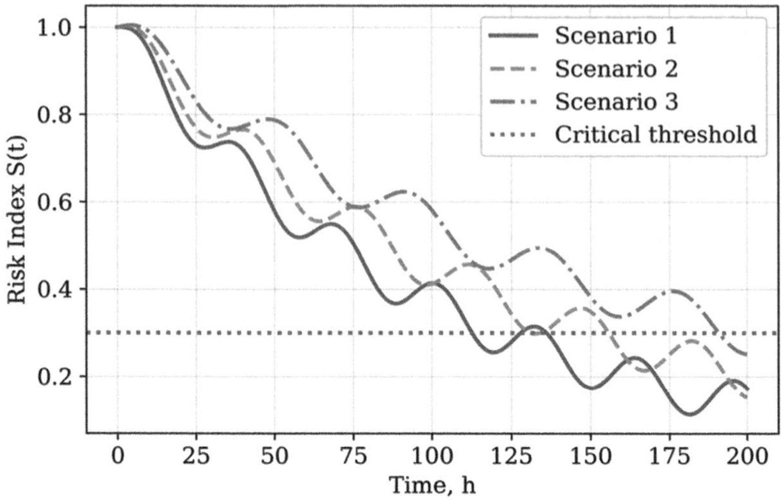

Figure 2.3 Temporal evolution of resonance exposure.

values measure between 0.10 and 0.30 mm, which indicates a move toward the critical level. The system is still stable; however, the system actually being at its maximum amplitude equal to the threshold level presents the risk of losing geometric stability and pressure on technical resource efficiency under prolonged loads or very heavy loads.

2.10 TIME SPENT IN THE RESONANCE ZONE $\tau r(t)$

Figure 2.3 shows the change in the time the system stays in the resonant vibration zone. For mooring operations, this is critical, as prolonged exposure to frequencies close to the natural frequency can cause excessive vibrations and damage to mooring lines or berth structures. The initial exceeding of the safe threshold of 2 seconds indicates a high risk, but the duration of the resonance impact decreases further. At the same time, periodic peaks signal the presence of a cyclic threat associated with repeated hydrometeorological disturbances. This requires predicting such effects and promptly adjusting mooring regimes to ensure safety.

2.11 COMPARISON WITH THE CLASSIC RISK MATRIX

To verify the adequacy of the results, a risk matrix (Figure 2.4), which is a classic tool of IMO and ISO standards, was built. The comparison shows that the proposed dynamic approach not only allows to take into account

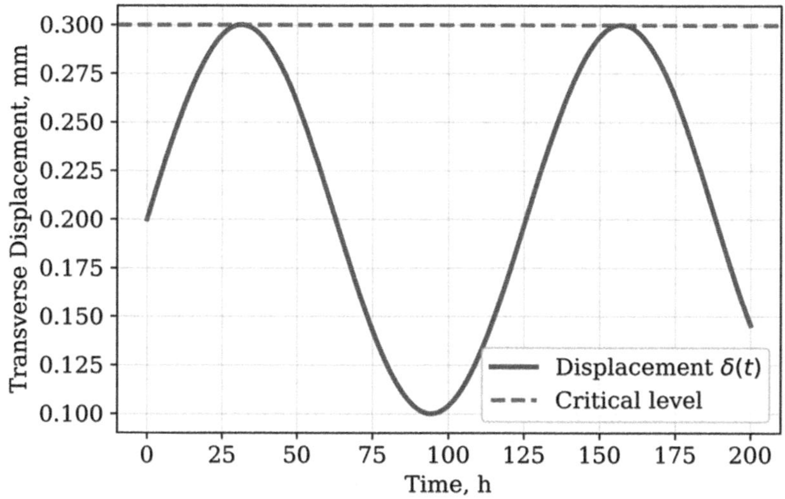

Figure 2.4 Integrated risk vs tug efficiency and vessel speed.

Table 2.2 Dependence of risk level on vessel speed and tug efficiency

Vessel speed (knots)	*Tug efficiency (0–1)*	*Risk level (approx.)*
Low (5–7)	High (0.8–1.0)	0.05–0.20 (very low)
Medium (10–15)	Medium (0.5–0.7)	0.30–0.50 (moderate)
High (18–20)	Low (0.1–0.3)	0.70–0.85 (high)

the real context and changes in the system but also provides greater flexibility in decision-making compared to discrete risk categorization.

The graph shows the dependence of the risk level on the vessel's speed and the efficiency of towing support. It is observed that an increase in speed with reduced tug efficiency significantly increases the risk (up to 0.8 and above), while with high efficiency, even at medium speeds, the risk is kept low. This indicates the critical role of tug support in risk management, especially in the context of complex maneuvers in port waters (Table 2.2).

2.12 DISCUSSION

The obtained results confirm the effectiveness of applying a dynamic risk assessment model in mooring operations under conditions of uncertainty. In contrast to classical approaches, such as static probability × consequence matrices, the proposed model allows not only to identify the moment when critical thresholds are exceeded but also to predict further developments in real time.

The integral risk index demonstrates stabilizing trends in scenarios with moderate or high management efficiency. At the same time, scenarios with high volatility require adaptive planning, including changes in strategy or automated attraction of reserve funds. The deviation is a sensitive indicator to changes in the environment and mechanical load. Its limit values can serve as triggers to prevent technical failures. The resonance time is especially relevant in the context of moorings in difficult hydrometeorological conditions, when structural elements may be overloaded due to harmonic oscillations. Detection of cyclic peaks allows for advanced optimization of operating modes.

Thus, the integration of dynamic analysis into decision support systems (DSSs) for port services significantly improves planning accuracy, reduces the likelihood of incidents, and ensures the sustainability of the maritime logistics network even in a changing environment.

2.13 CONCLUSIONS

This chapter proposes an approach to assessing the reliability of the maritime transport network based on risk-informed planning with consideration of operational uncertainties. The model combines the temporal dynamics of the integrated risk index, thermal impact, and resonant load, which allows for a more complete characterization of the change in the system state in real time. The modeling demonstrated that even under sinusoidal and exponential disturbances, early detection of exceeding critical risk thresholds is possible. Particularly dangerous are scenarios with high instability, where the risk remains in the critical zone for a long period of time. The resonance time analysis indicates cyclic periods of potential danger that are not taken into account by classical probability and consequence matrices. Comparative analysis with traditional approaches (IMO, ISO) proves the advantage of the dynamic model, which allows not only to assess the risk at the moment but also to predict its evolution, which is critical for planning adaptive management strategies.

The results obtained can be used to improve decision-making algorithms in port process management and safe operations planning under uncertainty, where further research could focus on the integration of stochastic models and multi-level risk scenarios.

REFERENCES

1. Kilic, K. I., Maity, S., Sung, I., & Nielsen, P. (2025). Challenges and AI-driven solutions in maritime search and rescue planning: A comprehensive literature review. *Marine Policy*, 178, 106692. https://doi.org/10.1016/j.marpol.2025.106692

2. Yang, L., Liu, J., Zhou, Q., Liu, Z., Chen, Y., Wang, Y., & Liu, Y. (2025). Enabling autonomous navigation: Adaptive multi-source risk quantification in maritime transportation. *Reliability Engineering & System Safety*, 261, 111118. https://doi.org/10.1016/j.ress.2025.111118
3. Li, Y., Yuen, K. F., & Zhou, Y. (2024). Risk assessment of achieving greenhouse gas emission reduction target in the maritime industry. *Transport Policy*, 155, 29–46. https://doi.org/10.1016/j.tranpol.2024.06.001
4. Lin, Y., Li, X., & Yuen, K. F. (2025). Machine learning applications for risk assessment in maritime transport: Current status and future directions. *Engineering Applications of Artificial Intelligence*, 155, 110959. https://doi.org/10.1016/j.engappai.2025.110959
5. Hausken, K., Welburn, J. W., & Zhuang, J. (2025). A review of game theory and risk and reliability analysis in infrastructures and networks. *Reliability Engineering & System Safety*, 261, 111123. https://doi.org/10.1016/j.ress.2025.111123
6. Elgharbi, S. E., Iturralde, M., Dupuis, Y., & Gaugue, A. (2025). Maritime monitoring through LoRaWAN: Resilient decentralised mesh networks for enhanced data transmission. *Computer Communications*, 241, 108276. https://doi.org/10.1016/j.comcom.2025.108276
7. He, Q., Liu, W., Liu, T., & Tian, Q. (2025). Robust coordinated path planning for unmanned aerial vehicles and unmanned surface vehicles in maritime monitoring with travel time uncertainty. *Transportation Research Part B: Methodological*, 199, 103284. https://doi.org/10.1016/j.trb.2025.103284
8. Deng, W., Ma, X., & Qiao, W. (2024). A novel methodology to quantify the impact of safety barriers on maritime operational risk based on a probabilistic network. *Reliability Engineering & System Safety*, 243, 109884. https://doi.org/10.1016/j.ress.2023.109884
9. Waskito, D. H., Kurt, R. E., Gusti, A. P., & Bowo, L. P. (2025). Assessing the risk of transporting battery electric vehicles through water transportation modes by integrating system theory and Bayesian Network approach. *Ocean Engineering*, 341, 122606. https://doi.org/10.1016/j.oceaneng.2025.122606
10. Dong, G., Zhang, J., & Dai, M. (2025). Evaluating the resilience impacts of emission reduction policies in green liner shipping network along the New Maritime Silk Road. *Regional Studies in Marine Science*, 86, 104209. https://doi.org/10.1016/j.rsma.2025.104209
11. Liu, Z., Deng, J., Shu, Y., Gan, L., Song, L., Li, H., & Yang, Z. (2025). Spatiotemporal prediction of offshore wind fields based on a hybrid deep learning model for maritime navigation. *Ocean & Coastal Management*, 269, 107841. https://doi.org/10.1016/j.ocecoaman.2025.107841
12. Liu, P., Dong, X., & Wang, P. (2025). An enhanced fuzzy cognitive map for human risk assessment in maritime transportation: Integrating causal mining and expert elicitation. *Advanced Engineering Informatics*, 68, 103624. https://doi.org/10.1016/j.aei.2025.103624
13. Chen, P., Zhang, A., Zhang, S., Dong, T., Zeng, X., Chen, S., Shi, P., Wong, Y. D., & Zhou, Q. (2025). Maritime near-miss prediction framework and model interpretation analysis method based on transformer neural network model with multi-task classification variables. *Reliability Engineering & System Safety*, 257, 110845. https://doi.org/10.1016/j.ress.2025.110845

14. Ay, C. (2025). Human reliability analysis in inert gas operations with fuzzy CREAM-based Bayesian networks. *Ocean Engineering*, 339, 122196. https://doi.org/10.1016/j.oceaneng.2025.122196
15. Animah, I. (2024). Application of Bayesian network in the maritime industry: Comprehensive literature review. *Ocean Engineering*, 302, 117610. https://doi.org/10.1016/j.oceaneng.2024.117610
16. Maidana, R. G., Kristensen, S. D., Utne, I. B., & Sørensen, A. J. (2023). Risk-based path planning for preventing collisions and groundings of maritime autonomous surface ships. *Ocean Engineering*, 290, 116417. https://doi.org/10.1016/j.oceaneng.2023.116417
17. Kong, D., Lin, Z., Li, W., & He, W. (2024). Development of an improved Bayesian network method for maritime accident safety assessment based on multiscale scenario analysis theory. *Reliability Engineering & System Safety*, 251, 110344. https://doi.org/10.1016/j.ress.2024.110344
18. Bokau, J. R. K., Samad, R., Park, Y., & Kim, D. (2025). From managing risk to reality: A case of maritime safety in Makassar Port, Indonesia using FRAM and AIS data analysis. *Ocean Engineering*, 339, 122012. https://doi.org/10.1016/j.oceaneng.2025.122012
19. Ledesma, O., & Lamo, P. (2025). Hybrid IoT network for real-time monitoring of maritime containers. *Computer Networks*, 271, 111627. https://doi.org/10.1016/j.comnet.2025.111627
20. Xu, Y., Peng, P., Claramunt, C., Qian, J., & Lu, F. (2025). Resilience of the global maritime LNG network: From static Bow-tie to dynamic SIR modeling. *Reliability Engineering & System Safety*, 265, 111605. https://doi.org/10.1016/j.ress.2025.111605
21. Fan, H., Jia, H., He, X., & Lyu, J. (2024). Navigating uncertainty: A dynamic Bayesian network-based risk assessment framework for maritime trade routes. *Reliability Engineering & System Safety*, 250, 110311. https://doi.org/10.1016/j.ress.2024.110311
22. Dong, Y., Ren, H., Tao, R., Sun, J., & Zhou, Y. (2025). Joint multi-objective optimization of maritime SAR equipment deployment using high-risk area recognition and SA-APSO. *Computers & Industrial Engineering*, 208, 111399. https://doi.org/10.1016/j.cie.2025.111399
23. Cooper-Baldock, Z., Turnock, S. R., & Sammut, K. (2025). Wake-informed 3D path planning for Autonomous Underwater Vehicles using A∗ and neural network approximations. *Ocean Engineering*, 332, 121353. https://doi.org/10.1016/j.oceaneng.2025.121353
24. Wang, W., Zhao, J., Chen, Y., Shao, P., & Jia, P. (2025). Process safety enhancement in maritime operations: A Bayesian network-based risk assessment framework for the RCEP region. *Process Safety and Environmental Protection*, 194, 1235–1256. https://doi.org/10.1016/j.psep.2024.12.072
25. Onyshchenko, S., & Melnyk, O. (2021). Probabilistic assessment method of hydrometeorological conditions and their impact on the efficiency of ship operation. *Journal of Engineering Science and Technology Review*, 14(6), 132–136. https://doi.org/10.25103/jestr.146.15
26. Melnyk, O., Onyshchenko, S., & Koryakin, K. (2021). Nature and origin of major security concerns and potential threats to the shipping industry. *Scientific Journal of Silesian University of Technology. Series Transport*, 113, 145–153. https://doi.org/10.20858/SJSUTST.2021.113.11

27. Melnyk, O., Bychkovsky, Y., Onishchenko, O., Onyshchenko, S., & Volianska, Y. (2023). Development the method of shipboard operations risk assessment quality evaluation based on experts review. In *Studies in Systems, Decision and Control* (Vol. 481, pp. 695–710). Springer. https://doi.org/10.1007/978-3-031-35088-7_40
28. Koskina, Y., Onyshenko, S., Drozhzhyn, O., & Melnyk, O. (2023). Efficiency of tramp fleet operating under the contracts of affreightment. *Scientific Journal of Silesian University of Technology. Series Transport*, 120, 137–149. https://doi.org/10.20858/sjsutst.2023.120.9
29. Shumylo, O., Yarovenko, V., Malaksiano, M., & Melnyk, O. (2023). Comprehensive assessment of hull geometry influence of a modernized ship on maneuvering performance and propulsion system parameters. *Pomorstvo*, 37(2), 314–325. https://doi.org/10.31217/p.37.2.13
30. Onishchenko, O., Bukaros, A., Melnyk, O., Yarovenko, V., Voloshyn, A., & Lohinov, O. (2023). Ship refrigeration system operating cycle efficiency assessment and identification of ways to reduce energy consumption of maritime transport. In *Studies in Systems, Decision and Control* (Vol. 481, pp. 641–652). Springer. https://doi.org/10.1007/978-3-031-35088-7_36
31. Zablotsky, Y. V., & Sagin, S. V. (2016). Enhancing fuel efficiency and environmental specifications of a marine diesel when using fuel additives. *Indian Journal of Science and Technology*, 9(46), 107516. https://doi.org/10.17485/ijst/2016/v9i46/107516
32. Sagin, S., Madey, V., Sagin, A., Stoliaryk, T., Fomin, O., & Kučera, P. (2022). Ensuring reliable and safe operation of trunk diesel engines of marine transport vessels. *Journal of Marine Science and Engineering*, 10(10), 1373. https://doi.org/10.3390/jmse10101373
33. Malaksiano, N. A. (2001). Exact inclusions of Gehring classes in Muckenhoupt classes. *Mathematical Notes*, 70(5–6), 673–681. https://doi.org/10.1023/a:1012983028054
34. Lapkina, I. O., Malaksiano, M. O., & Malaksiano, M. O. (2016). Optimization of the structure of sea port equipment fleet under unbalanced load. *Actual Problems of Economics*, 183(9), 364–371.
35. Chachra, A., Kumar, A., & Ram, M. (2024). A Markovian approach to reliability estimation of series-parallel system with Fermatean fuzzy sets. *Computers and Industrial Engineering*, 190, 110081. https://doi.org/10.1016/j.cie.2024.110081
36. Singh, K. U., Kumar, A., Singh, T., & Ram, M. (2022). Image-based decision making for reliable and proper diagnosing in NIFTI format using watermarking. *Multimedia Tools and Applications*, 81(27), 39577–39603. https://doi.org/10.1007/s11042-022-12192-9
37. Kumar, A., Bisht, S., Goyal, N., & Ram, M. (2021). Fuzzy reliability based on hesitant and dual hesitant fuzzy set evaluation. *International Journal of Mathematical, Engineering and Management Sciences*, 6(1), 166–179. https://doi.org/10.33889/IJMEMS.2021.6.1.010
38. Kormych, B., Malyarenko, T., & Wittke, C. (2023). Rescaling the legal dimensions of grey zones: Evidence from Ukraine. *Global Policy*, 14(3), 516–530. https://doi.org/10.1111/1758-5899.13233

39. Kormych, B., Averochkina, T., Demin, S., & Savych, O. (2019). Management in the field of innovative and import-substituting cluster in the modern Asian economy. *International Journal of Management and Business Research*, 9(4), 204–214.
40. Faizatama, A. M., Firdaus, N., Adiputra, R., Prabowo, A. R., Ghanbari-Ghazijahani, T., Fazeres-Ferradosa, T., Melnyk, O., & Salamena, G. G. (2025). Hydrodynamic impact of cold-water pipes and mooring systems on KVLCC2 for floating OTEC platforms. *Ocean Engineering*, 341, 122491. https://doi.org/10.1016/j.oceaneng.2025.122491
41. Zăgan, R., Chiţu, M.-G., & Manea, E. (2014). Ship manoeuvrability prediction using neural networks analysis. *Advanced Materials Research*, 1036, 946–951. https://doi.org/10.4028/www.scientific.net/AMR.1036.946
42. Popovici, D. M., Gerval, J. P., Hamza-Lup, F., Querrec, R., Polceanu, M., Popovici, N., & Zăgan, R. (2009). 3D virtual spaces supporting engineering learning activities. *International Journal of Computers, Communications and Control*, 4(4), 401–414. https://doi.org/10.15837/ijccc.2009.4.2456
43. Petculescu, P., Zăgan, R., & Prodan, G. (2006). Ultrasonic C-scan imaging for the bone sample. *Journal of Optoelectronics and Advanced Materials*, 8(1), 225–229.
44. Alwan, F. H. A., Prabowo, A. R., Muttaqie, T., Muhayat, N., Ridwan, R., & Laksono, F. B. (2022). Assessment of ballistic impact damage on aluminum and magnesium alloys against high velocity bullets by dynamic FE simulations. *Journal of the Mechanical Behavior of Materials*, 31(1), 595–616. https://doi.org/10.1515/jmbm-2022-0064
45. Prabowo, A. R., Bae, D. M., Sohn, J. M., Zakki, A. F., Cao, B., & Wang, Q. (2017). Analysis of structural behavior during collision event accounting for bow and side structure interaction. *Theoretical and Applied Mechanics Letters*, 7(1), 6–12. https://doi.org/10.1016/j.taml.2016.12.001
46. Prabowoputra, D. M., Hadi, S., Prabowo, A. R., & Sohn, J. M. (2020). Performance investigation of the Savonius horizontal water turbine accounting for stage rotor design. *International Journal of Mechanical Engineering and Robotics Research*, 9(2), 184–189. https://doi.org/10.18178/ijmerr.9.2.184-189

Chapter 3

Plyboard manufacturing plant with perfect switchover device

Reliability and availability analysis

Seema Saini, Arvind Kumar Lal, Sunil Kumar, Rajendra Kumar Sharma, and Jai Singh

3.1 INTRODUCTION

Reliability and availability (RAA) analysis leads to the betterment of the output that a manufacturer gets from his industry. This can further lead to significant profit in this age of technology, competition, and complexity. It also plays a significant role to get maximum output from the system. If we use standby component we can achieve maximum possible reliability. For this purpose, we should have the knowledge of the sensitivity of components in terms of their failure. By gaining the knowledge of such components, we can mend the system or a particular component. The management is also, in general, interested in the long run availability (LRA) of the system in order to have maximum profit. These parameters are often calculated for an industrial process that not only saves considerable efforts of maintenance personnel but also leads to better planning, better resource availability, and effective follow-up maintenance activities.

Mathematical analysis of the industrial processes might be helpful to understand their operational status. Several authors have studied the performance parameters of process industries and then carried out their behavioral analysis. Goel and Singh (1995) have done pioneering work in this field. They discussed the improvement in the reliability by introducing redundancy at various systems. Since then, several authors addressed the problem of reliability and LRA of process industries at various levels. In-depth discussions of the RAA analysis and system downtime involving repairable components are provided by Kiureghain et al. (2007). Dhillon et al. (1983) investigated power systems in the environment with fluctuation. The main focus of Singh (1984, 1989a, 1989b) was on various applications of reliability analysis (RA) on facilities that produce biogas, fertilizer, and sugar. Researchers have looked into availability analysis (AA) in the paper industries and paper manufacturing system with different repair policies (Kumar et al. 1989). Goel et al. (1995) covered the AA of dairy plants and the AA of a sophisticated standby (SB) system with an

DOI: 10.1201/9781003787525-3

imperfect switch (IS). The author also extended AA to the sugar industry's warm SB and SB switching failure (SF) system. Mahajan et al. (1999) discussed about the RA of the plant that makes utensils. Nonetheless, Singh (1984) used RA to analyze behaviors in the readymade garment industry Wang et al. (2006) investigated a comparative RAA analysis of four systems with warm SB SF and SB components. Kuo et al. (2009) conducted an AA between three systems with general repair rates (RR), reboot delays, and SF. Recently, Shakuuntala et al. (2011) studied the RA of the polytube industry using the supplementary variable technique. Garg et al. (2012) looked into the behavioral analysis in a paper mill using a Weibull fuzzy distribution. On the other hand, Tsai et al. (2004) highlighted the availability of multi-component system based on preventive maintenance. Misra et al. (2011) investigated the distribution of SB redundancy in parallel and series systems.

Most of these authors developed their models which govern a system of linear differential equations (SLDEs) and generally use analytical methods – Laplace transform and/or Lagrange's method to determine reliability of the system (ROTS). It is convenient and easy to apply Laplace transform method provided the systems under consideration are simple (i.e. having less number of subsystems (SS)) but problem becomes more complicated for complex systems (i.e. having several SS). We are unable to determine in this case the inverse Laplace transform of higher degree polynomials without which ROTS cannot be found. Even Lagrange's method may not be helpful for such a problem. In order to overcome this difficulty, we used numerical technique to calculate reliability of the complex systems.

Many authors discussed the behavior analysis of various manufacturing plants such as sugar industry, fertilizer industry, utensil industry, etc. However, certain manufacturing plants such as steel, plyboard, and woolen knitting have not been satisfactorily tackled for their behavioral analysis. In present study, we considered the plyboard manufacturing plant (PBMP) for its behavior analysis. After obtaining a system of linear differential equations (SLDEs) of PBMP using mnemonic rule, we applied Runge–Kutta fourth-order method to solve these equations in the same manner earlier discussed by Gupta et al. (2005, 2007). Lal et al. (2013) have also used the same numerical approach to discuss the behavioral analysis of piston manufacturing plant.

Section 3.1 presents introduction about the problem. A detailed description about PBMP and the symbols of the SS, required in the formulation of the problem, are described in Section 3.2. Assumptions for this model are also discussed in this section. The governing Chapman–Kolmogorov equations formulated to find out the reliability, LRA, and mean time before failure (MTBF) are discussed in Section 3.3. In Section 3.4, numerical results are presented for transient state as well as steady state. Finally, some conclusions based on the present study are drawn in Section 3.5.

3.2 SYSTEM, NOTATIONS, AND ASSUMPTIONS

We briefly discuss the manufacturing process of chosen PBMP. In this plant, wood logs of poplar (*Liriodendron tulipifera*), eucalyptus (*Eucalyptus globulus*), *etc.* are peeled in the wood-peeling machine to get the wood core. Wood core is then dried in the dryer. Simultaneously, logs of woods are cut into small pieces of specified size in band saw. These pieces are dried in the sizzling chamber and then cut into defined sizes using ripsaw. These pieces are now dazzled in the planner and converted into blocks. The thickness of the board depends upon the size of these blocks. Glue from the glue spreaders is applied on the two dry wood cores at the same time and pasted on the blocks of the pieces. Plyboard is now passed through the hot press to paste the cores on the board. With the help of cutting saw, the plyboard is divided into proper sizes. Sand paper machine is used to make the surface of the plyboard uniform. This process is shown in Figure 3.1. The PBMP consists of following nine SS:

1. **Band Saw A**: It is used to cut the big logs of wood into small pieces. It consists of a blade belt and a motor. Blade belt is supported by SB. This SS is subjected to major failure (MF) only.
2. **Sizzling Chamber B**: It is used to dry small wood pieces. It consists of three fans and three motors. If any one of the motors fails, then SS goes in reduced state (RS) and fails if more than one motor fails. The SS is subjected to MF or minor failures.
3. **Ripsaw C**: It is used to cut the pieces of wood into exact size. It consists of seven diamond cutters, one roller and one motor. SB units support cutters.

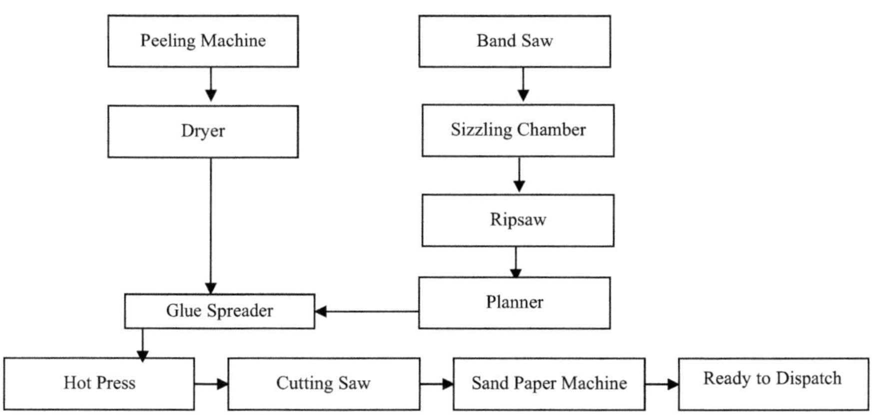

Figure 3.1 Flow chart of PBMP.

4. **Planner D:** It is used to dazzle the small pieces. It consists of four blades, one roller, and one motor. SB units here also support blades. The SS is subjected to MF only.
5. **Peeling Machine E:** It is used to get the core of wood. Big logs of wood are peeled into this system. It is also known as core veneer. It consists of one motor, six bearings, one roller, and one heavy gearbox. Based on the information given by the management, we have considered that the roller does not fail. The system is subjected to MF only.
6. **Dryer F:** It is used to dry the core. It consists of five motors, four fans, and one hydraulic pump (to increase the temperature). If one motor or one fan fails, the SS attains its RS, otherwise the SS fails. It has been taken that the hydraulic pump does never fail.
7. **Hot Press G:** It is used to paste the ply on the board. It consists of two motors, eight cylinders, one panel box, and one thermo-oil pump to rotate the oil. The SS is subjected to MF only.
8. **Cutting Saw H:** It is used to cut the plyboard in pre-defined sizes. It consists of two motors and two diamond cutters. SB units support diamond cutters. This SS is subjected to MF.
9. **Sand Paper Machine I:** It is used for dazzling the surface of the ply-board. It consists of one motor, one sandpaper belt, and one fan. SB units support sandpaper belt and the fan is considered to never fail. The SS is subjected to MF only.

All the SS in good states are symbolized as A, B, C, D, E, F, G, H, and I. $\overline{B}$ and $\overline{F}$ represent that SS B and F are working with less capacity, α_i $(i=1,2,\ldots,11)$ represent constant failure rates (FRs) of SS $A,C,D,E,G,H,I,\overline{B},\overline{F},B$ and F, respectively, β_i $(i=1,2,\ldots,11)$ represent constant RR of SS, $P_j(t)$ $(j=1,2,\ldots,34)$ represent the probability that system is in jthth state at time t and small letters a,b,c,d,e,f,g,h and i represent the failed states of the SS, A,B,C,D,E,F,G,H and I, respectively.

The mathematical formulations and further analyses of PBMP are also based on the following assumptions:

i. Failure and repair rates (FRRs) are measured in the unit per day and are independent of each other.
ii. FR of SS glue spreader is very small and we have considered that it never fails.
iii. No simultaneous failures among the SS are allowed.
iv. SS B and F can fail only through RS.
v. Repair of B or F in RS does not make these SS fully operational.
vi. Repaired components are like new components.
vii. Switching to SB is perfect.

In the light of the assumptions and symbols, the transition diagram of the PBMP is shown in Figure 3.2.

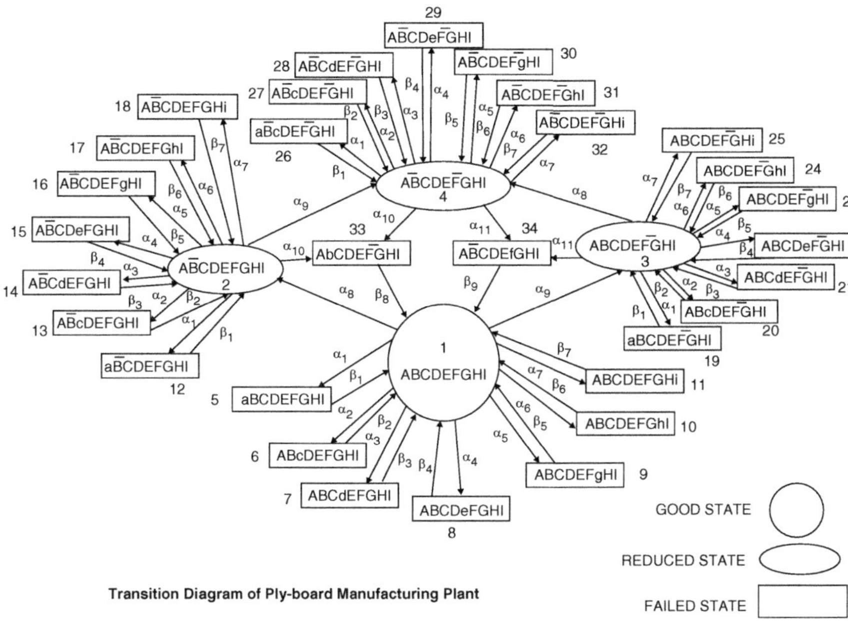

Figure 3.2 Transition of diagram of PBMP.

3.3 MATHEMATICAL MODEL OF THE PROBLEM

To determine RAA of a PBMP, mathematical formulation of the problem using transition diagram (Figure 3.2) is carried out following Gupta et al. (2005) for nine SS. For this plant, we will discuss the formulations of the problem for both transient and steady states, separately.

3.3.1 Transient state

Keeping in view the probability considerations of each state, we obtain the following first-order differential equations governing the PBMP at time $(t+\Delta t)$.

$$
\begin{aligned}
P_1(t+\Delta t) = {} & \left[1-(\alpha_1+\alpha_2+\alpha_3+\alpha_4+\alpha_5+\alpha_6+\alpha_7+\alpha_8+\alpha_9)\Delta t\right]P_1(t) \\
& +\big(\beta_1P_5(t)+\beta_2P_6(t)+\beta_3P_7(t)+\beta_4P_8(t)+\beta_5P_9(t)+\beta_6P_{10}(t) \\
& +\beta_7P_{11}(t)+\beta_8P_{33}(t)+\beta_9P_{34}(t))\Delta t\big)
\end{aligned}
$$

$$
\begin{aligned}
\frac{P_1(t+\Delta t)-P_1(t)}{\Delta t} = {} & \left[-(\alpha_1+\alpha_2+\alpha_3+\alpha_4+\alpha_5+\alpha_6+\alpha_7+\alpha_8+\alpha_9)\right]P_1(t) \\
& +\big(\beta_1P_5(t)+\beta_2P_6(t)+\beta_3P_7(t)+\beta_4P_8(t)+\beta_5P_9(t)+\beta_6P_{10}(t) \\
& +\beta_7P_{11}(t)+\beta_8P_{33}(t)+\beta_9P_{34}(t)\big)
\end{aligned}
$$

Taking $\Delta t \to 0$ on both sides, we get

$$P_1'(t) + \lambda_1 P_1(t) = \beta_1 P_5(t) + \beta_2 P_6(t) + \beta_3 P_7(t) + \beta_4 P_8(t) + \beta_5 P_9(t) + \beta_6 P_{10}(t) + \beta_7 P_{11}(t) + \beta_8 P_{33}(t) + \beta_9 P_{34}(t) \tag{3.1}$$

Similarly, for other states, we can write

$$P_2'(t) + \lambda_2 P_2(t) = \beta_1 P_{12}(t) + \beta_2 P_{13}(t) + \beta_3 P_{14}(t) + \beta_4 P_{15}(t) + \beta_5 P_{16}(t) + \beta_6 P_{17}(t) + \beta_7 P_{18}(t) + \alpha_8 P_1(t) \tag{3.2}$$

$$P_3'(t) + \lambda_3 P_3(t) = \beta_1 P_{19}(t) + \beta_2 P_{20}(t) + \beta_3 P_{21}(t) + \beta_4 P_{22}(t) + \beta_5 P_{23}(t) + \beta_6 P_{24}(t) + \beta_7 P_{25}(t) + \alpha_9 P_1(t) \tag{3.3}$$

$$P_4'(t) + \lambda_4 P_4(t) = \beta_1 P_{26}(t) + \beta_2 P_{27}(t) + \beta_3 P_{28}(t) + \beta_4 P_{29}(t) + \beta_5 P_{30}(t) + \beta_6 P_{31}(t) + \beta_7 P_{32}(t) + \alpha_8 P_3(t) + \alpha_9 P_2(t) \tag{3.4}$$

$$P_{4+i}'(t) + \beta_i P_{4+i}(t) = \alpha_i P_1(t), \quad i = 1,2,\ldots,7 \tag{3.5}$$

$$P_{11+i}'(t) + \beta_i P_{11+i}(t) = \alpha_i P_2(t), \quad i = 1,2,\ldots,7 \tag{3.6}$$

$$P_{18+i}'(t) + \beta_i P_{18+i}(t) = \alpha_i P_3(t), \quad i = 1,2,\ldots,7 \tag{3.7}$$

$$P_{25+i}'(t) + \beta_i P_{25+i}(t) = \alpha_i P_4(t), \quad i = 1,2,\ldots,7 \tag{3.8}$$

$$P_{33}'(t) + \beta_8 P_{33}(t) = \alpha_{10} P_2(t) + \alpha_{10} P_4(t) \tag{3.9}$$

$$P_{34}'(t) + \beta_9 P_{34}(t) = \alpha_{11} P_3(t) + \alpha_{11} P_4(t) \tag{3.10}$$

where

$$\lambda_1 = \alpha_1 + \alpha_2 + \alpha_3 + \alpha_4 + \alpha_5 + \alpha_6 + \alpha_7 + \alpha_8 + \alpha_9$$

$$\lambda_2 = \alpha_1 + \alpha_2 + \alpha_3 + \alpha_4 + \alpha_5 + \alpha_6 + \alpha_7 + \alpha_9 + \alpha_{10}$$

$$\lambda_3 = \alpha_1 + \alpha_2 + \alpha_3 + \alpha_4 + \alpha_5 + \alpha_6 + \alpha_7 + \alpha_8 + \alpha_{11}$$

$$\lambda_4 = \alpha_1 + \alpha_2 + \alpha_3 + \alpha_4 + \alpha_5 + \alpha_6 + \alpha_7 + \alpha_{10} + \alpha_{11}$$

with initial conditions

$$P_i(0) = \begin{cases} 1, & \text{if } 1(0) \\ 0, & \text{otherwise} \end{cases} \tag{3.11}$$

By finding the probability of each state in the system, reliability function $R(t)$ is computed by

$$R(t) = \sum_{i=1}^{4} P_i(t) \tag{3.12}$$

Once reliability at any time t is known, the MTBF is easily obtained by

$$MTBF = \int_0^{\infty} R(t)\, dt \tag{3.13}$$

3.3.2 Steady state

As management of an industry is always interested in LRA of the plant, we have also considered it. In this case, steady-state restriction must be taken into account. That is, $\left(\frac{d}{dt} \to 0 \quad as \quad t \to \infty\right)$. So, using $\mathrm{P_J}$ for $P_j(t \to \infty)$, $i = 1,2,\ldots,34$, SLDE (1–10) will be reduced in the following system of linear equations:

$$\lambda_1 P_1 = \beta_1 P_5 + \beta_2 P_6 + \beta_3 P_7 + \beta_4 P_8 + \beta_5 P_9 + \beta_6 P_{10} + \beta_7 P_{11} + \beta_8 P_{33} + \beta_9 P_{34} \tag{3.14}$$

$$\lambda_2 P_2 = \beta_1 P_{12} + \beta_2 P_{13} + \beta_3 P_{14} + \beta_4 P_{15} + \beta_5 P_{16} + \beta_6 P_{17} + \beta_7 P_{18} + \alpha_8 P_1 \tag{3.15}$$

$$\lambda_3 P_3 = \beta_1 P_{19} + \beta_2 P_{20} + \beta_3 P_{21} + \beta_4 P_{22} + \beta_5 P_{23} + \beta_6 P_{24} + \beta_7 P_{25} + \alpha_9 P_1 \tag{3.16}$$

$$\lambda_4 P_4 = \beta_1 P_{26} + \beta_2 P_{27} + \beta_3 P_{28} + \beta_4 P_{29} + \beta_5 P_{30} + \beta_6 P_{31} + \beta_7 P_{32} + \alpha_8 P_3 + \alpha_9 P_2 \tag{3.17}$$

$$\beta_i P_{4+i} = \alpha_i P_1, \quad i = 1,2,\ldots,7 \tag{3.18}$$

$$\beta_i P_{11+i} = \alpha_i P_2, \quad i = 1,2,\ldots,7 \tag{3.19}$$

$$\beta_i P_{18+i} = \alpha_i P_3, \quad i = 1,2,\ldots,7 \tag{3.20}$$

$$\beta_i P_{25+i} = \alpha_i P_4, \quad i = 1,2,\ldots,7 \tag{3.21}$$

$$\beta_8 P_{33} = \alpha_{10} P_2 + \alpha_{10} P_4 \tag{3.22}$$

$$\beta_9 P_{34} = \alpha_{11} P_3 + \alpha_{11} P_4 \tag{3.23}$$

Solving these equations recursively, we get

$$P_{4+i} = \frac{\alpha_i}{\beta_i} P_1, \quad i = 1,2,\ldots,7 \tag{3.24}$$

$$P_{11+i} = \frac{\alpha_i}{\beta_i} P_2, \quad i = 1,2,\ldots,7 \tag{3.25}$$

$$P_{18+i} = \frac{\alpha_i}{\beta_i} P_3, \quad i = 1,2,\ldots,7 \tag{3.26}$$

$$P_{25+i} = \frac{\alpha_i}{\beta_i} P_4, \quad i = 1,2,\ldots,7 \tag{3.27}$$

$$P_{33} = \frac{\alpha_{10}}{\beta_8} P_2 + \frac{\alpha_{10}}{\beta_8} P_4 \tag{3.28}$$

$$P_{34} = \frac{\alpha_{11}}{\beta_9} P_3 + \frac{\alpha_{11}}{\beta_9} P_4 \tag{3.29}$$

$$P_2 = K_1 P_1 \tag{3.30}$$

$$P_3 = K_2 P_1 \tag{3.31}$$

$$P_4 = K_3 P_1, \tag{3.32}$$

Since the sum of probability of all the states is one, that is, $\sum_{i=1}^{34} P_i = 1$, we get

$$P_1 = \left[(1 + K_1 + K_2 + K_3)\left(1 + \frac{\alpha_1}{\beta_1} + \frac{\alpha_2}{\beta_2} + \frac{\alpha_3}{\beta_3} + \frac{\alpha_4}{\beta_4} + \frac{\alpha_5}{\beta_5} + \frac{\alpha_6}{\beta_6} + \frac{\alpha_7}{\beta_7}\right) + \frac{\alpha_{10}}{\beta_8}(K_1 + K_3) + \frac{\alpha_{11}}{\beta_9}(K_2 + K_3) \right]^{-1} \tag{3.33}$$

where $K_1 = \dfrac{\alpha_8}{\alpha_9 + \alpha_{10}}$, $K_2 = \dfrac{\alpha_9}{\alpha_8 + \alpha_{11}}$, $K_3 = \dfrac{\alpha_8 K_2 + \alpha_9 K_1}{\alpha_{10} + \alpha_{11}}$

Thus, the LRA of the system denoted by$A(\infty)$ can finally be computed using

$$A(\infty) = \sum_{i=1}^{4} P_i = (1 + K_1 + K_2 + K_3) P_1 \tag{3.34}$$

3.4 BEHAVIOR ANALYSIS OF THE SYSTEM

3.4.1 Transient state

We have collected data of FRR for each SS of PBMP in the units of per day and hence computed ROTS for 360 days (i.e. 1 year) only. In order to compute ROTS, the SLDE (1–10) with initial conditions (3.11) has been solved using fourth-order Runge–Kutta method in the same way as used by Gupta et al. (2005, 2007). It is noted that the numerical integration is carried forward by assuming step size $h = 0.005$ and treating one iteration as 0.005 day. Thus, around 72,000 iterations are required to obtain the probabilities of each state for 360 days (i.e. 1 year) for various choices of FR of the SS. Finally, the ROTS using the equation (3.12) can be calculated efficiently. The computed results are finally shown in Tables 3.1a–h. The last row of each table shows the MTBF in days for the respective FR which is obtained by computing equation (3.13) using Simpson's one-third rule for 360 days only.

Table 3.1a Effect of FR of peeling machine on ROTS

α_4 *days*	*.0067*	*.0074*	*.010*	*.0102*	*.0113*
30	.976478	.976413	.976170	.976152	.976049
60	.974895	.974830	.974588	.974569	.974467
90	.974068	.974003	.973762	.973743	.973641
120	.973635	.973570	.973329	.973310	.973208
150	.973407	.973342	.973100	.973082	.972980
180	.973285	.973220	.972979	.972960	.972858
210	.973220	.973154	.972913	.972995	.972793
240	.973184	.973118	.972877	.972859	.972756
270	.973163	.973098	.972856	.972838	.972736
300	.973151	.973086	.972845	.972826	.972724
330	.973144	.973079	.972837	.972819	.972717
360	.973139	.973074	.972833	.972815	.972712
MTBF	350.832	350.811	350.727	350.724	350.685

Table 3.1b Effect of FR of hot press on ROTS

α_5 days	.0024	.0025	.0027	.0030	.0032
30	.976578	.976436	.976153	.975726	.975442
60	.974994	.974852	.974570	.974145	.973863
90	.974168	.974026	.973744	.973320	.973037
120	.973734	.973593	.973311	.972887	.972605
150	.973506	.973365	.973082	.972659	.972376
180	.973384	.973243	.972961	.972537	.972255
210	.973319	.973177	.972895	.972472	.972189
240	.973283	.973141	.972859	.972435	.972153
270	.973262	.973121	.972838	.972415	.972132
300	.973250	.973109	.972826	.972403	.972121
330	.973243	.973102	.972819	.972396	.972135
360	.973238	.973097	.972815	.972391	.972109
MTBF	350.868	350.808	350.720	350.571	350.472

Table 3.1c Effect of FR of sizzling chamber on ROTS

α_{10} days	.0041	.0048	.0055	.0067	.0083
30	.976819	.976482	.976152	.975599	.974890
60	.975552	.975053	.974570	.973778	.972791
90	.974859	.974290	.973744	.972864	.971790
120	.974477	.973878	.973111	.972403	.971311
150	.974264	.973656	.973082	.972171	.971082
180	.974144	.973524	.972961	.972053	.970971
210	.974075	.973467	.972895	.971992	.970918
240	.974034	.973428	.972859	.971960	.970892
270	.974009	.973405	.972838	.971943	.970880
300	.973993	.973391	.972826	.971934	.970873
330	.973982	.973382	.972819	.971928	.970870
360	.973975	.973377	.972815	.971925	.970868
MTBF	351.105	350.907	350.715	350.421	350.043

3.4.1.1 *Effect of FR of peeling machine (α_4) on the ROTS*

The impact of the SS peeling machine's FR on ROTS is investigated by varying its values as follows: $\alpha_4 = 0.0113, 0.0102, 0.0100, 0.0074$, and 0.0067, while other parameters have fixed values: $\alpha_1 = 0.0027$, $\alpha_2 = 0.0018$, $\alpha_3 = 0.0018$, $\alpha_5 = 0.0027$, $\alpha_6 = 0.0027$, $\alpha_7 = 0.0027$, $\alpha_8 = 0.017$, $\alpha_9 = 0.0027$, $\alpha_{10} = 0.0055$, $\alpha_{11} = 0.00054$, $\beta_1 = 0.67$, $\beta_2 = 0.83$, $\beta_3 = 0.83$, $\beta_4 = 10.20$, $\beta_5 = 0.67$, $\beta_6 = 0.67$, $\beta_7 = 0.67$, $\beta_8 = 0.67$ and $\beta_9 = 0.50$. The calculated outcomes are displayed in Table 3.1a. Days have been taken into consideration while calculating the

Table 3.1d Effect of FR of dryer on ROTS

α_{11} *days*	*.00050*	*.00052*	*.00054*	*.00057*	*.0006*
30	.976157	.976154	.976152	.976148	.976144
60	.974580	.974575	.974570	.974562	.974455
90	.973756	.973750	.973744	.973734	.973725
120	.973325	.973318	.973311	.973299	.973289
150	.973098	.973091	.973082	.973070	.973058
180	.972977	.972970	.972961	.972947	.972935
210	.972913	.972904	.972895	.972881	.972869
240	.972877	.972868	.972859	.972845	.972832
270	.972857	.972848	.972838	.972824	.972811
300	.972845	.972836	.972826	.972812	.972798
330	.972838	.972829	.972819	.972804	.972791
360	.972837	.972824	.972815	.972800	.972786
MTBF	350.725	350.723	350.721	350.715	350.709

Table 3.1e Effect of RR of peeling machine on ROTS

β_4 *days*	*8.2*	*9.2*	*10.2*	*11.2*	*12.2*
30	.975920	.976048	.976152	.976237	.976308
60	.974338	.974466	.974569	.974654	.974725
90	.973512	.973640	.973743	.973828	.973898
120	.973079	.973207	.973310	.973395	.973465
150	.972851	.972979	.973082	.973166	.973237
180	.972729	.972857	.972960	.973045	.973115
210	.972664	.972792	.972895	.972979	.973050
240	.972628	.972756	.972858	.972943	.973014
270	.972607	.972735	.972838	.972922	.972993
300	.972595	.972723	.972826	.972910	.972981
330	.972588	.972716	.972819	.972903	.972974
360	.972583	.972711	.972814	.972899	.972969
MTBF	350.639	350.684	350.720	350.749	350.776

values of α_4. When the duration is increased from 30 to 210 days, the ROTS drop by about 0.34%, and thereafter it drops gradually. Nonetheless, when the peeling machine's FR rises, it falls by about 0.05%. As the FR rises from 0.0067 to 0.0113, the MTBF in days falls by about 0.042%.

3.4.1.2 Effect of FR of hot press (α_5) on the ROTS

The following FRR combinations are used to examine the ROTS. The following parameters of hot press are taken into consideration: $\alpha_5 = 0.0032$,

Table 3.1f Effect of RR of hot press on ROTS

β_5 *days*	*.52*	*.60*	*.67*	*.80*	*1.0*
30	.975047	.975705	.976152	.976775	.977418
60	.973469	.974124	.974569	.975190	.975831
90	.972644	.973298	.973743	.974363	.975003
120	.972211	.972865	.973310	.973930	.974570
150	.971983	.972637	.973082	.973702	.974342
180	.971862	.972516	.972960	.973580	.974220
210	.971796	.972450	.972895	.973515	.974154
240	.971760	.972414	.972858	.973478	.974118
270	.971739	.972393	.972838	.973458	.974098
300	.971727	.972381	.972826	.973446	.974086
330	.971720	.972374	.972819	.973439	.974079
360	.971715	.972369	.972814	.973434	.974074
MTBF	350.335	350.564	350.720	350.945	351.161

Table 3.1g Effect of RR of sizzling chamber on ROTS

β_8 *days*	*.52*	*.60*	*.67*	*.80*	*1.0*
30	.975394	.975843	.976125	.976587	.977041
60	.973357	.974078	.974569	.975257	.975970
90	.972300	.973159	.973743	.974560	.975404
120	.971750	.972678	.973310	.974192	.975103
150	.971463	.972426	.973082	.973996	.974941
180	.971312	.972293	.972960	.973891	.974853
210	.971232	.972221	.972895	.973834	.974804
240	.971189	.972182	.972858	.973801	.974776
270	.971165	.972161	.972838	.973783	.974759
300	.971152	.972148	.972826	.973772	.974748
330	.971144	.972141	.972819	.973765	.974742
360	.971139	.972136	.972814	.973760	.974737
MTBF	350.195	350.506	350.718	351.017	351.324

0.0030, 0.0027, 0.0025, 0.0024 and other parameters have been considered as: $\alpha_1 = 0.0027$, $\alpha_2 = 0.0018$, $\alpha_3 = 0.0018$, $\alpha_4 = 0.0102$, $\alpha_6 = 0.0027$, $\alpha_7 = 0.0027$, $\alpha_8 = 0.017$, $\alpha_9 = 0.0027$, $\alpha_{10} = 0.0055$, $\alpha_{11} = 0.00054$, $\beta_1 = 0.67$, $\beta_2 = 0.83$, $\beta_3 = 0.83$, $\beta_4 = 10.20$, $\beta_5 = 0.67$, $\beta_6 = 0.67$, $\beta_7 = 0.67$, $\beta_8 = 0.67$, $\beta_9 = 0.50$. Table 3.1b displays the calculated value of ROTS using these FRR values. In this instance, extending the period from 30 to 360 days results in a roughly 0.34% decrease in ROTS. However, as the FR of hot press increases, reliability and MTBF both drop by about 0.11%.

Table 3.1h Effect of RR of dryer on ROTS

β_9 *days*	*.40*	*.44*	*.50*	*.57*	*.67*
30	.976135	.976143	.976152	.976160	.976169
60	.974537	.974551	.974569	.974586	.974603
90	.973698	.973718	.973743	.973765	.973789
120	.973255	.973280	.973310	.973337	.973366
150	.973020	.973048	.973082	.973112	.973145
180	.972893	.972923	.972960	.972993	.973029
210	.972823	.972856	.972895	.972930	.972967
240	.972784	.972818	.972858	.972895	.972934
270	.972761	.972796	.972838	.972876	.972916
300	.972747	.972783	.972826	.972865	.972906
330	.972739	.972775	.972819	.972858	.972900
360	.972733	.972770	.972814	.972854	.972896
MTBF	350.679	350.708	350.720	350.730	350.741

3.4.1.3 Effect of FR of sizzling chamber (α_{10}) on the ROTS

Using the five level of sizzling chamber as α_{10} = 0.0083, 0.0067, 0.0055, 0.0048, 0.0041 and α_1 = 0.0027, α_2 = 0.0018, α_3 = 0.0018, α_4 = 0.0102, α_5 = 0.0027, α_6 = 0.0027, α_7 = 0.0027, α_8 = 0.017, α_9 = 0.0027, α_{11} = 0.00054, β_1 = 0.67, β_2 = 0.83, β_3 = 0.83, β_4 = 10.20, β_5 = 0.67, β_6 = 0.67, β_7 = 0.67, β_8 = 0.67, β_9 = 0.50, we examine the impact of the chamber's FR on the ROTS. Table 3.1c displays the ROTS, which has been calculated for these FRR selections. According to this table, ROTS drops by roughly 0.41% as the duration increases from 30 to 360 days. As the sizzling chamber's FR increases, reliability drops by 0.24%–0.31% and MTBF drops by 0.30%.

3.4.1.4 Effect of FR of dryer (α_{11}) on the ROTS

FRR of all the SS are now kept fixed at the levels: α_1 = 0.0027, α_2 = 0.0018, α_3 = 0.0018, α_4 = 0.0102, α_5 = 0.0027, α_6 = 0.0027, α_7 = 0.0027, α_8 = 0.017, α_9 = 0.0027, α_{10} = 0.0055, β_1 = 0.67, β_2 = 0.83, β_3 = 0.83, β_4 = 10.20, β_5 = 0.67, β_6 = 0.67, β_7 = 0.67, β_8 = 0.67, β_9 = 0.50 and five levels of FR of dryer: α_{11} = 0.0006, 0.00057, 0.00054, 0.00052, and 0.00050 have been taken into account to calculate ROTS and the results are shown in Table 3.1d. This table shows that the ROTS drops by roughly 0.34% at the start with the increase from 30 to 210 days but after that rate of decrease is extremely tiny and reliability decreases from 0.001% to 0.005% with the increase of FR from 0.00050 to 0.00060. However, MTBF is decreased by approximately 0.0046% with the increase in FR of dryer.

3.4.1.5 Effect of RR of peeling machine (β_4) on the ROTS

Next, consider the effect of RR of peeling machine on the ROTS. We have used constant FRR of all the SS other than peeling machine. Following data have been used, $\beta_4 = 8.20, 9.20, 10.20, 11.20, 12.20$, and $\alpha_1 = 0.0027$, $\alpha_2 = 0.0018$, $\alpha_3 = 0.0018$, $\alpha_4 = 0.0102$, $\alpha_5 = 0.0027$, $\alpha_6 = 0.0027$, $\alpha_7 = 0.0027$, $\alpha_8 = 0.017$, $\alpha_9 = 0.0027$, $\alpha_{10} = 0.0055$, $\alpha_{11} = 0.00054$, $\beta_1 = 0.67$, $\beta_2 = 0.83$, $\beta_3 = 0.83$, $\beta_5 = 0.67$, $\beta_6 = 0.67$, $\beta_7 = 0.67$, $\beta_8 = 0.67$, $\beta_9 = 0.50$. This effect on reliability is shown in Table 3.1e. This table reveals that the ROTS decreases by approximately 0.34% with the increase in time from 30 to 360 days. However, reliability and MTBF both increase by approximately 0.04% with the increase in the RR of peeling machine from 8.20 to 12.20.

3.4.1.6 Effect of RR of hot press (β_5) on the ROTS

We have also computed the ROTS after varying the RR of hot press and taking other FRR as fixed. With the following values of various FRR: $\beta_5 = 0.52$, 0.60, 0.67, 0.80, 1.00 and $\alpha_1 = 0.0027$, $\alpha_2 = 0.0018$, $\alpha_3 = 0.0018$, $\alpha_4 = 0.0102$, $\alpha_5 = 0.0027$, $\alpha_6 = 0.0027$, $\alpha_7 = 0.0027$, $\alpha_8 = 0.017$, $\alpha_9 = 0.0027$, $\alpha_{10} = 0.0055$, $\alpha_{11} = 0.00054$, $\beta_1 = 0.67$, $\beta_2 = 0.83$, $\beta_3 = 0.83$, $\beta_4 = 10.20$, $\beta_6 = 0.67$, $\beta_7 = 0.67$, $\beta_8 = 0.67$, $\beta_9 = 0.50$ computation is carried out and the results are given in Table 3.1f. The ROTS decreases by approximately 0.34% at the start with the increase in time from 30 to 210 days but after that rate of decrease is very small. MTBF and reliability both increase by approximately 0.24% with the increase in the RR of the SS from 0.52 to 1.00.

3.4.1.7 Effect of RR of sizzling chamber (β_8) on ROTS

We consider the effect of RR of sizzling chamber on the ROTS. Following value combinations of FRR of various SS, $\alpha_1 = 0.0027$, $\alpha_2 = 0.0018$, $\alpha_3 = 0.0018$, $\alpha_4 = 0.0102$, $\alpha_5 = 0.0027$, $\alpha_6 = 0.0027$, $\alpha_7 = 0.0027$, $\alpha_8 = 0.017$, $\alpha_9 = 0.0027$, $\alpha_{10} = 0.0055$, $\alpha_{11} = 0.00054$, $\beta_1 = 0.67$, $\beta_2 = 0.83$, $\beta_3 = 0.83$, $\beta_4 = 10.2$, $\beta_5 = 0.67$, $\beta_6 = 0.67$, $\beta_7 = 0.67$, $\beta_9 = 0.5$ and five levels of $\beta_8 = 0.52$, 0.60, 0.67, 0.8, 1.00, we have computed ROTS and the values are shown in Table 3.1g. This clearly indicates that ROTS decreases by approximately 0.24% with the increase in time from 30 to 360 days and it increases by approximately 0.17%–0.37% with the increase in the RR of sizzling chamber from 0.52 to 1.00. However, MTBF increases by 0.32% with the increase in RR.

3.4.1.8 Effect of RR of dryer (β_9) on the ROTS

The following combinations of FRR of various SS have now been used to calculate the ROTS: $\alpha_1 = 0.0027$, $\alpha_2 = 0.0018$, $\alpha_3 = 0.0018$, $\alpha_4 = 0.0102$, $\alpha_5 = 0.0027$, $\alpha_6 = 0.0027$, $\alpha_7 = 0.0027$, $\alpha_8 = 0.017$, $\alpha_9 = 0.0027$, $\alpha_{10} = 0.0055$, $\alpha_{11} = 0.00054$, $\beta_5 = 0.67$, $\beta_6 = 0.67$, $\beta_7 = 0.67$, $\beta_8 = 0.67$, and $\beta_9 = 0.40$, 0.44,

0.5, 0.57, 0.67. The calculated outcomes are displayed in Table 3.1h. It is determined that dependability rises by roughly 0.004%–0.02% with an increase in the dryer's RR from 0.40 to 0.67, while ROTS falls by roughly 0.34% with an increase in time from 30 to 360 days. When the dryer's RR rises, the MTBF in days increases by 0.018%.

3.4.2 Steady state

The LRA of the plant, as defined in (3.34), has been calculated for various choices of the FRR of the SS and is tabulated in Tables 3.2a–d.

3.4.2.1 *Effect of FR of band saw and peeling machine on LRA*

Here, effect of FR of SS *A* and *E* on LRA of the system is studied by varying their values as: $\alpha_1 = 0.0025, 0.0027, 0.0030, 0.0032$, and $\alpha_4 = 0.0074, 0.010,$

Table 3.2a LRA of the system for the combination of band saw and peeling machine

α_1 α_4	*.0025*	*.0027*	*.0030*	*.0032*
.0074	.9733458	.9730631	.9726393	.9723570
.010	.9731044	.9728218	.9723980	.9721161
.0102	.9730858	.9728032	.9723797	.9720975
.0113	.9729837	.9727012	.9722777	.9719956

Table 3.2b LRA of the system for the combination of FR of sizzling chamber and dryer

α_{10} α_{11}	*.0048*	*.0055*	*.0067*	*.0083*
.00052	.9733714	.9728131	.9719279	.9708726
.00054	.9733607	.9728032	.9719192	.9708650
.00057	.9733447	.9727886	.9719062	.9708536
.00060	.9733290	.9727740	.9718933	.9708423

Table 3.2c LRA of the system for the combination of band saw and peeling machine

β_1 β_4	*.60*	*.67*	*.8*	*1*
8.2	.9721280	.9725725	.9731923	.9738320
9.2	.9722558	.9727004	.9733204	.9739603
10.2	.9723585	.9728032	.9734233	.9740634
11.2	.9724429	.9728877	.9735080	.9741481

Table 3.2d LRA of the system for the combination of RR of sizzling chamber and dryer

β_8 / β_9	.60	.67	.8	1.0
.44	.9720794	.9727572	.9737029	.9774590
.5	.9722397	.9729178	.9738637	.9776211
.57	.9722671	.9729451	.9738911	.9776487
.67	.9722962	.9729743	.9739204	.9776782

0.0102, 0.0113. The FRRs of other SS have been taken as: $\alpha_2 = 0.0018$, $\alpha_3 = 0.0018$, $\alpha_5 = 0.0027$, $\alpha_6 = 0.0027$, $\alpha_7 = 0.0027$, $\alpha_8 = 0.017$, $\alpha_9 = 0.0027$, $\alpha_{10} = 0.0055$, $\alpha_{11} = 0.00054$, $\beta_1 = 0.67$, $\beta_2 = 0.83$, $\beta_3 = 0.83$, $\beta_4 = 10.2$, $\beta_5 = 0.67$, $\beta_6 = 0.67$, $\beta_7 = 0.67$, $\beta_8 = 0.67$, $\beta_9 = 0.5$. The LRA of the system has been calculated and is given in Table 3.2a. This table exhibits 0.1% increase in FR of band saw (α_1) and has negative impact on the LRA of the plant, whereas FR of peeling machine (α_4) has negative impact of approximately 0.037%.

3.4.2.2 Effect of FR of sizzling chamber and dryer on LRA

We have fixed FR of all the SS other than sizzling chamber and dryer at the levels as: $\alpha_1 = 0.0027$, $\alpha_2 = 0.0018$, $\alpha_3 = 0.0018$, $\alpha_4 = 0.0102$, $\alpha_5 = 0.0027$, $\alpha_6 = 0.0027$, $\alpha_7 = 0.0027$, $\alpha_8 = 0.017$, $\alpha_9 = 0.0027$, $\beta_1 = 0.67$, $\beta_2 = 0.83$, $\beta_3 = 0.83$, $\beta_4 = 10.2$, $\beta_5 = 0.67$, $\beta_6 = 0.67$, $\beta_7 = 0.67$, $\beta_8 = 0.67$, $\beta_9 = 0.5$. The values of FR of sizzling chamber and dryer have been considered as: $\alpha_{10} = 0.0048$, 0.0055, 0.0067, 0.0083, and $\alpha_{11} = 0.00052$, 0.00054, 0.00057, 0.0060. Based on these values, LRA of the plant is computed and is given in Table 3.2b. This table exhibits that increase in FR of sizzling chamber (α_{10}) affects LRA by approximately 0.26% and FR of dryer (α_{11}) affects it by approximately 0.004%. As such, FR of sizzling chamber affects the LRA of plant more than the FR of dryer.

3.4.2.3 Effect of RR of band saw and peeling machine on LRA of the plant

We have also calculated the LRA of the system by varying the RR of band saw and peeling machine and computed results are shown in Table 3.2c. The RRs of band saw and peeling machine have been varied as: $\beta_1 = 0.60$, 0.67, 0.80, 1.00 and $\beta_4 = 8.20$, 9.20, 10.20, 11.20. The FRRs of other SS have been fixed as: $\alpha_1 = 0.0027$, $\alpha_2 = 0.0018$, $\alpha_3 = 0.0018$, $\alpha_4 = 0.0102$, $\alpha_5 = 0.0027$, $\alpha_6 = 0.0027$, $\alpha_7 = 0.0027$, $\alpha_8 = 0.017$, $\alpha_9 = 0.0027$, $\alpha_{10} = 0.0055$, $\alpha_{11} = 0.00054$, $\beta_2 = 0.83$, $\beta_3 = 0.83$, $\beta_5 = 0.67$, $\beta_6 = 0.67$, $\beta_7 = 0.67$, $\beta_8 = 0.67$, $\beta_9 = 0.50$. Table 3.2c reveals that increase in RR of band saw (β_1) increases

LRA of the system by approximately 0.19% and RR of peeling machine (β_4) increases it by approximately 0.032%. So, β_1 affects the LRA of the plant more as compared with the RR of peeling machine (β_4).

3.4.2.4 Effect of RR sizzling chamber and dryer on LRA

The RR of sizzling chamber and dryer is varied as: β_8 = 0.60, 0.67, 0.80, 1.00 and β_9 = 0.44, 0.50, 0.57, 0.67 keeping other parameters fixed such as: $\alpha_1 = 0.0027$, $\alpha_2 = 0.0018$, $\alpha_3 = 0.0018$, $\alpha_4 = 0.0102$, $\alpha_5 = 0.0027$, $\alpha_6 = 0.0027$, $\alpha_7 = 0.0027$, $\alpha_8 = 0.017$, $\alpha_9 = 0.0027$, $\alpha_{10} = 0.0055$, $\alpha_{11} = 0.00054$, $\beta_1 = 0.67$, $\beta_2 = 0.83$, $\beta_3 = 0.83$, $\beta_4 = 10.20$, $\beta_5 = 0.67$, $\beta_6 = 0.67$, $\beta_7 = 0.67$. Using these data, LRA of the plant is calculated and presented in Table 3.2d. We conclude from this table that increase in RR of sizzling chamber (β_8) increases LRA of the system and so does the increase in RR of melting vats, but increasing the RR of sizzling chamber affects the LRA by approximately 0.55% and increasing the RR of dryer increases it by approximately 0.022%.

3.5 CONCLUSIONS

The results of this work are based on actual FRR data for the SS collected from the relevant PBMP. These results have been analyzed, and they are compatible with what is happening in the concerned manufacturing plant. A further examination of Tables 3.1a–h and 3.2a–d demonstrates that SS, namely sizzling chamber, has the greatest impact on the reliability and LRA of the entire plant when compared to other SS. The tables show that the other SS are approximately similarly effective. As a result, management should take the best possible care of this SS in order to improve the plant's overall performance. Our suggestions to the concerned management may be useful in lowering maintenance costs and enhancing plant availability.

ACKNOWLEDGMENTS

The authors are grateful to Graphic Era Deemed to be University, Dehradun, Uttarakhand, India, for support. This chapter was written in good faith using information available at the time of publication. The views, findings, and conclusion articulated are those of the author's.

REFERENCES

Agnihotri, R. K., Khare, A. and Jain, S. (2008). Reliability analysis of a system of boiler used in readymade garment industry. *Journal of Reliability and Statistical Methods*. 1(1): 33–41.

Dhillon, B. S. and Nateson, J. (1983). Stochastic analysis of outdoors power system in fluctuating environment. *Microelectron Reliability.* 23: 867–881.

Flehinger, B. J. (1958). Reliability improvement through redundancy at various system levels. *IBM Journal of Research and Development.* 2(2): 148–158.

Garg, H., Sharma, S. P. and Rani, M. (2012). Behavioral analysis of pulping unit in a paper mill with Weibull fuzzy distribution function using ANCBLT technique. *International Journal of Applied Mathematics and Mechanics.* 8(4): 86–96.

Goel, P. and Singh, J. (1995). Reliability analysis of a standby complex system having imperfect switch over device. *Microelectron Reliability.* 35: 285–288.

Gupta, P., Lal, A. K., Sharma, R. K. and Singh, J. (2005). Numerical analysis of reliability and availability of the serial processes in butter-oil processing plant. *International Journal of Quality and Reliability Management.* 22(3): 303–316.

Gupta, P., Lal, A. K., Sharma, R. K. and Singh, J. (2007). Analysis of reliability and availability of serial processes of plastic-pipe manufacturing plant: A case study. *International Journal of Quality and Reliability Management.* 24: 404–419.

Kiureghain, A. D. and Ditlevson, O. D. (2007). Availability, reliability and downtime system with repairable components. *Reliability Engineering and System Safety.* 92(2): 66–72.

Kumar, D. and Singh, J. (1989). Availability of washing system in the paper industries. *Microelectron Reliability.* 27: 775–778.

Kuo-Hslung Wang, K. H. and Chen, Y. J. (2009). Comparative analysis of availability between three systems with general repair rates, reboot delay and switching failures. *Applied Mathematics and Computations.* 2159(1): 384–394.

Lal, A. K., Kaur, M. and Lata, S. (2013). Behavioral study of Piston manufacturing plant through stochastic models. *Journal of Industrial Engineering International.* 9: 24.

Mahajan, P. and Singh, J. (1999). Reliability of utensils manufacturing plant – A case study. *Opsearch.* 36: 260–271.

Misra, N., Misra, A. K. and Dhariyal, I. D. (2011). Standby redundancy allocations in series and parallel system. *Journal of Applied Probability.* 48: 43–55.

Shakuntala, S., Lal, A. K., Bhatia, S. S. and Singh, J. (2011). Reliability analysis of polytube industry using supplementary variable technique. *Applied Mathematics and Computations.* 218(8): 3981–3992.

Singh, J. (1984). Reliability of fertilizer production supply problem. *Proceedings of ISPTA*, Wiley Eastern Ltd., New Delhi, pp. 95–98.

Singh, J. (1989a). A warm standby redundant system with common causes failures. *Reliability Engineering and System Safety.* 26: 135–141.

Singh, J. (1989b). Reliability analysis of biogas plant having two dissimilar units. *Microelectronics Reliability.* 29: 779–781.

Tsai, Y. and Wang, L. T. (2004). A study of availability – Centered preventive maintenance for multi-component systems. *Reliability Engineering System Safety.* 84: 261–270.

Wang, K. H, Wein, L. D. and Jyh, B. K. (2006). Comparison of reliability and the availability between four systems with warm stand by components and stand by switching failures. *Applied Mathematics and Computations.* 183(2): 1310–1322.

Chapter 4

Bayesian availability analysis of repairable *k*-out-of-*n* systems using Weighted Exponential-Lindley model

Sunita Sharma

4.1 INTRODUCTION

In practical engineering, the judgment and prior knowledge of an engineer play a crucial role when dealing with the parameters of basic life distribution in Bayesian reliability/availability calculations. When the sample size is small, this becomes much more crucial. There is a significant incentive to rely on operational data about the device's components when information about the device as a whole is either expensive or unavailable. The parameter is viewed as a random variable in the Bayesian approach, and its probability density function (PDF) or probability mass function is calculated using the engineer's experience. By applying prior knowledge, the engineer may efficiently address difficult situations and minimize possible problems.

The reliability characteristics of systems must be evaluated in order to achieve dependability objectives. Availability is an important metric for a well-maintained item since it takes repair capability into account. The assessment of a repairable system's steady-state availability is crucial in this context. The evaluation of system availability has been thoroughly studied in a variety of system configurations and at various phases, from the design phase to the operational phase. Series, parallel, k-out-of-n, stand-by, multi-state, and mixed architectures are some of these setups. There are two primary ways that are frequently used to assess system availability: analytical techniques and simulation methods. Moreover, a k-out-of-n: F system only fails when k or more of the n components malfunction.

More precisely, an n-out-of-n: F system can be thought of as a parallel system configuration, and a 1-out-of-n: F system can be thought of as a series system. The k-out-of-n structure is a frequently used type of redundancy in fault-tolerance systems. A thorough investigation of applying the Bayesian technique to reliability/availability issues was provided by Tillman et al. [1]. In his own research, Mazumdar [2] also performed a comparative examination of different availability estimations.

Many times, prior knowledge about system parameters can be easily obtained from experience with similar systems, either past or present.

DOI: 10.1201/9781003787525-4

Because engineers who are familiar with certain systems can informally incorporate this knowledge, there is typically a strong incentive to use it. However, by defining some or all of the additional data as prior distributions for system parameters and using Bayesian inference techniques, a more thorough examination might be advantageous. The availability of previous data in the conditional distribution is crucial to the Bayesian technique in reliability/availability analysis. This approach represents the increasing trend of using all-encompassing probabilistic techniques to address the uncertainties in modern engineering challenges. Engineers can more robustly and intelligently handle and negotiate the obstacles presented by these uncertainties by using Bayesian approaches. The operational system availability has been the subject of numerous studies in the body of current literature.

In 1968, Brender [3] introduced the concept of Bayesian availability and predicted system reliability through Bayesian treatment. Subsequently, Brender [4] further expanded the Bayesian evaluation findings of system availability across various categories. Moreover, in their study, Hajeeh [5] provided analytical formulations for estimating the mean time to failure and steady-state availability of a system, considering both random and common cause failures. Jain and Gupta [6] obtained mathematical formulations for system reliability and availability across various configurations in their study. Also, Ke and Chu [7] conducted an analysis of the steady-state availability of a repairable system. Islam and Khan [8] utilized geometric failure and repair time distribution to estimate the Bayesian point and availability for a k-out-of-n system. Additionally, Khan and Islam [9] derived different Bayesian point estimates considering a half-normal lifetime. Madhumitha and Vijayalakshmi [10] examined Bayesian estimate of system availability for consecutive k-out-of-n: F system. Yaghoubi et al. [11] obtained a closed-form expression for the steady-state availability of a cold standby repairable k-out-of-n system. Further, Sharma and Bhutani [12] developed an efficient approach to acquire the Bayes point estimator and its corresponding confidence interval for system availability. Gonzalez-Lopez et al. [13] studied E-Bayesian estimation for reliability and availability analysis of k-out-of-m systems based on exponential distribution. Gholizadeh et al. [14] discussed Bayesian system reliability and availability estimation under a vague environment based on exponential distribution in their study.

The remainder of this chapter is organized as follows: Section 4.2 presents the model description. Section 4.3 introduces the notations and assumptions adopted throughout the study. Section 4.4 outlines the proposed methodology. Section 4.5 focuses on the Bayesian estimation of availability. Section 4.6 provides numerical illustrations, followed by the results and discussion in Section 4.7. Finally, Section 4.8 concludes the chapter and highlights future research directions.

4.2 MODEL DESCRIPTION

This section contains the description of the WXLD [15] model having the PDF,

$$f(t) = \frac{4\theta^3 t(2+\theta+t)e^{-2\theta t}}{(1+\theta)^2}; \quad t \geq 0, \ \theta \geq 0 \tag{4.1}$$

where t is a random variable following WXLD and θ is the parameter of the distribution.

Moreover, the cumulative distribution function (CDF) of the distribution is

$$F(t) = 1 - e^{-2\theta t}\left(\frac{2\theta^2 t^2}{(1+\theta)^2} + 2\theta t + 1\right), \quad t \geq 0, \ \theta \geq 0 \tag{4.2}$$

Additionally, two unknown reliability time indices of WXLD are also explored in this study, namely: reliability function (RF) and hazard rate function (HRF) at a time $t > 0$, which are provided by

$$R(t,\lambda) = e^{-2\lambda t}\left(\frac{2\lambda^2 t^2}{(1+\lambda)^2} + 2\lambda t + 1\right) \tag{4.3}$$

and

$$h(t,\lambda) = \frac{4\lambda^3 t(2+\lambda+t)}{2\lambda^2 t^2 + 2\lambda t(1+\lambda)^2 + (1+\lambda)^2} \tag{4.4}$$

The graphical representations of the PDF, CDF, HRF, and RF for various values of θ are displayed in Figure 4.1.

4.3 NOTATIONS AND ASSUMPTIONS

4.3.1 Notations

θ	Failure rate of component
μ	Repair rate of component
MTBF	Mean time between failure
MTTR	Mean time to repair
SELF	Squared error loss function
PDF	Probability density function
A_c	Steady-state availability of component

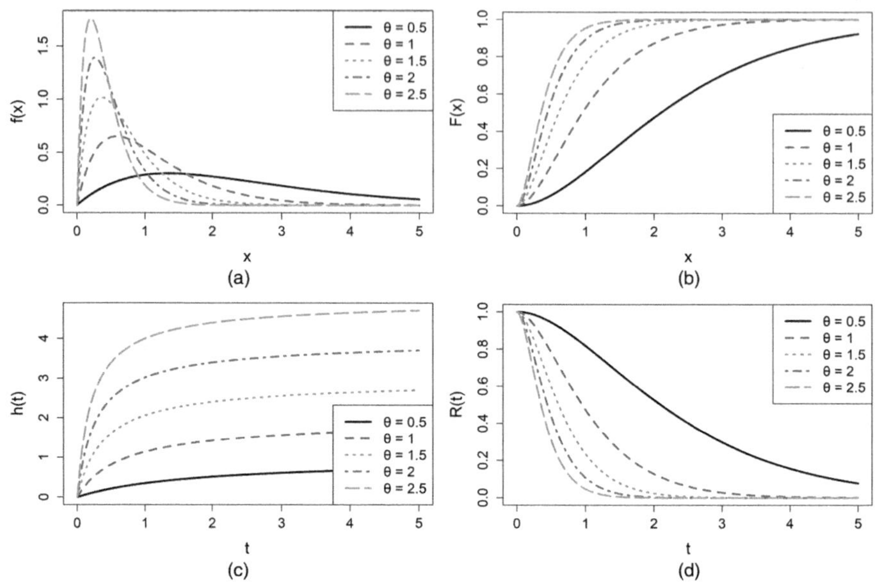

Figure 4.1 Plots of the (a) PDF, (b) CDF, (c) HRF, and (d) RF for various values of θ.

A_s	Steady-state availability of system
T	Total testing time
r	Number of failures in $(0, T)$
s	Number of repairs in $(0, T)$
A_{s1}	A_s of a series system
A_{s2}	A_s of a parallel system
θ^*	Bayes estimate of θ
μ^*	Bayes estimate of μ
A_c^*	Bayes estimate of A_c
A_s^*	Bayes estimate of A_s
A_{s1}^*	Bayes estimate of A_{s1}
A_{s2}^*	Bayes estimate of A_{s2}

4.3.2 Assumptions

1. The k-out-of-n: F system under consideration is homogeneous, where all n components are uniform and operate independently of the system.
2. At time $t = 0$, all components are in a good condition and functioning properly.
3. The component is characterized by a dichotomous state, where it can be either in a working state or a failed state.
4. The system fails only if k or more out of n components become faulty.

5. The components exhibit independence and identical distribution (i.i.d) characteristics.
6. The failure time and repair time of each component follow WXLD.
7. There is a single repairman available who immediately begins the repair process when a component fails.
8. The repair tasks are prioritized based on a first-come, first-served basis.
9. The repaired system is as good as new.

4.4 METHODOLOGY

Assume that the failure time X of each component follows WXLD with PDF specified as

$$f(x,\theta)=\frac{4\theta^3 x(2+\theta+x)e^{-2\theta x}}{(1+\theta)^2}, \quad x\geq 0,\ \theta>0 \tag{4.5}$$

The MTBF is $1/\theta$.

Similarly, the repair time Y of each component to be distributed as WXLD with PDF is,

$$f(y,\mu)=\frac{4\mu^3 y(2+\mu+y)e^{-2\mu y}}{(1+\mu)^2}, \quad y\geq 0,\ \mu>0 \tag{4.6}$$

Here, the MTTR is $1/\mu$.

Now, the steady-state component availability A_c, which represents the long-term probability of the component being available for operation, is expressed as,

$$A_c=\frac{\text{MTBF}}{\text{MTBF}+\text{MTTR}}$$

$$A_c=\frac{1/\theta}{1/\theta+1/\mu}=\frac{\mu}{\mu+\theta} \tag{4.7}$$

By employing simple probabilistic reasoning, the availability of a k-out-of-n system A_s can be determined as

$$A_s=\sum_{i=k}^{n}\binom{n}{i}(A_c)^i(1-A_c)^{n-i} \tag{4.8}$$

For a series system the availability is given as

$$A_{s1} = (A_c)^n = \left(\frac{\mu}{\mu+\theta}\right)^n \tag{4.9}$$

Similarly, by substituting $k = 1$ in equation (4.8) we obtained the parallel system availability as:

$$A_{s2} = 1-(1-A_c)^n = 1-\left(1-\frac{\mu}{\mu+\theta}\right)^n \tag{4.10}$$

For Bayesian availability analysis, the prior distribution of the failure parameter θ is assumed to be gamma (α_1, β_1) with PDF

$$g_1(\theta) = \frac{\beta_1^{\alpha_1} e^{-\theta\beta_1} \theta^{\alpha_1-1}}{\Gamma\alpha_1} \tag{4.11}$$

Similarly, the prior distribution of μ is assumed to be another gamma (α_2, β_2) given as

$$g_2(\mu) = \frac{\beta_2^{\alpha_2} e^{-\mu\beta_2} \mu^{\alpha_2-1}}{\Gamma\alpha_2} \tag{4.12}$$

Let T denote the overall testing duration. Throughout the operational process, let r represent the count of failures and s denote the count of repairs recorded within $(0, T)$. It is reasonable to assume that the random variables r and $s$are independent of each other during the course of operation.

4.4.1 Posterior analysis of system availability

For calculating the probability of observing r failures during $(0, T)$ given that the failure rate distribution for a component is WXLD with MTBF $1/\theta$, we need to calculate the conditional probability $P_1(r \mid \theta)$.

The PDF of the failure time is given in equation (4.5). Now, we can calculate the probability of observing r failures during the total testing time T as follows:

$$P_1(r \mid \theta) = \int_0^\infty f(x,\theta)\,dx = \int_0^\infty \frac{4\theta^3 x(2+\theta+x)e^{-2\theta x}}{(1+\theta)^2}\,dx \tag{4.13}$$

However, the integration involved in finding this probability can be complex and may not have a closed-form solution. In such cases, numerical methods or approximation techniques can be used. One common method is numerical integration. Further, to calculate the probability of observing s repairs during the total testing time T, given that the repair rate distribution for a component is WXLD, we can use a similar approach as before and compute the probability as

$$P_2(s \mid \mu) = \int_0^{\infty} \frac{4\mu^3 y(2+\mu+y)e^{-2\mu y}}{(1+\mu)^2} dy \tag{4.14}$$

Now, by incorporating the observed data of r failures in $(0, T)$ and updating the prior distribution given in equation (4.11), the posterior distribution of θ can be obtained as:

$$\pi_1(\theta \mid r) = \frac{P_1(r \mid \theta) g_1(\theta)}{\int_0^{\infty} P_1(r \mid \theta) g_1(\theta) d\theta} \tag{4.15}$$

$$= \frac{\int_0^{\infty} \frac{4\theta^3 x(2+\theta+x)e^{-2\theta x}}{(1+\theta)^2} \frac{\beta_1^{\alpha_1} e^{-\theta\beta_1} \theta^{\alpha_1-1}}{\Gamma\alpha_1} dx}{\int_0^{\infty}\int_0^{\infty} \frac{4\theta^3 x(2+\theta+x)e^{-2\theta x}}{(1+\theta)^2} \frac{\beta_1^{\alpha_1} e^{-\theta\beta_1} \theta^{\alpha_1-1}}{\Gamma\alpha_1} d\theta dx}$$

$$= \frac{\frac{\beta_1^{\alpha_1}}{\Gamma\alpha_1} \int_0^{\infty} \frac{4x\theta^{3+\alpha_1-1}(2+\theta+x)e^{-\theta(2x+\beta_1)}}{(1+\theta)^2} dx}{\frac{\beta_1^{\alpha_1}}{\Gamma\alpha_1} \int_0^{\infty}\int_0^{\infty} \frac{4x\theta^{3+\alpha_1-1}(2+\theta+x)e^{-\theta(2x+\beta_1)}}{(1+\theta)^2} d\theta dx}$$

The integration in equation (4.15) can be quite complex and may not have a closed-form solution. In such cases, numerical methods can be used to approximate the integral. One common approach is to use numerical integration technique.

Similarly, the posterior distribution of s repairs in $(0, T)$ with respect to prior equation (4.12) is given by:

$$\Pi_2(\mu \mid s) = \frac{P_2(s \mid \mu) g_2(\mu)}{\int_0^{\infty} P_2(s \mid \mu) g_2(\mu) d\mu} \tag{4.16}$$

$$= \frac{\int_0^\infty \frac{4\mu^3 y(2+\mu+y)e^{-2\mu y}}{(1+\mu)^2} \frac{\beta_2^{\alpha_2} e^{-\mu\beta_2} \mu^{\alpha_2-1}}{\Gamma\alpha_2} dy}{\int_0^\infty\int_0^\infty \frac{4\mu^3 y(2+\mu+y)e^{-2\mu y}}{(1+\mu)^2} \frac{\beta_2^{\alpha_2} e^{-\mu\beta_2} \mu^{\alpha_2-1}}{\Gamma\alpha_2} d\mu dy}$$

$$= \frac{\frac{\beta_2^{\alpha_2}}{\Gamma\alpha_2}\int_0^\infty \left[\frac{4y\mu^{3+\alpha_2-1}(2+\mu+y)e^{-\mu(2y+\beta_2)}}{(1+\mu)^2}\right]dy}{\frac{\beta_2^{\alpha_2}}{\Gamma\alpha_2}\int_0^\infty\int_0^\infty \left[\frac{4y\mu^{3+\alpha_2-1}(2+\mu+y)e^{-\mu(2y+\beta_2)}}{(1+\mu)^2}\right]d\mu dy}$$

Since a closed-form solution is unavailable for equation (4.16), numerical integration techniques are employed to handle the integration.

The posterior distribution of A_c given the observed values of θ and μ, which are obtained from equations (4.15) and (4.16) respectively, can be determined by applying the relationship $A_c = \frac{\mu}{\mu+\theta}$ as follows:

$$f(A_c \mid r,s) = P(A_c \mid r, s)P(r, s) \tag{4.17}$$

$$= \frac{\int_0^\infty\int_0^\infty \frac{4\theta^3 x(2+\theta+x)e^{-2\theta x}}{(1+\theta)^2} \cdot \frac{4\mu^3 y(2+\mu+y)e^{-2\mu y}}{(1+\mu)^2} I\left(A_c = \frac{\mu}{\mu+\theta}\right) dxdy^* \frac{1}{B(\alpha',\beta')} A_c^{a_1-1}(1-A_c)^{b_2-1}}{P(r,s)}$$

where

$$P(r,s) = \int_0^1 \left[\int_0^\infty\int_0^\infty \frac{4\theta^3 x(2+\theta+x)e^{-2\theta x}}{(1+\theta)^2} \cdot \frac{4\mu^3 y(2+\mu+y)e^{-2\mu y}}{(1+\mu)^2} / \left(A_c = \frac{\mu}{\mu+\theta}\right) dxdy^* \frac{1}{B(\alpha',\beta')} A_c^{a_1-1}(1-A_c)^{b_1-1}\right] dA_c \tag{4.18}$$

where $a_1 = r + \beta$, $b_1 = s + v$ and $I(\cdot)$ is the indicator function that evaluates to 1 when the condition inside the parentheses is true and 0 otherwise. Again, the integration process for equation (4.17) using equation (4.18) can be complex and may require numerical integration methods.

4.5 BAYESIAN ESTIMATION

The Bayes estimators for θ and μ, namely θ^* and μ^* respectively, can be obtained by calculating the posterior expectations of their respective distributions under SELF as:

$$\theta^* = E(\theta \mid r') = \int_0^\infty \theta \Pi_1(\theta \mid r') d\theta \tag{4.19}$$

$$= \frac{\int_0^\infty \int_0^\infty \frac{4x\theta^{3+\alpha}(2+\theta+x)e^{-\theta(2x+\beta)}}{(1+\theta)^2} dx d\theta}{\frac{\beta_1^{\alpha_1}}{\Gamma\alpha_1} \int_0^\infty \int_0^\infty \frac{4x\theta^{3+\alpha-1}(2+\theta+x)e^{-\theta(2x+\beta)}}{(1+\theta)^2} d\theta dx}$$

The given equation (4.19) represents the ratio of two double integrals. However, due to the complexity of the integrands, it is not possible to provide a closed-form solution for this equation. Evaluating this expression would require numerical methods such as numerical integration techniques.

Similarly,

$$\mu^* = E(\mu \mid s) = \int_0^\infty \mu \Pi_2(\mu \mid s) d\mu \tag{4.20}$$

$$= \frac{\frac{\beta_2^{\alpha_2}}{\Gamma\alpha_2} \int_0^\infty \left[\frac{4y\mu^{3+\alpha_2}(2+\mu+y)e^{-\mu(2y+\beta_2)}}{(1+\mu)^2} \right] dy d\mu}{\frac{\beta_2^{\alpha_2}}{\Gamma\alpha_2} \int_0^\infty \int_0^\infty \left[\frac{4y\mu^{3+\alpha_2-1}(2+\mu+y)e^{-\mu(2y+\beta_2)}}{(1+\mu)^2} \right] d\mu dy}$$

Again, we use numerical techniques for finding the solution of equation (4.20). Therefore, based on the relationship given by equation (4.7), the Bayes estimator for A_c can be calculated as follows:

$$A_c^* = \frac{\mu^*}{\mu^* + \theta^*} \tag{4.21}$$

where θ^* and μ^* are the values obtained from the expressions given in equations (4.19) and (4.20), respectively.

4.5.1 Bayesian availability analysis of *k*-out-of-*n* repairable system

Given the posterior distribution of A_c obtained from equation (4.17) and assuming SELF, we can derive the Bayes estimator for *k*-out-of-*n*: F system, A_s, defined in equation (4.8) as,

$$A_s^* = E[A_s \mid r, s] \tag{4.22}$$

$$= \int_0^\infty A_s f(A_c \mid r, s)] dA_c = \int_0^\infty \sum_{i=k}^{n} \binom{n}{i} \left(A_c^*\right)^i \left(1 - A_c^*\right)^{n-i} f(A_c \mid r, s)] dA_c$$

where, $f(A_c \mid r, s)$ is expressed in equation (4.17) and A_c^* is expressed in equation (4.21).

To find the solution, we use numerical methods or approximation techniques, such as numerical integration techniques.

4.5.2 Particular cases

4.5.2.1 *Case 1. Series system*

By substituting $k = n$ in equation (4.22), we get the Bayes estimate of availability of a series system, denoted by A_{s1}^*. The exact expression for A_{s1}^* will depend on the specific form of the posterior distribution of A_c obtained from equation (4.17).

4.5.2.2 *Case 2. Parallel system*

Similarly, the Bayes estimate of the availability of a parallel system, denoted as A_{s2}^*, is obtained by substituting $k = 1$ into equation (4.22).

4.6 NUMERICAL ILLUSTRATION

Equation (4.21) provides the Bayes point estimator for the steady-state availability of the component of *k*-out-of-*n*: F system. Additionally, equation (4.22) offers an alternative formulation of the Bayes point estimator for the steady-state availability of the same system, using SELF. Moreover, in order to analyze the results, the estimates of system availability can be examined by holding certain parameters constant while varying others. This allows for a systematic evaluation of the impact of different parameters on the availability estimates. As an illustration, we set n as 6 and consider three scenarios: 1-out-of-6 (parallel), 3-out-of 6, and 6-out-of-6 (series) systems. Tables 4.1 and 4.2 present the changes in system availability for varying values of r, considering different parameter settings such as ($\beta_1 = 1, \beta_2 = 1, s = 2$) and

($\beta_1 = 2, \beta_2 = 2, s = 2$), respectively. Similarly, Tables 4.3 and 4.4 illustrate the trend in system availability for different values of s, while considering ($\beta_1 = 1, \beta_2 = 1, r = 2$) and ($\beta_1 = 2, \beta_2 = 2, r = 2$) parameter configurations.

Table 4.1 Bayesian estimator of availability for variation in $r(\beta_1 = 1, \beta_2 = 1, s = 2)$

System	*Number of failures r in (0,T)*	A_c^*	A_s^*
	1	0.46871	0.66314
	2	0.35562	0.62533
1-out-of-6 system	3	0.30587	0.57231
	4	0.28045	0.54326
	5	0.22671	0.51233
	1	0.48211	0.69227
	2	0.40302	0.62678
3-out-of-6 system	3	0.30987	0.58892
	4	0.30015	0.57362
	5	0.25321	0.51674
	1	0.48857	0.70156
	2	0.41321	0.63395
6-out-of-6 system	3	0.31876	0.59871
	4	0.31089	0.58218
	5	0.25763	0.53762

Table 4.2 Bayesian estimator of availability for variation in $r(\beta_1 = 2, \beta_2 = 2, s = 2)$

System	*Number of failures r in (0,T)*	A_c^*	A_s^*
	1	0.45279	0.71125
	2	0.36521	0.70235
1-out-of-6 system	3	0.31785	0.68452
	4	0.30116	0.65576
	5	0.22625	0.61045
	1	0.48633	0.72245
	2	0.41268	0.71374
3-out-of-6 system	3	0.38657	0.69114
	4	0.32214	0.65879
	5	0.25074	0.62032
	1	0.48621	0.72459
	2	0.40125	0.70985
6-out-of-6 system	3	0.38332	0.70123
	4	0.35501	0.68765
	5	0.26142	0.64018

Table 4.3 Bayesian estimator of availability for variation in s($\beta_1 = 1, \beta_2 = 1, r = 2$)

System	*Number of repairs s in* $(0, T)$	A_c^*	A_s^*
1-out-of-6 system	1	0.42632	0.49624
	2	0.50702	0.53578
	3	0.51415	0.56201
	4	0.56217	0.58045
	5	0.58525	0.60187
3-out-of-6 system	1	0.47411	0.55847
	2	0.50854	0.60802
	3	0.52974	0.62142
	4	0.55104	0.65458
	5	0.58245	0.66622
6-out-of-6 system	1	0.48744	0.51278
	2	0.50447	0.54974
	3	0.52123	0.56256
	4	0.55385	0.61251
	5	0.59302	0.61962

Table 4.4 Bayesian estimator of availability for variation in s($\beta_1 = 2, \beta_2 = 2, r = 2$)

System	*Number of repairs s in* $(0, T)$	A_c^*	A_s^*
1-out-of-6 system	1	0.36622	0.38249
	2	0.42815	0.44663
	3	0.45545	0.45923
	4	0.50023	0.51852
	5	0.52545	0.53125
3-out-of-6 system	1	0.41241	0.45445
	2	0.47752	0.51857
	3	0.50147	0.54254
	4	0.55565	0.58257
	5	0.60788	0.61987
6-out-of-6 system	1	0.50125	0.58145
	2	0.54174	0.63566
	3	0.58402	0.67458
	4	0.61102	0.69852
	5	0.63754	0.70875

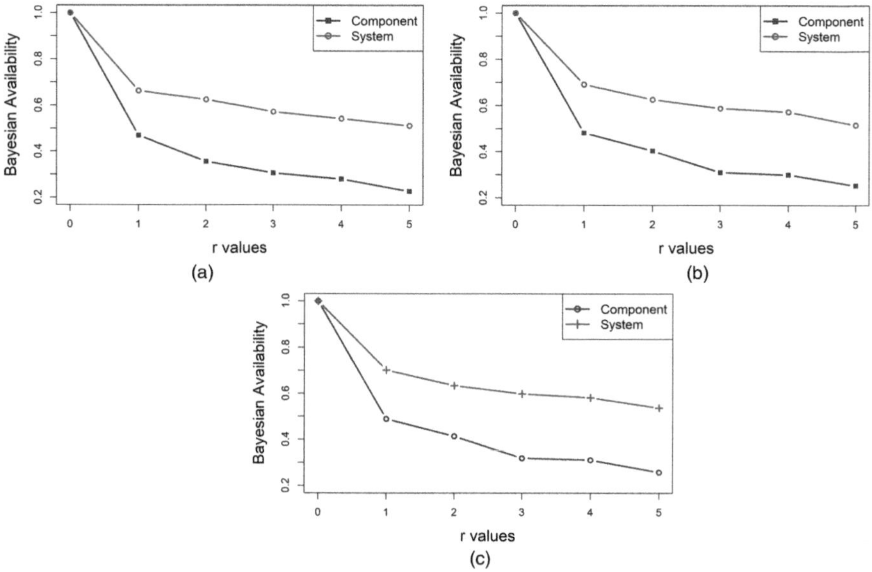

Figure 4.2 Bayesian availability v/s r values for (a) 1-out-of-6 system, (b) 3-out-of-6 system, and (c) 6-out-of-6 system when $\beta_1 = 1$, $\beta_2 = 1$, $s = 2$, $\theta = 1$, and $\mu = 2$.

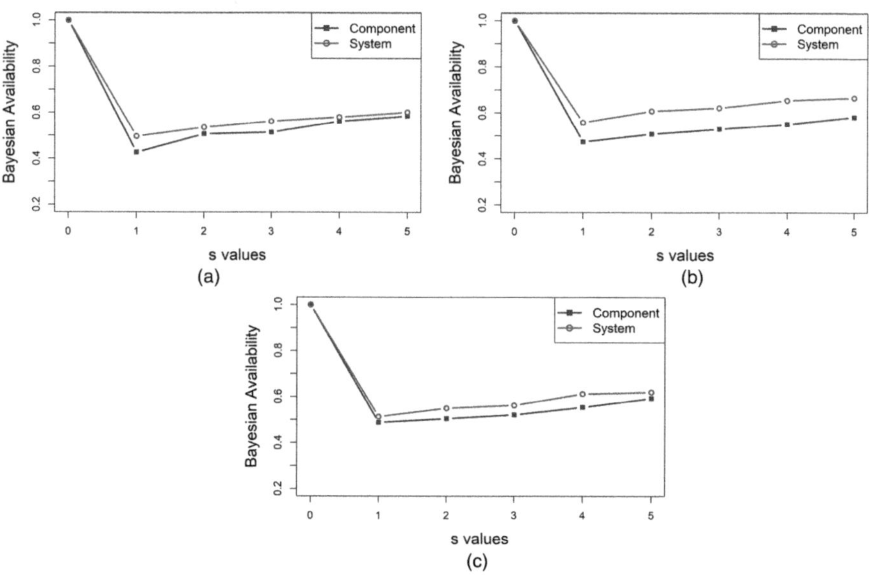

Figure 4.3 Bayesian availability v/s s values for (a) 1-out-of-6 system, (b) 3-out-of-6 system, and (c) 6-out-of-6 system when $\beta_1 = 1$, $\beta_2 = 1$, $r = 2$, $\theta = 1$, and $\mu = 2$.

4.7 RESULTS AND DISCUSSION

In this study, we focused on the Bayesian estimation of k-out-of-n systems, specifically examining cases where k takes values of 1, 3, and 6. We conducted computations using the R software to obtain Bayesian estimates for the availability of the system as a whole as well as the availability of each individual component. Tables 4.1 and 4.2 show the results of this experiment, which examined the effects of changing the number of failures (r) in various systems. Surprisingly, we found a comparable pattern in all the systems under consideration: a discernible decline in the Bayesian availability estimates was correlated with an increase in failure rates. This result highlights the crucial connection between system availability and failures, highlighting the fact that more failures result in poorer overall availability. Furthermore, as shown in Tables 4.3 and 4.4, we investigated how Bayesian component and system availability behaved in response to variations in the number of repairs (s).

This investigation allowed us to comprehend how Bayesian availability responds to different levels of repair activity. Interestingly, our findings consistently indicated that as the number of repairs (s) increased, both component and system availabilities exhibited an upward trend. This observation suggests that increased repair efforts positively influence the overall availability of the system. To visually reinforce these patterns, we included Figure 4.2 which graphically illustrate the decreasing trend of Bayesian availability estimates as failures (r) increase. Correspondingly, Figure 4.3 visually underscores the increasing trend of Bayesian availability as the number of repairs (s) rises. These figures not only provide a clear visualization of the results of this study but also serve as important visual aids aiming to grasp the observed trends more intuitively. These visual aids can aid maintenance professionals in identifying critical points of failure and prioritizing repair efforts accordingly.

4.8 CONCLUSIONS AND FUTURE SCOPE

In order to estimate the steady-state availability of the suggested system, this chapter presents and explains the Bayesian approach. Because Bayesian approach can make well-informed decisions and improve the testing process, it has the potential to reduce test time and material costs. Efficient resource allocation is made possible by Bayesian approaches, which include prior knowledge, update beliefs based on fresh facts, and quantify uncertainty. Additionally, a particular case in which $n = 6$ and k takes the values 1, 3, and 6 is subjected to numerical analysis. This topic looks at how the quantity of repairs and failures affects the Bayesian estimation of steady-state availability. The fundamental goal of system maintenance is to increase the system's availability, which is accomplished by performing regular maintenance and repair of its equipment while it is operational. The numerical findings show that the system's availability decreases as the number of failures rises. On the other hand, the availability of the system improves as the number of fixes rises.

The information presented in this chapter opens up a number of interesting directions for future study. First, by investigating several loss functions other than the squared error loss function, researchers may be able to determine which is best for particular situations. Second, the method would be relevant to a greater variety of real-world systems and their distinctive features if the analysis were expanded to include a larger range of failure and repair time distributions. Another interesting approach is to include covariates or external factors that affect failure and repair processes. This can result in more accurate calculations when external variables are important. Additionally, opportunities for analyzing and optimizing availability in changing environments are provided by dynamic systems and Bayesian models that take time-varying characteristics into consideration.

REFERENCES

1. Tillman F. A., Kuo W., Hwang C. L. and Grosh D. L. (1982). Bayesian Reliability and Availability - A Review. *IEEE Transactions on Reliability*, R-31(4), 362–372.
2. Mazumdar M. (1982). A Comparison of Several Estimates of Availability. *Naval Research Logistics Quarterly*, 29(3), 411–418.
3. Brender D. M. (1968). The Prediction and Measurement of System Availability: A Bayesian Treatment. *IEEE Transactions on Reliability*, R-17(3), 127–138.
4. Brender D. M. (1968). The Bayesian Assessment of System Availability: Advanced Applications and Techniques. *IEEE Transactions on Reliability*, R-17(3), 138–147.
5. Hajeeh M. A. (2011). Reliability and Availability of a Standby System with Common Cause Failure. *International Journal of Operational Research*, 11(3), 343–363.
6. Jain M. and Gupta R. (2014). Availability Analysis of Repairable Redundant System with Three Types of Failures Subject to Common Cause Failure. *International Journal of Mathematics in Operational* Research, 6(3), 271–296.
7. Ke J. C. and Chu Y. K. (2007). Comparative Analysis of Availability for a Redundant Repairable System. *Applied Mathematics and Computation*, 188(1), 332–338.
8. Islam H. M. and Khan M. A. (2010). Bayesian Analysis of System Availability with Geometric Failure Law in Life Testing. *Journal of Quality in Maintenance Engineering*, 16(2), 214–221.

9. Khan M. A. and Islam H. M. (2012). Bayesian Analysis of System Availability with Half-Normal Lifetime. *Quality Technology and Quantitative Management*, 9(2), 203–209.
10. Madhumitha J. and Vijayalakshmi G. (2021). Bayesian Estimate of System Availability for Consecutive *k*-out-of-*n*: F System. *Indonesian Journal of Electrical Engineering and Computer Science*, 24(3), 1692–1699.
11. Yaghoubi, A., Niaki, S. T. A. and Rostamzadeh H. (2019). A Closed-Form Equation for Steady-State Availability of Cold Standby Repairable *k*-out-of-*n*: G Systems. *International Journal of Quality and Reliability Management*, 37(1), 145–155.
12. Sharma, K. K. and Bhutani, R. K. (1993). Bayesian Analysis of System Availability. *Microelectronics Reliability*, 33(6), 809–811.
13. Gonzalez-Lopez, V. A., Gholizadeh, R. and Galarza, C. E. (2017). E-Bayesian Estimation for System Reliability and Availability Analysis Based on Exponential Distribution. *Communications in Statistics - Simulation and Computation*, 46(8), 6221–6241.
14. Gholizadeh, R., Shirazi, A. M. and Hadian, M. (2012). Bayesian System Reliability and Availability Analysis under the Vague Environment Based on Exponential Distribution. *International Journal on Soft Computing*, 3(1), 25–44.
15. Sharma, S. and Kumar, V. (2023). Bayesian Analysis of *k*-out-of-*n* System Using Weighted Exponential-Lindley Distribution. *International Journal of Reliability, Quality and Safety Engineering*, 30(6), 2350028.

Chapter 5

Reliability assessment of cloud security system using Markov model

Nupur Goyal, Nikhil Nawani, Mangey Ram, Akansha Gupta, and Neeraj Sengar

5.1 INTRODUCTION

The cloud has become an essential part of various industries, revolutionizing the way businesses operate and transforming the way we store, access, and analyse the data. With its scalability, flexibility, and cost-efficiency, cloud computing has become an attractive option for businesses and empowered across sectors to streamline their operations, boost productivity, and drive innovation. Despite the considerable benefits of cloud computing environments, there are also concerns around reliability and security. As more organizations move their operations and data to the cloud, the risk of malicious attacks, data breaches, and other security challenges also increases. Therefore, it is important for the organizations to prioritize the security of their cloud computing environments, as failing to do so can result in significant financial and reputational harm. That's why before implementing cloud computing, organizations should conduct a comprehensive evaluation of the risks and benefits involved [1–3].

According to the U.S. National Institute of Standards and Technology (NIST) definition: "Cloud computing is a model for enabling convenient, on-demand network access to share a pool of configurable computing resources (for example servers, networks, storage, services, and applications) that can be quickly provisioned and released with least management effort or service provider interaction" [4].

There are three service models in cloud computing which are SaaS, Paas, and IaaS. In SaaS (*Software as a service*), it is the provider's application that is available to consumers via a thin client but the end-user itself does not operate or govern the backend cloud infrastructure [5]. In the case of PaaS (*Platform as a Service*), consumers are able to deploy their self-developed or third-party applications on a cloud infrastructure through specific programming languages, library options, available services and tools offered by the service provider. Additionally, the customer has no responsibility in terms of supervising or administering the cloud infrastructure [6,7]. While in IaaS (*Infrastructure as a Service*) cloud computing model, the provider provides the basic computing resources that are virtualized, such as

DOI: 10.1201/9781003787525-5

processing power, storage, and a network and any fundamental computing resource. In this model, the consumer controls the operating system (OS), applications, and storage [7].

The cloud has three deployment models: private, public, and hybrid as well as three service models. Private cloud is for a single organization that is managed and maintained by either of the organization, a third party or both of them, on or away from the organization premises. Public clouds belong to various groups and companies, but they are freely available for public sharing and usage. A hybrid cloud brings together two or more separate cloud infrastructures including private, community, or public clouds and facilitates easy movement of data and applications among these clouds [4].

The term "cloud security" refers to different mechanisms and procedures established and applied with an aim to secure data, applications, and virtual infrastructures on cloud. It is essential for providing protection of sensitive information, maintaining privacy, integrity, and availability of cloud-based resources. With the increasing adoption of cloud services by organizations and individuals, cloud security has become a top priority in the digital age [8,9].

5.1.1 Factors affecting cloud security

Misconfigured Resources: Misconfiguration of cloud resources, such as storage buckets or database, can expose sensitive data to unauthorized access. Addressing these misconfigurations, from tightening access controls to implementing automated configuration management, is crucial in securing your cloud environment [10].

Insecure API: Almost every cloud application exposes its services in the form of APIs or UIs. Since they act as front door of the system, they are attacked and scanned continuously. This can include attacks such as cross site scripting (XSS) or SQL injection attacks [11]. These exploits can hijack user sessions, steal sensitive data, or even manipulate application functionality. Maintaining robust API security through rigorous validation, encryption, and access controls is crucial for the cloud security.

Poor Patching: It is an important part of cloud security as it ensures that software components are kept up to date with the latest security fixes. Outdated software harbours vulnerabilities, and these vulnerable can be exploited and could compromise sensitive data, disrupt operations, and damage reputation.

Identity Mismanagement: It involves the process and procedures for managing user identities accessing privileges and authentication mechanism (IAM) policies. Weak passwords, unclear roles, and outdated IAM policies create vulnerabilities. Tightening identity controls, implementing multi-factor authentication, and regularly reviewing access levels are crucial steps for securing cloud.

Distributed Denial of Service Attacks (DDoS): They are a type of cyber-attack where the attackers flood a cloud-based service or application with an overwhelming amount of traffic requests or commands to render it inaccessible to legitimate users. Protection against DDoS requires robust security measures, vigilant monitoring, and swift response plans to ensure your cloud service is up and running.

In the first section of the chapter cloud security system has been introduced. In this portion, authors try to explain the basic concept of cloud and cloud security. Which type of failures can affect the cloud security system is also discussed in this part of the chapter. The related literature of cloud security system and reliability issues in various networks, software, etc. are discussed in Section 5.2. The third section contains all the basic details of mathematical modelling of cloud security system such as system description, state description, assumptions, and the notations utilized in the model. The fourth section depicts the mathematical formulation of the model. In the fifth section, authors determined the performance indices and explain with the help of numerical example. In the sixth section of this research work, the obtained results of availability, reliability, and MTTF with the effect of failure parameters are discussed. The last section that is the seventh portion examines the conclusions acquired from the model.

5.2 LITERATURE REVIEW

Cloud computing is a critical technology driving the fourth industrial revolution, enabling industries such as Internet of Things (IoT), C4T, Industry 4.0, and Smart Cities to share and obtain resources on a pay-per-use and elastic provisioning model. However, cloud-based applications are vulnerable to the services provided by the cloud platforms. Any possible outage can result in significant loss of reputation and money for the application provider. With this in mind, researchers have explored the security and reliability issues related to cloud computing and have proposed various solutions to mitigate them.

One of the primary security concerns in cloud computing is security misconfiguration vulnerabilities. Researchers have found that default security configuration settings for Apache, MySQL, and PHP packages differ significantly from the recommended settings. Secure configuration requires controls for provisioning, administration, monitoring, validation, and management. Insecure configurations resulting from small human errors can lead to costly security breaches and reduce user confidence [12]. Eshete et al. [13] have developed a tool that can analyse the existing security configuration settings and notify users of any necessary changes to achieve recommended security configurations for deploying web applications. This tool is believed to be an innovative and reliable solution for addressing security misconfiguration vulnerabilities, which are a primary cause of security breaches in web applications. Furthermore, secure configuration within cloud environments is of utmost importance.

Cloud environments are also susceptible to threats beyond those faced by other distributed systems due to their multi-tenancy and virtualization architecture. Cryptographic solutions have been proposed to address these risks, but most cryptosystems rely on public key infrastructure (PKI), which has some deficiencies when applied to cloud security. Therefore, Ahamed et al. [14] have suggested that newer, cryptography-based solutions such as homomorphic encryption techniques and data-centric approaches are better suited to address risks encountered in cloud environments. Another approach will be to use innovative solutions such as blockchain and AI/ML to improve cloud computing security in the future [15].

One can group security issues into architecture-related issues, service delivery model-related issues, cloud characteristic-related issues, and cloud stakeholder-related issues. This will help cloud providers and security vendors to have a better understanding of the problem. It also helps researchers to be aware of the existing problem dimensions and gaps [16]. To mitigate security risks and protect resources, enterprises can take steps such as proper planning and understanding of emerging risks, threats, and possible countermeasures. Additionally, appropriate security measures such as load balancing, caching, parallel processing, encryption, authentication, and intrusion detection systems can be implemented to protect data and resources [17]. Ghobadi et al. [18] mention few common causes of failure including hardware and software errors, network failures, and power outages. To address these issues, they have proposed various fault tolerance techniques, such as redundancy, checkpointing, and self-healing systems.

A generalized view of these issues is required to be present here to enhance the understanding of the security flaws of the cloud computing framework and to devise suitable countermeasures for them. Finally, various cloud security schemes need to be discussed on a comparative framework. Also, a comprehensive overview of the current status and potential future developments in cloud security is also important [19].

To design efficient and effective fault tolerance algorithms, an overview of cloud computing concepts, components, service, and deployment models is described by Kumari and Kaur [20]. In addition to understanding data centre network topologies that can impact the design of these algorithms, various fault tolerance techniques and algorithms, including heuristics, meta-heuristics, and clustering approaches have been introduced. The use of fault tolerance mechanisms such as FTCloudSim and MCS for recovering and evaluating failures in cloud computing environments is also advised. However, it will be still impossible to design a cloud that is completely free from failures [21].

To recover from the reliability-related issues in this type of systems, researchers gave many techniques and studied various issues in different systems. Nachiappan et al. [22] have discussed challenges in enhancing data reliability for Big Data applications in cloud computing; they have also delved into replication and erasure coding techniques, analysing their trade-offs. To address these limitations, a novel hybrid technique is proposed. To achieve optimal cloud service performance, Cui et al. [23] analysed comprehensive reliability using Markov-based method. This multi-objective optimization challenge was solved with genetic algorithm-based chaotic ant swarm (GA-CAS) algorithm. This innovative approach demonstrably outperforms other meta-heuristics in terms of convergence speed and solution quality. Mesbahi et al. [24] have outlined a valuable reference roadmap for future works. The roadmap covers all aspects of the problem and proposed specific measure for each step. It explored the strategies for ensuring high availability and reliability within cloud computing data centres.

Recognizing the impact of slow network connections on cloud service reliability and the key role of failure characteristic analysis in addressing disruptions, Kaur and Kumar [21] have highlighted the potential of fault tolerance mechanisms like FTCloudSim and MCS. These mechanisms offer a powerful solution for cloud environments by enabling both failure recovery and evaluation, ultimately strengthening service resilience. Vishwanath and Nagappan [25] observed hardware failures and repairs in large data centres. They explored the intricate relationship between failure occurrences and key factors like machine age and hard disk configurations. By deciphering these failure characteristics, authors aim to pave the way for improved operational experiences and reduced hardware costs, ultimately enhancing the overall efficiency of data centre operations. Building on the observation that cloud storage systems face unique reliability challenges, Zhang et al. [26] proposed a novel modelling approach. This technique demonstrably enhanced replication performance and system reliability by identifying the optimal strategy for balancing both factors and showed the effectiveness and superiority of this approach compared to traditional static methods.

Pokorn [27] has identified challenges in comprehensively assessing the reliability and availability of the IoT using the traditional MIL-HDBK 217 approach. Recognizing the limitations of the classical method, which solely focuses on hardware reliability, they proposed a broader framework encompassing hardware, software, human factors, protocols, standards, and logistics, all of which significantly influence the reliability of an IoT system. Sharma et al. [28] suggested the critical intersection of reliability and energy efficiency that provide a comprehensive review of existing techniques and their inherent trade-offs, offering a solid foundation for further advancements. Recognizing the complexities of resource management, they classified resource failures, fault tolerance mechanisms, and energy management approaches. The detailed analysis paves the way for the development of optimized resource provisioning policies that are both energy-aware and failure-aware.

In this research work, authors have taken into consideration internal failure as well as hacking attacks in a cloud security system to analyse the performance. They modelled the system Markov process and the obtained differential equation is solved by Laplace transformation. In this study, authors examined the effect of considered failures on various performance indices like availability, reliability, and MTTF of the cloud security system.

5.3 MATHEMATICAL MODEL DETAILS

5.3.1 System description

The cloud has become indispensable across various industries. However, despite its numerous benefits, the cloud security is susceptible to failure. Cloud security is vulnerable to both partial and complete failure due to a range of issues. While issues like poor patching, misconfiguration, and insecure APIs, as well as DDoS attacks, can result in complete failure, other issues like a misconfigured identity and access management (IAM) policy can lead to unauthorized access to resources, which can reduce efficiency and partially compromise the cloud security (Figure 5.1). Overall Cloud security demands a multi-layered approach to address diverse vulnerabilities to ensure the confidentiality, accuracy, and accessibility of cloud resources.

The state of the cloud security notified in Figure 5.1 is described as follows:

S_0 Good working condition of the cloud security.

S_1 The cloud security is working with reduced efficiency due to *poor patching*.

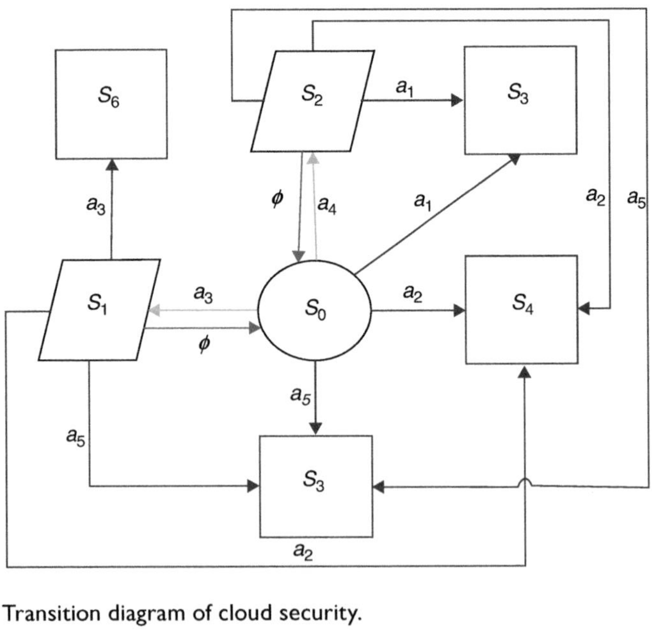

Figure 5.1 Transition diagram of cloud security.

S_2 The cloud security is working partially due to *identity mismanagement.*
S_3 The cloud security has breakdown due to *misconfiguration.*
S_4 Breakdown state of the cloud security due to *insecure APIs.*
S_5 The cloud security failed completely due to *DDoS attacks.*
S_6 The cloud security is in breakdown state due to *lack of patching.*

5.3.2 Assumptions

The designed cloud security model is associated with the following assumptions:

- Initially, the system is working properly without any failure.
- The system covered three types of states: good, degraded, and failed.
- Failure rates of the system have constant distribution.
- The sufficient repair facility is always available for the degraded state of the system.
- Repair rates follow constant distribution according to modelling.
- The system repairs automatically without waiting time to repair, after repair it works like a new one.

5.3.3 Notations

Table 5.1 lists all the key terms and their description used throughout this research.

5.4 MATHEMATICAL FORMULATION AND SOLUTION

The following differential equations describe the mathematical model of cloud security, which has been derived using the Markov process:

$$\left[\frac{\partial}{\partial t}+\sum_{i=1}^{5}a_i\right]P_0(t)=\varnothing\sum_{i=1}^{2}P_i(t) \tag{5.1}$$

Table 5.1 Notations

Notation	*Description*
t/s	Time scale/Laplace transform variable.
$\bar{P}(s)$	Laplace transformation of $P(t)$.
$P_i(t)$	The probability of state S_i at time t; $i = 0, 1, 2, \ldots, 6$.
$a_1/a_2/a_3/a_4/a_5$	Failure rate due to misconfigured resources/insecure API/poor patching/ identity mismanagement/DDoS attacks occurring
	Repair rate for degraded state of the system.
$P_{up}(t)/P_{down}(t)$	probability of upstate/downstate of cloud security system at time t.
Av(t)/Rl(t)	Availability/reliability of the cloud security system at time t.

$$\left[\frac{\partial}{\partial t}+\sum_{i=1}^{3}a_i+a_5+\varnothing\right]P_1(t)=a_3P_0(t) \tag{5.2}$$

$$\left[\frac{\partial}{\partial t}+a_1+a_2+a_5+\varnothing\right]P_2(t)=a_4P_0(t) \tag{5.3}$$

$$\frac{\partial}{\partial t}P_3(t)=a_1\sum_{i=0}^{2}P_i(t) \tag{5.4}$$

$$\frac{\partial}{\partial t}P_4(t)=a_2\sum_{i=0}^{2}P_i(t) \tag{5.5}$$

$$\frac{\partial}{\partial t}P_5(t)=a_5\sum_{i=0}^{2}P_i(t) \tag{5.6}$$

$$\frac{\partial}{\partial t}P_6(t)=a_3P_1(t) \tag{5.7}$$

Initial condition

$$P_0(0)=1 \text{ and other state probabilities are zero at } t=0 \tag{5.8}$$

Taking the Laplace transform of equations (5.1–5.7) using equation (5.8),

$$\left[s+\sum_{i=1}^{5}a_i\right]\bar{P}_0(s)=\varnothing\sum_{i=1}^{2}\bar{P}_i(s) \tag{5.9}$$

$$\left[s+\sum_{i=1}^{3}a_i+a_5+\varnothing\right]\bar{P}_1(s)=a_3\bar{P}_0(s) \tag{5.10}$$

$$\left[s+a_1+a_2+a_5+\varnothing\right]\bar{P}_2(s)=a_4\bar{P}_0(s) \tag{5.11}$$

$$s\bar{P}_3(s)=a_1\sum_{i=0}^{2}\bar{P}_i(s) \tag{5.12}$$

$$s\overline{P}_4(s) = a_2 \sum_{i=0}^{2} \overline{P}_i(s) \tag{5.13}$$

$$s\overline{P}_5(s) = a_5 \sum_{i=0}^{2} \overline{P}_i(s) \tag{5.14}$$

$$s\overline{P}_6(s) = a_3 \overline{P}_1(s) \tag{5.15}$$

After solving equations (5.9–5.15), we get probability of each state as

$$\overline{P_0}(s) = \frac{1}{S + K_1 - \varnothing K_4}$$

$$\overline{P_1}(s) = \frac{a_3}{S + K_2} \overline{P_0}(s)$$

$$\overline{P_2}(s) = \frac{a_4}{S + K_3} \overline{P_0}(s)$$

$$\overline{P_3}(s) = \frac{a_1}{S}[1 + K_4]\overline{P_0}(s)$$

$$\overline{P_4}(s) = \frac{a_2}{S}[1 + K_4]\overline{P_0}(s)$$

$$\overline{P_5}(s) = \frac{a_5}{S}[1 + K_4]\overline{P_0}(s)$$

$$\overline{P_6}(s) = \frac{(a_3)^2}{S(S + K_2)} \overline{P_0}(s)$$

where

$$K_1 = a_1 + a_2 + a_3 + a_4 + a_5$$

$$K_2 = a_1 + a_2 + a_3 + a_5 + \varnothing$$

$$K_3 = a_1 + a_2 + a_5 + \varnothing$$

$$K_4 = \frac{a_3}{S + K_2} + \frac{a_4}{S + K_3}$$

Laplace transformation of working state probability of the cloud security is

$$\overline{P_{\mathbf{up}}}(s) = \sum_{i=0}^{2} \overline{P_i}(s) \tag{5.16}$$

$$= (1 + K_4)\overline{P_0}(s)$$

Probability of failed state of the cloud security system is

$$\overline{P_{\text{down}}}(s) = \sum_{i=3}^{6} \overline{P_i}(s) \tag{5.17}$$

$$= \frac{1}{S}\left[(1 + K_4)(a_1 + a_2 + a_3) + \frac{a_3{}^2}{S + K_2}\right]\overline{P_0}(s)$$

It is also noticed that

$$\overline{P_{\text{up}}}(s) + \overline{P_{\text{down}}}(s) = \frac{1}{s}$$

5.5 PERFORMANCE INDICES AND NUMERICAL ANALYSIS

To explore system behaviour, some performance indices, reliability, mean time to failure (MTTF), steady-state availability, and some other measures such as recovery state probability, reboot state probability, and failure frequency have been investigated.

5.5.1 Availability analysis

Availability in cloud security refers to the measure of how often a cloud service is accessible and operational. It is a crucial aspect of cloud computing, as it directly impacts user experience, business productivity, and overall trust in cloud services. Availability of the designed cloud security model can be examined by taking the inverse Laplace transformation of equation (5.16) in various cases-

Case 1. In a comprehensive case, putting the value of failure as $a_1 = 0.32$, $a_2 = 0.20$, $a_3 = 0.28$, $a_4 = 0.12$, $a_5 = 0.08$, $u = 1$, we get

$$Av(t) = -0.03670e^{(-2.17t)} + 0.00881e^{(-1.66t)} + 1.02788e^{(-0.65t)}$$

Case 2. When cloud security has perfect configured resources, then availability is

$$Av(t) = -0.03670e^{(-1.85t)} + 0.00881e^{(-1.34t)} + 1.02788e^{(-0.33t)}$$

Case 3. Assume that each API is secure, then

$$Av(t) = -0.03670e^{(-1.97t)} + 0.00881e^{(-1.46t)} + 1.02788e^{(-0.45t)}$$

Case 4. Cloud security system has good patching, then system availability is

$$Av(t) = e^{(-0.60t)}$$

Case 5. When there is no identity mismanagement in cloud security system, then availability is

$$Av(t) = -0.03571e^{(-2.11t)} + 1.03571e^{(-0.65t)}$$

Case 6. When no DDoS attacks occur in cloud security system, then

$$Av(t) = -0.03670e^{(-2.09t)} + 0.00881e^{(-1.58t)} + 1.02788e^{(-0.57t)}$$

After varying the time unit from 0 to 15, one can obtain security system's availability in each case which is calculated in Table 5.2 and its graphical behaviour with respect to each failure parameter is demonstrated in Figure 5.2.

5.5.2 Reliability analysis

Cloud security's reliability refers to how cloud infrastructure and services provide secure and dependable operation consistently. It encompasses the following key aspects: availability, resilient to failures, data integrity and protection, threat prevention detection, and continuous improvements. Reliability of the designed cloud security model can be examined by taking the inverse Laplace transformation of equation (5.16) in various cases.

Case 1. In a comprehensive case, putting the value of failure as $a_1 = 0.32$, $a_2 = 0.20$, $a_3 = 0.28$, $a_4 = 0.12$, $a_5 = 0.08$, $u = 0$, we get the system's reliability as

$$Av(t) = 2.33333e^{(-0.88t)} + 0.30000e^{(-0.60t)} - 1.63333e^{(-t)}$$

Table 5.2 Availability of cloud security

	Availability of the system					
Time	*Case 1*	*Case 2*	*Case 3*	*Case 4*	*Case 5*	*Case 6*
0	1.00000	1.00000	1.00000	1.00000	1.00000	1.00000
1	0.53571	0.73773	0.65431	0.54881	0.53527	0.58032
2	0.28166	0.53416	0.42019	0.30120	0.28062	0.33053
3	0.14757	0.38541	0.26890	0.16530	0.14641	0.18760
4	0.07727	0.27792	0.17197	0.09072	0.07631	0.10641
5	0.04046	0.20040	0.10998	0.04979	0.03976	0.06036
6	0.02119	0.14450	0.07034	0.02732	0.02072	0.03424
7	0.01109	0.10420	0.04498	0.01500	0.01079	0.01942
8	0.00581	0.07514	0.02877	0.00823	0.00562	0.01102
9	0.00304	0.05418	0.01840	0.00452	0.00293	0.00625
10	0.00159	0.03907	0.01177	0.00248	0.00153	0.00354
11	0.00083	0.02817	0.00752	0.00136	0.00080	0.00201
12	0.00044	0.02031	0.00481	0.00075	0.00041	0.00114
13	0.00023	0.01465	0.00308	0.00041	0.00022	0.00065
14	0.00012	0.01056	0.00197	0.00022	0.00011	0.00037
15	0.00006	0.00762	0.00126	0.00012	0.00006	0.00021

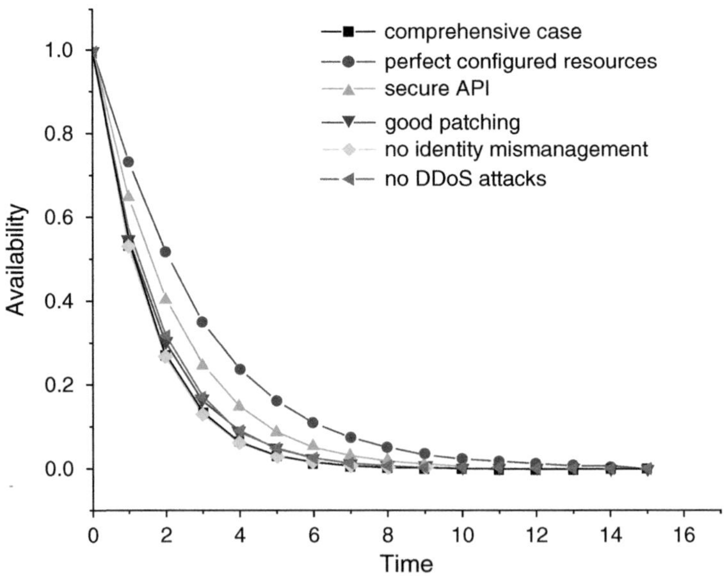

Figure 5.2 Availability of cloud security system with respect to time.

Case 2. When cloud security has perfect configured resources, then cloud security system's reliability is

$$Av(t) = 2.33333e^{(-0.56t)} + 0.30000e^{(-0.28t)} - 1.63333e^{(-0.68t)}$$

Case 3. Assume that each API is secure, then the system's reliability is

$$Av(t) = 2.33333e^{(-0.68t)} + 0.30000e^{(-0.40t)} - 1.63333e^{(-0.80t)}$$

Case 4. Cloud security system has good patching, then the system reliability is

$$Av(t) = e^{(-0.60t)}$$

Case 5. When there is no identity mismanagement in cloud security system, then reliability is

$$Av(t) = (0.28000\ t + 1)e^{(-0.88t)}$$

Case 6. When no DDoS attacks occur in the cloud security system, then system reliability is

$$Av(t) = 2.33333e^{(-0.80t)} + 0.30000e^{(-0.52t)} - 1.63333e^{(-0.92t)}$$

After varying the time unit from 0 to 15, one can obtain cloud security system's reliability in each case which is computed in Table 5.3 and its graphical behaviour with respect to each failure parameter is explained in Figure 5.3.

5.5.3 MTTF analysis

MTTF is a reliability metric that measures the average time between failures of a system or component. It is calculated by dividing the total operating time of a system by the number of failures that occur during that time. MTTF is an important metric for both availability and reliability, as it provides an indication of how often a system is likely to fail. Higher MTTF values indicate greater reliability and availability. MTTF is used for the non-repairable systems. So, by considering no repair, one can evaluate MTTF as the function of failure rates as

$$\begin{aligned} \text{MTTF} &= \lim_{s \to 0} \overline{P_{\text{up}}}(s) \\ &= \frac{1}{a_1 + a_2 + a_3 + a_4 + a_5}\left[1 + \frac{a_3}{a_1 + a_2 + a_3 + a_5} + \frac{a_4}{a_1 + a_2 + a_5}\right] \end{aligned} \tag{5.18}$$

Table 5.3 Reliability of cloud security system

	Reliability of the system					
Time	*Case 1*	*Case 2*	*Case 3*	*Case 4*	*Case 5*	*Case 6*
0	1.00000	1.00000	1.00000	1.00000	1.00000	1.00000
1	0.53160	0.73208	0.64930	0.54881	0.53092	0.57588
2	0.27075	0.51347	0.40391	0.30120	0.26839	0.31773
3	0.13478	0.35201	0.24559	0.16530	0.13130	0.17134
4	0.06637	0.23869	0.14770	0.09072	0.06275	0.09139
5	0.03258	0.16136	0.08856	0.04979	0.02947	0.04860
6	0.01603	0.10935	0.05322	0.02732	0.01365	0.02591
7	0.00794	0.07456	0.03219	0.01500	0.00625	0.01390
8	0.00397	0.05129	0.01964	0.00823	0.00284	0.00752
9	0.00200	0.03565	0.01211	0.00452	0.00128	0.00411
10	0.00102	0.02505	0.00755	0.00248	0.00057	0.00227
11	0.00053	0.01779	0.00476	0.00136	0.00026	0.00127
12	0.00027	0.01277	0.00302	0.00075	0.00011	0.00072
13	0.00014	0.00925	0.00194	0.00041	0.00005	0.00041
14	0.00008	0.00675	0.00126	0.00022	0.00002	0.00023
15	0.00004	0.00496	0.00082	0.00012	0.00000	0.00014

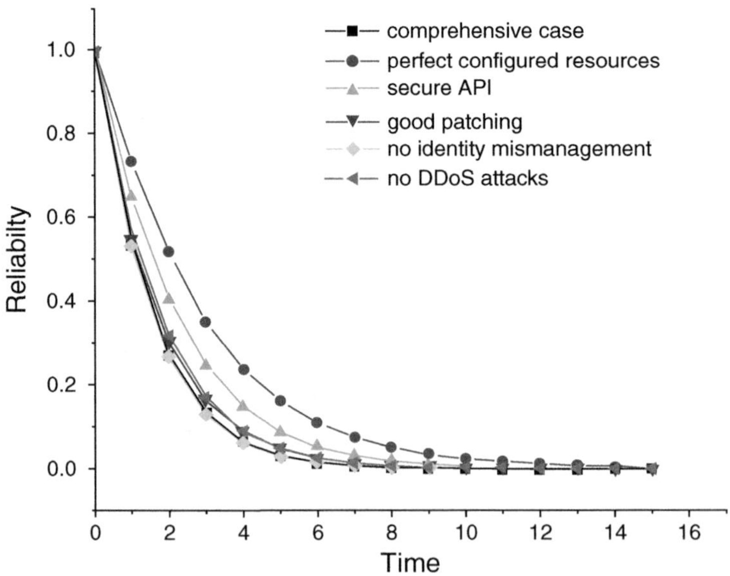

Figure 5.3 Reliability of cloud security system with respect to time.

Putting $a_1 = 0.32$, $a_2 = 0.20$, $a_3 = 0.28$, $a_4 = 0.12$, $a_5 = 0.08$ and varying the failure rates as 0.00, 0.05, 0.10, 0.15, 0.20, ..., 0.90, respectively, in equation (5.18), one obtains the variation in MTTF with respect to failure rates as shown in Table 5.4 and Figure 5.4.

5.5.4 Neural network model performance evaluation

To further confirm the prediction of the Markov-based reliability model for cloud security systems, a neural network was trained to transform system inputs into their corresponding reliability outputs. This was done using the Levenberg-Marquardt training algorithm, which is a popular algorithm for nonlinear regression problems because it has a fast convergence.

Data Specifications

Predictors (Input): 16 × 1 double array (16 observations, 1 feature)

Responses (Output): 16 × 6 double array (16 observations, 6 outputs)

Neural Network Parameters

Training Algorithm: Levenberg-Marquardt

Performance Measure: Mean Squared Error (MSE)

Neural network model training and validation results in terms of MSE and R value are given in Table 5.5 and graph is demonstrated in Figure 5.5.

Table 5.4 MTTF as a function of failure rate

Variation in failure rates	*MTTF*				
	a_1	a_2	a_3	a_4	a_5
0.00	2.83613	2.13971	1.66667	1.49793	1.71823
0.05	2.49678	1.94145	1.65618	1.50701	1.58757
0.10	2.23081	1.77664	1.63763	1.51524	1.47516
0.15	2.01618	1.63740	1.60920	1.52251	1.37742
0.20	1.83911	1.51818	1.57609	1.52918	1.29167
0.25	1.69042	1.41494	1.54033	1.53526	1.21582
0.30	1.56375	1.32468	1.50327	1.54083	1.14827
0.35	1.45453	1.24508	1.46581	1.54595	1.08772
0.40	1.35939	1.17438	1.42857	1.55066	1.03316
0.45	1.27576	1.11117	1.39194	1.55502	0.98373
0.50	1.20169	1.05432	1.35618	1.55907	0.93875
0.55	1.13563	1.00293	1.32147	1.56283	0.89766
0.60	1.07636	0.95625	1.28788	1.56634	0.85996
0.65	1.02289	0.91366	1.25547	1.56962	0.82527
0.70	0.97440	0.87466	1.22427	1.57269	0.79323
0.75	0.93024	0.83881	1.19426	1.57557	0.76355
0.80	0.88986	0.80574	1.16541	1.57828	0.73599
0.85	0.85279	0.77516	1.13771	1.58084	0.71033
0.90	0.81866	0.74678	1.11111	1.58325	0.68638

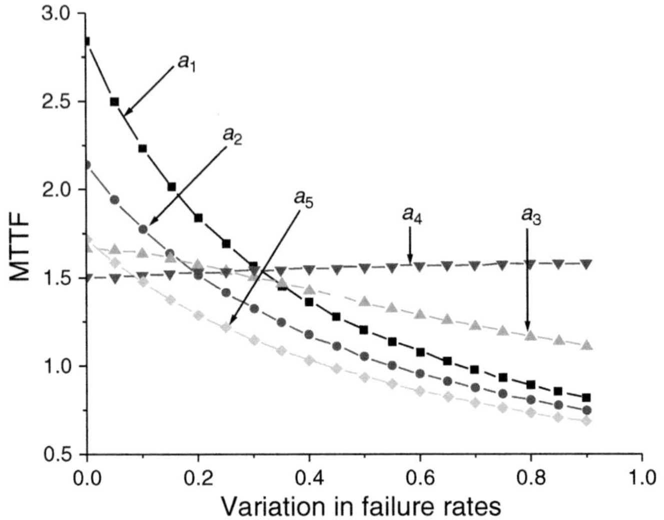

Figure 5.4 MTTF with respect to failure rates.

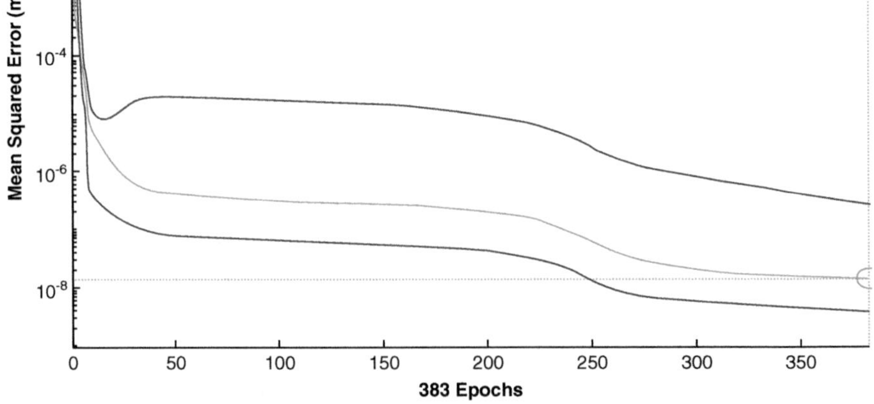

Figure 5.5 Mean squared error with respect to epochs.

Table 5.5 Model performance metrics

	Observations	*MSE*	*R*
Training	12	3.7172e-09	1.0000
Validation	2	1.4015e-08	1.0000
Test	2	2.7093e-07	1.0000

5.6 RESULTS AND DISCUSSION

The research highlights the importance of maintaining a robust and resilient cloud infrastructure to ensure security, reliability, and availability. By understanding the probabilities of transitioning between good, degraded, and failed states, we can design and implement appropriate security measures to minimize the likelihood of breaches and maximize system uptime. On the basis of the numerical example of availability, reliability, and MTTF of cloud security system, one can obtain the following results:

The graph of availability versus time shows an exponential decline in availability over time for all the cases such as "comprehensive case," "perfect configured resources," "secure API," "good patching," "no identity mismanagement," and "no DDoS attacks." In each case, initially system has full availability. From the graphs shown in Figure 5.2, one can see that the availability of the designed system is approximately the same in comprehensive case, case of no identity mismanagement, no DDoS attacks, and good patching. In the case of perfect configuration and secure APIs, high availability of the cloud security can be achieved. From the critical examination of the graph, one can see that after eight units of time, the availability will be constant.

The graph of reliability vs time (Figure 5.3) demonstrates the decrement in system reliability as time increases. Misconfigured resources significantly contribute to reduced reliability as we can see from the graph. DDoS attacks have a relatively smaller impact on reliability compared to other factors. Security measures such as secure APIs, good patching, and identity management help mitigate declines in reliability but don't eliminate attacks entirely as we can see by the persistent gap between the "comprehensive case" line and the lines of the remaining cases in the graph.

From the study of the MTTF graph given in Figure 5.4, one can see that the MTTF decreases with respect to each failure rate except good patching. In the case of perfect configured resources and No Identity Mismanagement, the MTTF of cloud security is lowest, whereas in the case of good patching, it is highest.

Figure 5.5 illustrates the performance of the neural network on 383 training epochs, with MSE plotted on a logarithmic scale for training set (blue), validation set (green), and test set (red). The performance curve shows a steep drop initially, which is characteristic of fast learning, followed by saturation of both training and validation errors to very low values of around 10^{-9} and 10^{-8}, respectively. The optimal validation performance occurs at epoch 383 with an MSE of 1.4015e-08, thereby validating the model's superior generalization ability.

While the testing error has risen slightly, it is still low and indicates a consistent drop, which shows strength and no overfitting. A correlation coefficient (R) of 1.0000 at all levels shows a nearly perfect relationship between predicted and measured values. Overall, these results strongly indicate that the trained network can consistently link input conditions to their related reliability states within the Markov-based cloud security system model.

5.7 CONCLUSION

From this study of failures effect on the performance of cloud security system, the authors conclude that the cloud security is most sensitive with respect to misconfigured resources and API security. So, by controlling these failures, one can attain a highly reliable cloud security. Form the critical examination of Figure 5.4, one can find that failure rate of identity mismanagement has not much effect on system MTTF while when cloud security has up to 30% misconfigured resources, then MTTF is highest. But due to DDoS attacks, MTTF of the cloud security is lowest. So, through the overall study, authors conclude that configured resources give the maximum reliability and DDoS attacks give lowest reliability.

The findings of this research can assist cloud service providers in making informed decisions regarding resource allocation and risk management. By quantifying the impact of security events on reliability and availability, providers can prioritize investments in security measures and allocate resources effectively to minimize disruptions and maintain service quality. The obtained results is validated by MSE and R value. This study provides more comprehensive information for engineers in developing highly reliable cloud security.

DECLARATION OF STATEMENT

Ethical approval and consent to participate

The current research totally belongs to the contributors and related works are properly cited in the article. There are no human- and animal-related research involved in this work.

CONSENT FOR PUBLICATION

Here, authors did not use any third-party material. All tables and figures were made by own study.

COMPETING INTEREST

The authors confirm that there is no conflict of interest to declare for this publication.

REFERENCES

1. Kulkarni, G., Chavan, N., Chandorkar, R., Waghmare, R., & Palwe, R. (2012, October). Cloud security challenges. In *2012 7th International Conference on Telecommunication Systems, Services, and Applications (TSSA)* (pp. 88–91). IEEE. Denpasar-Bali, Indonesia.
2. Behl, A. (2011, December). Emerging security challenges in cloud computing: An insight to cloud security challenges and their mitigation. In *2011 World Congress on Information and Communication Technologies* (pp. 217–222). IEEE. Mumbai, India.
3. Cinque, M., Russo, S., Esposito, C., Choo, K. K. R., Free-Nelson, F., & Kamhoua, C. A. (2018). Cloud reliability: Possible sources of security and legal issues? *IEEE Cloud Computing*, *5*(3), 31–38.
4. Mell, P., & Grance, T. (2011). *The NIST Definition of Cloud Computing*. NIST.
5. Tsai, W., Bai, X., & Huang, Y. (2014). Software-as-a-service (SaaS): Perspectives and challenges. *Science China Information Sciences*, *57*, 1–15.
6. Qian, L., Luo, Z., Du, Y., & Guo, L. (2009). Cloud computing: An overview. In *Cloud Computing: First International Conference, CloudCom 2009*, Beijing, China, *December 1–4, 2009*. Proceedings 1 (pp. 626–631). Springer. Berlin, Heidelberg.
7. Rani, D., & Ranjan, R. K. (2014). A comparative study of SaaS, PaaS and IaaS in cloud computing. *International Journal of Advanced Research in Computer Science and Software Engineering*, 4(6), 458–461.
8. Singh, A., & Chatterjee, K. (2017). Cloud security issues and challenges: A survey. *Journal of Network and Computer Applications*, 79, 88–115.
9. Sun, P. (2020). Security and privacy protection in cloud computing: Discussions and challenges. *Journal of Network and Computer Applications*, 160, 102642.
10. Alatawi, S., Alhasani, A., Alfaidi, S., Albalawi, M., & Almutairi, S. M. (2020, September). A survey on cloud security issues and solution. In *2020 International Conference on Computing and Information Technology (ICCIT-1441)* (pp. 1–5). IEEE. Tabuk, Saudi Arabia.
11. Qazi, F. (2022). Application programming interface (API) security in cloud applications. *EAI Endorsed Transactions on Cloud Systems*, 7(23), e1.

12. Wood, K., & Pereira, E. (2011). Impact of misconfiguration in cloud–investigation into security challenges. *International Journal Multimedia and Image Processing*, 1(1), 17–25.
13. Eshete, B., Villafiorita, A., & Weldemariam, K. (2011, August). Early detection of security misconfiguration vulnerabilities in web applications. In *2011 Sixth International Conference on Availability, Reliability and Security* (pp. 169–174). IEEE. Vienna, Austria.
14. Ahamed, F., Shahrestani, S., & Ginige, A. (2013). Cloud computing: Security and reliability issues. *Communications of the IBIMA, 2013*, 1–12.
15. Pandey, S., & Farik, M. (2015). Cloud computing security: Latest issues & countermeasures. *International Journal of Scientific & Technology Research*, 4(11), 3–6.
16. Bouayad, A., Blilat, A., Mejhed, N. E. H., & El Ghazi, M. (2012, October). Cloud computing: Security challenges. In *2012 Colloquium in Information Science and Technology* (pp. 26–31). IEEE. Fez, Morocco.
17. Tiwari, P. K., & Mishra, B. (2012). Cloud computing security issues, challenges and solution. *International Journal of Emerging Technology and Advanced Engineering*, 2(8), 306–310.
18. Ghobadi, A., Karimi, R., Heidari, F., & Samadi, M. (2014, February). Cloud computing, reliability and security issue. In *16th International Conference on Advanced Communication Technology* (pp. 504–511). IEEE. Pyeongchang, Korea (South).
19. Basu, S., Bardhan, A., Gupta, K., Saha, P., Pal, M., Bose, M., Basu, K., Chaudhury, S., & Sarkar, P. (2018, January). Cloud computing security challenges & solutions – A survey. In *2018 IEEE 8th Annual Computing and Communication Workshop and Conference (CCWC)* (pp. 347–356). IEEE. Las Vegas, NV.
20. Kumari, P., & Kaur, P. (2021). A survey of fault tolerance in cloud computing. *Journal of King Saud University-Computer and Information Sciences*, 33(10), 1159–1176.
21. Kaur, G., & Kumar, R. (2015). A review on reliability issues in cloud service. *International Journal of Computer Applications*, 975, 8887.
22. Nachiappan, R., Javadi, B., Calheiros, R. N., & Matawie, K. M. (2017). Cloud storage reliability for big data applications: A state of the art survey. *Journal of Network and Computer Applications*, 97, 35–47.
23. Cui, H., Li, Y., Liu, X., Ansari, N., & Liu, Y. (2017). Cloud service reliability modelling and optimal task scheduling. *IET Communications*, 11(2), 161–167.
24. Mesbahi, M. R., Rahmani, A. M., & Hosseinzadeh, M. (2018). Reliability and high availability in cloud computing environments: A reference roadmap. *Human-Centric Computing and Information Sciences*, 8, 1–31.
25. Vishwanath, K. V., & Nagappan, N. (2010, June). Characterizing cloud computing hardware reliability. In *Proceedings of the 1st ACM symposium on Cloud computing* (pp. 193–204). Association for Computing Machinery, Indianapolis, IN.

26. Zhang, R., Lin, C., Meng, K., & Zhu, L. (2013, November). A modeling reliability analysis technique for cloud storage system. In *2013 15th IEEE International Conference on Communication Technology* (pp. 32–36). IEEE, Guilin, China.
27. Pokorni, S. J. (2019). Reliability and availability of the Internet of things. *Vojnotehnicki glasnik/Military Technical Courier*, 67(3), 588–600.
28. Sharma, Y., Javadi, B., Si, W., & Sun, D. (2016). Reliability and energy efficiency in cloud computing systems: Survey and taxonomy. *Journal of Network and Computer Applications*, 74, 66–85.

Chapter 6

Multistate network reliability – an overview

Partha Chakrabarti and Neeraj Kumar Goyal

6.1 INTRODUCTION

Modern infrastructure systems such as power transmission grids (Jia et al., 2020), production systems (Chang, 2019; Chang & Lin, 2015), computer networks (Lin & Huang, 2014), and logistics and supply chains (Golpra, 2020; Wu et al., 2008) must deliver a required level of service despite degradations, congestion, maintenance outages, and random disruptions. In such systems, components rarely behave in a purely binary way; power transmission lines may operate under temporary thermal limits, routers may throttle throughput under load, and logistics links/facilities may continue operating at reduced capacity rather than failing outright. Reliability assessment for these networks therefore needs models that can represent graded performance and quantify the probability that a target demand level can still be delivered from source to destination.

As stated above, real-world infrastructure failures repeatedly demonstrate that network components rarely fail outright but instead undergo progressive degradation that severely affects system performance. In 2021, the massive Ever Given grounding in the Suez Canal reduced the channel's operational capacity to zero for 6 days, halting nearly 12% of global maritime trade and triggering downstream congestion across road, rail, and warehousing networks worldwide. The 2022 Rogers Communications outage in Canada began as a software update issue that degraded core routing performance before cascading into a near-national blackout of mobile, broadband, and payment networks, highlighting how partial performance drops in backbone systems propagate across dependent layers. In power systems, the 2003 Northeast Blackout in the U.S.-Canada region was initiated by a set of transmission lines operating under heavy thermal stress; as their capacity limits were exceeded, line sag and protective trips triggered sequential deratings and outages, ultimately collapsing a grid serving 55 million customers. Similar capacity level degradations were observed during European heatwaves (2019–2022), where thermal derating of transmission lines forced utilities to operate below nominal capacity, reducing deliverable power despite no component being fully "failed." In cloud and

DOI: 10.1201/9781003787525-6

computing infrastructures, the 2020 Gmail and Google Cloud outage was caused by a quota misconfiguration that throttled authentication servers, degrading throughput until service became inaccessible for millions of users. In supply chains, the COVID-19 period exposed how partial shutdowns that reduced staffing, constrained port operations, and limited warehouse throughput can cause systemic delays even without complete facility failure.

A natural framework for this purpose is the Multistate Flow Network (MFN) model, where each component (edge and/or node) can take one of several discrete capacity states, with an associated probability distribution (Lin, 2001). The system performance is typically defined through a flow-based measure which is the maximum transferable flow between a source and a sink. The demand (d) two-terminal reliability is the probability that the realized network can support at least d units of flow from source to the destination (Bai et al., 2015b; Lin & Huang, 2014). While MFNs provide a far more realistic abstraction than binary networks for many applications, their reliability evaluation becomes computationally challenging as the number of components and capacity levels grows. A key difficulty is that exact MFN reliability evaluation is non-deterministic polynomial-time hard (NP-hard) in general, largely due to the rapid growth of the feasible state space and the combinatorial explosion in the number of d-minimal paths (d-MPs). Consequently, much of the literature has focused on two intertwined tasks: (1) generating the relevant d-MPs/d-minimal cuts (MCs) efficiently and (2) computing the probability of the resulting union of success events using exact (e.g., inclusion–exclusion, sum-of-disjoint-products (SDP)/factoring) or approximate (e.g., sampling and bounding) strategies. In practice, even when a complete d-MP set is available, the efficiency of recursive disjointing/factoring procedures can vary substantially depending on how these paths are processed.

6.2 RELATED LITERATURE

An established theoretical formulation defines system reliability as the probability of a success event over the vector of component states, wherein system performance is quantified by the maximum flow attainable under the given state. Conceptually, the most direct exact technique is state-space enumeration (Bai et al., 2018), which sums probabilities over all state vectors that satisfy the demand. To reduce this combinatorial explosion, early exact approaches exploited decomposition and max-flow computations within a structured enumeration. Classic work by Doulliez and Jamoulle (1972) introduced a decomposition-style reliability evaluation using the Ford–Fulkerson framework (Fulkerson, 1966), establishing a template for combining flow computations with probabilistic state reasoning. Lin et al. (2001) proposed a pivotal d-MP enumeration method that reduced redundant enumerations compared with earlier formulations, and subsequent enhancements addressed unreliable nodes, feasibility checks, and all-demand-level d-MP generation. As early works used inclusion–exclusion methods which were computationally intensive, the new SDP technique traces to fast reliability evaluation algorithms initially for binary networks, including Aggarwal et al.'s (1975) fast evaluation approach and Abraham's improved algorithm (1979). Locks (2009) refined Abraham's approach by introducing an alphanumeric ordering scheme for both MPs and generated disjoint terms, leading to more efficient computations.

Zuo et al. (2007) introduced a recursive formulation for evaluating reliability for MFNs grounded in the SDP framework, which they termed the Recursive SDP (RSDP) method. This approach demonstrated notable computational advantages over earlier SDP-based techniques, particularly for networks with a large number of components. Subsequently, Yeh (2015) proposed an enhanced version, referred to as the improved SDP (iSDP), which applies algebraic simplification strategies to reduce the required summations and multiplications, thereby increasing the overall efficiency of RSDP when generating disjoint d-MPs. Building on these developments, Datta and Goyal (2023) proposed a single-variable inversion SDP technique by introducing three rules that directly identify redundant and shared capacity states among d-MPs and eliminate them efficiently, resulting in a notable reduction in redundant computations.

More recently, the Chakrabarti and Goyal (2024) proposed a multi-variable inversion (MVI) technique which removes shared and redundant capacity states more efficiently during disjointing, achieving substantial reductions in the generation of number of disjoint vectors compared to the well-established SDP algorithms available in the literature.

Another class of reliability evaluation method is the universal generating functions (UGFs) proposed by the authors (Levitin, 2005; Lisnianski & Levitin, 2003; Ushakov, 1986), which represent discrete performance distributions and combined subsystem UGFs through structure-dependent composition operators, while multi-valued decision diagrams (MDDs) encode the multistate structure function as a layered Directed Acyclic Graph (DAG) enabling dynamic-programming probability propagation (Shrestha et al., 2010; Xing & Dai, 2008; Zang et al., 2003). These approaches are well-established for multistate systems and have been adapted to network contexts where the structure function is induced by max-flow feasibility. Although UGFs/MDDs are exact and efficient for certain topologies, their scalability depends strongly on diagram size and the chosen variable ordering.

It was observed that pre-ordering the d-MPs significantly simplifies the subsequent SDP disjointing process. In this context, the ordering heuristics proposed by Bai et al. (2015a) provide early improvements by restructuring d-MP vectors using generalized length and score-based criteria to reduce the growth of disjoint product terms within RSDP. However, the pre-ordered

MP method introduced in Chakrabarti and Goyal (2024) offers a stronger and more simplified advantage by prioritizing d-MPs according to the number of component capacity states they eliminate, enabling the SDP algorithm to remove the largest shared regions early in the disjointing process.

For large MFNs where exact d-MP/d-MC enumeration is impractical, approximate and simulation-based methods play a decisive role. Monte Carlo simulation (Cancela et al., 2013; Ramirez-Marquez & Coit, 2005) estimates reliability by repeatedly sampling component capacities and checking if the realized network can route at least demand d.

Finally, bounding and partial-enumeration approaches (Butler, 1982; Hudson & Kapur, 1985; Niu & Shao, 2011) provide a middle ground. Bounding methods estimate lower/upper reliability values by avoiding complete enumeration; early bounding strategies can be fast but sometimes loose, motivating later work on tighter bounds via smarter decomposition. Heuristic truncation of minimal sets is also effective, generating only a subset of higher-impact MPs/MCs and pruning low-impact objects that can yield near-exact estimates with large speedups, with accuracy tightening as more MPs are included (Forghani-elahabad et al., 2019).

6.3 BASIC CONCEPTS AND DEFINITIONS

In a network context, components are typically nodes or edges of a graph. Let $G=(V,E)$ be a directed or undirected graph where each edge $e \in E$ has a multistate capacity c_e, and s and t are specified source and terminal nodes (two-terminal reliability), or more general sets of origin and destination nodes (multi-terminal networks). For a given realization of edge capacities $c=(c_e \ : e \in E)$, define the system performance as the maximum flow $F(c)$ from s to t capturing total served demand, connectivity, or other quality measure. Such a network (G) is considered to satisfy the following assumptions further:

1. The flow throughout the network (G) obeys the flow conservation law, meaning that for every intermediate node the total incoming flow equals the total outgoing flow except the source and the terminal node.
2. All demands are exclusively transferred from the source node (s) to the terminal node (t).
3. All nodes are perfectly reliable, i.e., do not experience capacity degradation or failure.
4. The capacity states of all components are mutually independent.
5. Each arc possesses an integer-valued capacity defined by a specified discrete probability distribution that characterizes its multistate performance.

The network is deemed operational at demand d if $F(c) \geq d$ and the network reliability at level d is $R(d) = P(F(C) \geq d)$. When the performance measure is maximum flow, the model is often called a MFN or limited-flow network. A central contribution of multistate network reliability theory is the generalization of binary MP/MC concepts to d-MPs and d-MCs. A state vector x is a d-MP if the system performance under x is at least d : $\varphi(x) \geq d$ (or $F(c) \geq d$).

A state vector x is a d-MP if for any state vector $y \leq x$ with $y \neq x$, i.e., $\varphi(y) < d$. In other words, a d-MP is a minimal sufficient combination of component performance levels to meet demand d. To verify whether a candidate vector is a true d-MP, we apply **Theorem 1**.

Theorem 1

ϕ_{min} denotes the minimal set of lower-boundary flow vectors (d-MPs) required to satisfy a given demand d; a lower-boundary vector is included in ϕ_{min} only if it is not dominated component-wise by another vector in the set.

Proof

Let $\phi = \{X_1, X_2, \ldots, X_k\}$ be the set of lower-boundary flow vectors where each $X_j = (c_{j,1}, c_{j,2}, \ldots, c_{j,m})$. Consider two distinct vectors X_o and X_p such that $c_{o,i} \geq c_{p,i}$ for all $i = 1, 2, \ldots, m$. In this situation, every feasible system state in the range $[X_o, M]$ is already contained within the range $[X_p, M]$, meaning X_o does not contribute any additional success states. Hence, X_o is a redundant lower-boundary vector and belongs to $\phi_{\text{redundant}}$, leading to the minimal set $\phi_{\text{min}} = \phi - \phi_{\text{redundant}}$.

For example, if $X_o^l = (4,4,4,3,3,4,4,6,6)$ and $X_p^l = (4,3,1,0,3,4,4,4)$ with maximal capacities $M = (4,4,4,4,4,4,6,6)$, then since $c_{o,i} \geq c_{p,i}$ holds for all components, the interval $[X_o^l, M]$ is entirely contained within $[X_p^l, M]$; thus, X_o^l is redundant and excluded from ϕ_{min}.

Given a set of all d-MPs, $\{x(1), \ldots, x(K)\}$, the event that the system meets level d is the union of the events that the actual state vector dominates at least one d-MP as stated in equation (6.1).

$$\{\text{ system} \geq \text{d}\} = \bigcup_{k=1}^{K} \{\mathbf{X} \geq \mathbf{x}^{(k)}\} \tag{6.1}$$

The computation of $R(d)$ then reduces to the union probability of these monotone events, commonly addressed through inclusion–exclusion formulas, SDP representations, and recursive factoring algorithms.

The bottleneck is often not only the union probability but the prior enumeration of all d-MPs or d-MCs. Finding all d-MPs is itself an NP-hard problem; extensive research has gone into developing more efficient enumeration algorithms, including branch-and-bound, dynamic programming, candidate pruning via dominance relations, and specialized testing of candidate state vectors.

6.4 EXACT EVALUATION METHODS

Exact reliability evaluation aims to compute $R(d)$ without approximation. For small or moderately sized networks, it is feasible but for large, dense, or high-dimensional multistate networks, it often becomes intractable due to the underlying NP-hardness. The exact reliability evaluation methods available in the literature are reviewed and discussed in the following subsections.

6.4.1 State-space enumeration

The most direct approach is to enumerate all possible component state vectors $x \in X_1 \times \cdots \times X_n$, compute $\varphi(x)$, and sum the probabilities of those meeting the demand as shown in equations (6.2) and (6.3).

$$R(d) = \sum_{\mathbf{x}:\phi(\mathbf{x}) \geq d} P(\mathbf{X} = \mathbf{x}) \tag{6.2}$$

If components are independent and $P(X_i = k) = p_{ik}$, then

$$P(\mathbf{X} = \mathbf{x}) = \prod_{i=1}^{n} p_{i,x_i} \tag{6.3}$$

This naive enumeration has complexity proportional to $\prod_{i=1}^{n}(m_i + 1)$, which explodes quickly as the number of components and states increases. Nonetheless, it serves as a conceptual benchmark and may be practical for small networks.

6.4.2 UGF and decision diagrams

Another powerful class of methods is based on UGFs and MDDs. The UGF is a compact mathematical representation of the performance distribution of a component, subsystem, or network element. For a subsystem whose performance can take several discrete levels $g_1, g_2, \ldots, g_K$ with respective probabilities $p_1, p_2, \ldots, p_K$, the UGF is shown in equation (6.4).

$$u(z) = \sum_{k=1}^{K} p_k z^{g_k} \tag{6.4}$$

where p_k is the probability of subsystem performance level g_k. Parallel, series, and more complex structures have corresponding composition operators that combine UGFs of subsystems to obtain the UGF of the whole system. Reliability at level d can then be obtained from the UGF by summing the probabilities associated with performance levels $g_k \geq d$ as shown in equation (6.5).

$$R(d) = \sum_{k:g_k \geq d} p_k \tag{6.5}$$

In the MDD-based approach, the multistate system (or network) is described by a multistate structure function $\phi(x_1, x_2, \ldots, x_m)$, where x_i denotes the performance state (e.g., capacity level) of component i, typically taking values in a finite set $\{0, 1, \ldots, c_i^{\max}\}$. This structure function is encoded as a MDD, which is a layered DAG with one layer for each component. At the layer corresponding to component i, each outgoing arc from a node is labeled by a possible state value x_i and the probability $P(X_i = x_i)$ of that state. A complete path from the root node to a terminal node represents one specific combination of component states $(x_1, x_2, \ldots, x_m)$. Assuming statistical independence, the probability of that global state is obtained by multiplying the probabilities along the arcs of the path. The terminal nodes are associated with the resulting system performance level $g = \phi(x_1, x_2, \ldots, x_m)$.

To compute the reliability at demand level d, one traverses the MDD and accumulates the probabilities of all paths that end in terminal nodes with performance level $g \geq d$. In practice, this is done via dynamic programming along the MDD where probability masses are propagated from the root to the terminals, combining contributions that reach the same node, and finally summing the total probability that ends in terminals satisfying $\phi(\mathbf{x}) \geq d$. This yields an exact value of the multistate system (or network) reliability while avoiding explicit enumeration of all possible state vectors.

6.4.3 Recursive SDP and factoring methods

Once all d-MPs are known, exact reliability evaluation often proceeds via SDP-based or recursive factoring methods. The key ideas are:

- Represent the union of events in disjoint (mutually exclusive) form so that probabilities can be summed directly.
- Exploit network structure to recursively decompose the problem through conditioning on component states.

Building on this foundation, the RSDP method extends classical SDP principles to multistate networks by systematically decomposing the success event into disjoint product terms through a recursive expansion that eliminates overlap among d-MPs. To improve computational efficiency, further iSDP formulations incorporate some proposed rules based on GCF (Greatest Common Factors) to decrease the number of calculations in the final reliability expression. A more algebraically compact alternative, the MVI method, avoids explicit disjoint decomposition and instead derives a compact reliability expression through MVI of component capacities over path sets during the disjointing procedure. We will be presenting an illustrative example to find network reliability using MVI which is the most efficient method available in the literature.

6.4.4 Approximate and simulation-based methods

For large multistate networks, exact methods become infeasible due to the size of the state space and the explosion in the number of d-MPs/d-MCs. Approximate and simulation-based methods then play a crucial role. Monte Carlo simulation estimates $R(d)$ by repeatedly sampling component states according to their distributions, computing the system performance $F(\mathbf{C})$, and counting the fraction of trials where $F(\mathbf{C}) \geq d$. Given N samples, the estimator $\hat{R}(d) = \frac{1}{N}\sum_{j=1}^{N} I\left(F\left(\mathbf{C}^{(j)}\right) \geq d\right)$ is unbiased, with standard error approximately $\sqrt{\frac{R(d)\left(1 - R(d)\right)}{N}}$. However, when the network is highly reliable (i.e., $R(d)$ close to 1) or when the event of interest is rare, crude Monte Carlo becomes inefficient as an enormous number of samples are needed to achieve acceptable accuracy. To mitigate this, variance reduction techniques are employed:

- Importance sampling, which biases sampling toward more "critical" configurations (e.g., near the demand boundary) and then reweights samples.
- Splitting methods, which split trajectories that approach the failure (or success) region.
- Control variates and stratified sampling, leveraging analytical approximations or decompositions.

These methods have been adapted to multistate networks by identifying suitable importance measures and critical regions in the high-dimensional component state space.

6.4.5 Bounding and approximation via partial enumeration

In many practical situations especially for large multistate networks where full state-space enumeration is computationally prohibitive, obtaining tight and computationally efficient bounds on $R(d)$ is often sufficient. Alternatively enumerating all d-MPs/d-MCs is NP-hard, various heuristic and metaheuristic methods have been proposed to generate good subsets of d-MPs quickly as follows:

- Prioritizing d-MPs according to the degree of overlap among their corresponding capacity state vectors, arranging them so that successive d-MPs exhibit the highest overlap.
- Greedy search guided by edge importance or capacity contribution.
- Genetic algorithms and evolutionary search in the space of paths/cuts.
- Local search and tabu search strategies.
- Learning-based estimators that approximate the mapping from component reliabilities and network topology to $R(d)$.

By restricting attention to this subset, one can derive conservative lower and upper bounds on the reliability without requiring complete identification of all MPs/MCs. Since each d-MP represents a minimal sufficient configuration for achieving demand d, the union of the corresponding success regions forms a subset of the true success space. Therefore, evaluating the probability of this truncated union yields $R_{\text{LB}}(d) \le R(d)$, where the bound becomes tighter as additional d-MPs are included.

6.5 SYSTEM RELIABILITY EVALUATION (MVI-BASED APPROACH)

This section outlines the workflow for computing the level-d reliability using the MVI approach as discussed in this section including the generation of MPs and ending with the final reliability expression.

6.5.1 Step 1: Generate feasible and valid flow allocations from MPs

Let the set of MPs between the source node(s) and terminal node (t) be $P_1, P_2, \ldots, P_k$. A flow allocation (flow vector) is denoted by

$$\mathrm{F} = (f_1, f_2, \ldots, f_k),$$

where f_i is the amount of demand routed through MP P_i. A flow vector is feasible if it satisfies:

1. $f_i \geq 0$ for all $i = 1, 2, \ldots, k$, and
2. $\sum_{i-1}^{k} f_i = d$, where d is the required demand level.

However, feasibility in this sense does not guarantee that the allocation can be simultaneously supported by shared components. Because different MPs may overlap, the total flow through a common component may exceed its capacity. For instance, consider $\mathrm{F} = (4, 3, 0, 0)$ for $d = 7$ with four MPs. If a component c_1 is shared by both paths P_1 and P_2 and its maximum capacity is 4 units, then routing 4 units on P_1 and 3 units on P_2 would require more than 4 units through c_i, making this allocation infeasible in practice. Such vectors are said to be infeasible and removed from consideration. The remaining allocations are termed as valid flow vectors (VFVs).

6.5.2 Step 2: Convert each VFV to a boundary flow vector

For every VFV we construct a boundary flow vector in the component state space. This boundary vector represents a component capacity state in which the assigned path flows in VFV can pass from s to t exactly as allocated. Since reliability evaluation requires component capacity states (and their probabilities), this step maps each valid allocation into the corresponding capacity state requirement on every component.

6.5.3 Step 3: Obtain lower-boundary flow vectors (d-MPs)

The boundary flow vectors are then reduced to lower-boundary flow vectors, i.e., the d-MPs. Each d-MP is a minimal component capacity state vector that is sufficient to support demand level d; reducing any component state in a d-MP would violate the ability to transmit d.

6.5.4 Step 4: Compute point reliability

Let the d-MP set be $\{X_1, X_2, \ldots, X_L\}$, where each X_i is a component-wise state vector. The following preliminaries and rules are required for applying the MVI technique and calculating the final reliability.

For a network with m multistate elements $c_1, \ldots, c_m$ and maximum capacity vector M, each disjoint term is stored as an MVI container with three length-m arrays: a lower-bound flow vector $X_{j'}^{l}$ an upper-bound flow vector X_j^u, and a group-label vector G_j. Ungrouped elements are those with $G_j(i) = 0$; their admissible "success" states lie between their lower and upper bounds. Grouped elements share the same positive group label and contribute through a difference of products that subtracts the overlap block defined by the group's upper bounds. Let $A = \{i : G_j(i) = 0\}$ and $B_g = \{i : G_j(i) = g\}$ for groups $g = 1,\ldots,r$. Then the probability of an MVI container is evaluated using equation (6.6).

$$\Pr(\text{MVI}_j) = \prod_{i \subset A} \Pr(c_i^l : c_i^u) \prod_{g=1}^{r} \left(\prod_{i \subset B_g} \Pr(c_i^l) - \prod_{i \subset B_g} \Pr(c_i^u) \right) \tag{6.6}$$

where $\Pr(c_i^l)$ and $\Pr(c_i^u)$ are taken directly from the cumulative/survival capacity tables (e.g., $\Pr(X_i \geq c_i)$). In practice, $\Pr(c_i^l : c_i^u)$ is computed from the same table as the probability mass assigned to the allowed interval between the lower and upper bounds (with the standard convention that when $c_i^u = M_i$, this reduces to the survival value at c_i^l).

6.5.4.1 Rule 1. Disjointing two lower-boundary flow vectors

Given two lower-boundary flow vectors X_j (already processed) and X_{j+1} (current), overlap occurs whenever they share one or more component capacity states. The basic MVI disjointing rule removes this overlap by truncating the current container's upper bounds at the earlier lower bounds and recording the truncation via grouping. Concretely, for each element i such that $c_{i,j+1}^l < c_{i,j}^l$, the upper bound $c_{i,j+1}^u$ is replaced by $c_{i,j}^l$, and the corresponding group label is updated from 0 to the index of the earlier term being subtracted.

6.5.4.2 Rule 2. Disjointness check

To avoid unnecessary operations, two checks are applied repeatedly during disjointing. First, two terms are already disjoint if (1) among ungrouped elements, at least one element's upper bound is strictly below the compared lower bound, or (2) for grouped elements, at least one full group has all its upper bounds less than or equal to the compared lower bounds.

6.5.4.3 Rule 3. Redundancy check

Two lower-boundary flow vectors are redundant if its lower bounds are component-wise greater than or equal to those of an earlier term (so it does not contribute any new success states); such terms are discarded immediately.

6.5.4.4 Rule 4.Group breaking w.r.t. common elements

During disjointing, a case arises when an element that needs truncation is already inside a group (a "common element"), meaning its admissible region is coupled with other grouped elements. In this case, the MVI method performs a group-breaking operation: the current container is split into two containers so that one part becomes automatically disjoint (by restricting the common element's range below the compared lower bound), while the remaining part retains the non-overlapping region and continues disjointing.

When multiple common elements occur within the same group, the same split is applied to each such element in that group; when common elements occur across multiple groups, groups are opened sequentially, yielding at most 2^n containers if n groups contain common elements, followed by disjointness and redundancy checks on each resulting container.

A further mixed case occurs when, within a grouped set of common elements, some have upper bounds already below the compared lower bound while others do not; the container is again split so that the portion with "small" upper bounds becomes disjoint, and only the remaining portion requires continued disjointing.

6.6 OVERALL WORKFLOW AND RELIABILITY COMPUTATION

Starting from the precomputed set of d-MP (lower-bound) vectors $\{X_1, X_2 \ldots, X_k\}$, the MVI procedure initializes the first disjoint term as a single container with $LB = X_1$, $UB = M$, and all group labels zero. For each subsequent X_i, a new container is created with $LB = X_i$, $UB = M$, $G = 0$, and then it is disjointed sequentially with respect to all earlier $\{X_1, X_2 \ldots, X_{k-1}\}$. At each comparison, the method applies: (1) the disjointness test to skip computation whenever possible, (2) the truncation and grouping rule to remove overlap, (3) group breaking when common elements are encountered, and (4) redundancy pruning to discard containers that add no new success states.

Because the resulting MVI containers are mutually exclusive by construction, the point reliability at demand d is obtained by summing container probabilities as stated in equation (6.7).

$$R_d = \Pr\left(\bigcup_{j=1}^{\delta} \mathrm{MVI}_j\right) = \sum_{j=1}^{\delta} \Pr\left(\mathrm{MVI}_j\right) \tag{6.7}$$

An illustrative example

To demonstrate the application of the MVI technique for multistate network reliability evaluation, we consider a simple yet widely used benchmark bridge network taken from the established literature. The network, shown in Figure 6.1, consists of one source node S, two intermediate nodes $N_1$and N_2, and a terminal node T. The connectivity between these nodes is provided by six distinct arcs (denoted c_1–c_6), forming the classic bridge or diamond network topology commonly used for illustrating flow-based reliability models. Each component operates in multiple capacity states and may experience partial degradation or failure. To reduce the computational effort required during the MVI-based reliability evaluation, the cumulative probability distributions of the component capacity states are employed, as provided in Table 6.1.

The supervisor would like to know the reliability of the network to pass 3 units of demand from source (s) to terminal (T) (i.e. R_3). We obtain the set of three 3-MPs using Theorem 1 which are $X_1 = (3, 2, 1, 0, 0, 1)$, $X_2 = (2, 2, 0, 0, 1, 1)$, $X_3 = (2, 1, 1, 0, 1, 2)$. The upper boundary vector for all the d-MPs is $X_i^u = (3, 2, 1, 1, 1, 2)$. The reliability index (R_3) can be calculated by the step-by-step procedure as given below.

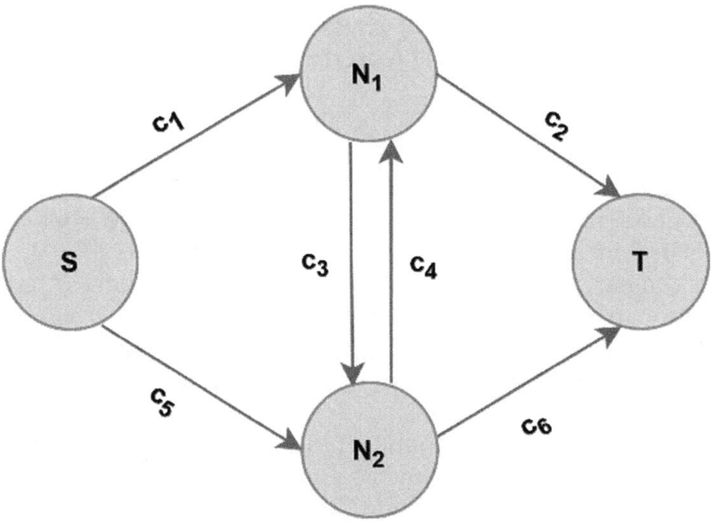

Figure 6.1 A benchmark bridge network.

Table 6.1 Cumulative probabilities of element capacities in Figure 6.1

Component	c_1	c_2	c_3	c_4	c_5	c_6
Capacity 0	1.00	1.00	1.00	1.00	1.00	1.00
Capacity 1	0.90	0.90	0.90	0.90	0.90	0.95
Capacity 2	0.85	0.60				0.70
Capacity 3	0.60					

STEP 0: Initialize $\text{MVI}_1\left[X_i^l = X_1, X_i^u = M \text{ and } G = 0\right]$

$$\text{MVI}_1 = \left[X_1^l = (3,2,1,0,0,1), X_1^u = (3,2,1,1,1,2), G_1 = (0,0,0,0,0,0)\right]$$

Probability (all ungrouped components) can be calculated directly using equation (6.6) and Table 6.1 as follows:

$$\begin{aligned}\Pr(\text{MVI}_1) &= c_{1,1}(3)c_{2,1}(2)c_{3,1}(1)c_{4,1}(0)c_{5,1}(0)c_{6,1}(1)\\ &= 0.6\times0.6\times0.9\times1\times1\times0.95\\ &= 0.3078\end{aligned}$$

STEP 1: Disjoint X_2 w.r.t. X_1 using **Rule 1** which says to find elements where $c_{i,2}^l < c_{i,1}^l$ and replace the upper bound of $c_{i,2}^l$ by the lower bound of $c_{i,1}^l$ and assign a group number.

Compare $X_2 = (2,2,0,0,1,1)$ with $X_1 = (3,2,1,0,0,1)$ where $c_{1,2}^l(2) < c_{1,1}^l(3)$, $c_{3,2}^l(0) < c_{3,1}^l(1)$ and all other components in $X_2(c_2, c_4, c_5, c_6)$ are greater than or equal to the elements in X_1 so they will remain ungrouped. Hence, the second MVI container is given below:

$$\text{MVI}_2 = \left[X_{2,1}^l(2,2,0,0,1,1), X_{2,1}^u(3,2,1,1,1,2), G_{2,1}(1,0,1,0,0,0)\right]$$

$$\Pr(\text{MVI}_2) = (0.85\times1-0.6\times0.9)(0.6)(1)(0.9)(0.95) = 0.15903$$

STEP 2: Disjoint X_3 w.r.t. X_1, then w.r.t. X_2

2 (a) Disjoint X_3 w.r.t. X_1 using **Rule 1**

Compare $X_3 = (2,1,1,0,1,2)$ with $X_1 = (3,2,1,0,0,1)$ where $c_{1,3}^l(2) < c_{1,1}^l(3)$, $c_{2,3}^l(0) < c_{2,1}^l(1)$, and all other components in $X_2(c_3, c_4, c_5, c_6)$ are greater than or equal to the elements in X_1 so they will remain ungrouped. Hence, the third MVI container is given below:

$$\text{MVI}_3 = \left[X_{3,1}^l(2,1,1,0,1,2), X_{3,1}^u(3,2,1,1,1,2), G_{3,1} = (1,1,0,0,0,0)\right]$$

2(**b**) Now again we need to disjoint this with X_2. While trying to disjoint using Rule 1 we find that in $X_{3,1}^l$ the second component $c_{2,3}^l(1)$ is less than the second component of X_2^l which is $c_{2,2}^l(2)$. As $c_{2,3}^l(1)$ in $X_{3,1}^l$ already belongs to group ($G=1$) we split MVI_3 into two parts based on **Rule 4** of group breaking as follows:

$$\text{MVI}_{3(1)} = \left[X_{3,1}^{l}(2,1,1,0,1,2), X_{3,1}^{u}(3,2,1,1,1,2), G_{3,1}=(0,1,0,0,0,0)\right]$$

and

$$\text{MVI}_{3(2)} = \left[X_{3,1}^{l}(2,2,1,0,1,2), X_{3,1}^{u}(3,2,1,1,1,2), G_{3,1}=(1,0,0,0,0,0)\right]$$

As $\text{MVI}_{3(1)}$ is already disjoint according to **Rule 2** and $\text{MVI}_{3(2)}$ is redundant according to **Rule 3** we will store only $\text{MVI}_{3(1)}$ and remove redundant $\text{MVI}_{3(2)}$.

$$\text{MVI}_3 = \left[X_{3,2,1}^{l}(2,1,1,0,1,2), X_{3,2,1}^{u}(3,2,1,1,1,2), G_{3,2,1}=(0,1,0,0,0,0)\right]$$

$$\Pr(\text{MVI}_3) = 0.85\times(0.9-0.6)\times0.9\times1\times0.9\times0.7 = 0.144585$$

$$\begin{aligned} R_d &= \sum_{j=1}^{\delta}\Pr(\text{MVI}_j) \\ &= \Pr(\text{MVI}_1)+\Pr(\text{MVI}_2)+\Pr(\text{MVI}_3) \\ &= 0.3078+0.15903+0.144585 = 0.611415 \end{aligned}$$

6.7 APPLICATIONS

Multistate network reliability is used wherever flows (power, data, people, goods) move through networks whose components can work at several capacity levels rather than simply "up" or "down." Nodes and arcs are modeled with discrete capacity states (as in the distribution tables), and performance is typically measured by total deliverable flow, satisfied demand, or achieved service level. Below are representative examples of realistic networks with possible format of capacity distribution tables for its different components. These distributions can be refined further by incorporating real-time failure and degradation data collected from field operating conditions.

6.7.1 Power generation, transmission, and distribution

In power systems, generators and transmission lines operate at several output or rating levels, from full capacity down to complete outage, as represented in the generator and line capacity distributions. Network performance is usually defined as the maximum MW deliverable to a load area, the fraction of demand served, or the amount of load shed. Using these multistate models, one can (1) compute the probability that at least dMW is deliverable, (2) evaluate how line uprating or new generation changes adequacy indices, and (3) study cascading and derating effects as components move between partial and full-capacity states rather than jumping directly from "in service" to "failed." We can treat each transmission line and generator in the power distribution network as having its own discrete capacity distribution of this form as presented in Tables 6.2 and 6.3.

6.7.2 Communication and computing networks

For communication networks, each link in the capacity table represents different bandwidth or QoS levels (e.g., from full throughput to link unavailable), while computing nodes have multistate processing capacities due to failures, throttling, or power-saving modes. Performance measures include the probability that a target data rate is achievable between terminals, that delay and loss remain within QoS thresholds, or that enough processing capacity is available to serve current demand as shown in Tables 6.4 and 6.5. These models support design and planning (sizing capacity and redundancy), resilience studies (impact of partial degradations on end-to-end service), and cloud/data center architecture evaluation (probability that application-level performance targets are met given fluctuating node and link capacities).

Table 6.2 Generator unit capacity distribution (single thermal unit)

State	*Description*	*Output (MW)*	*Probability*	*Comment*
0	Forced outage	0	0.05	Unit tripped or unavailable
1	Severe derating	40	0.10	Major fault or severe constraint
2	Partial capacity	80	0.20	Minor faults/high ambient temp
3	Rated capacity	100	0.65	Normal operating condition

Table 6.3 Transmission line capacity distribution (thermal rating)

State	*Description*	*Capacity (MVA)*	*Probability*	*Comment*
0	Out of service	0	0.03	Outage or protection trip
1	Emergency rating	60	0.07	Short-term overload allowed
2	Reduced rating	80	0.25	High ambient temperature/ maintenance
3	Nominal rating	100	0.65	Normal condition

Table 6.4 Communication link capacity distribution (backbone link)

State	*Description*	*Bandwidth (Mbps)*	*Probability*	*Comment*
0	Link down	0	0.02	Hardware/software failure
1	Severely degraded	50	0.08	Heavy congestion/multiple faults
2	Moderately degraded	150	0.25	Normal traffic with some congestion
3	Full capacity	300	0.65	Light load, QoS guarantees maintained

Table 6.5 Compute node/server cluster capacity distribution

State	*Description*	*Processing capacity (requests/s)*	*Probability*	*Comment*
0	Node unavailable	0	0.03	Hardware failure/ maintenance
1	Minimal service	1,000	0.12	Running on backup/ degraded mode
2	Partial capacity	3,000	0.25	High load, throttling, or power saving
3	Full capacity	5,000	0.60	Nominal operating condition

6.7.3 Transportation and logistics networks

In transportation networks, road or rail links have multiple effective capacities (from free flow to blocked), as shown in Table 6.6, while in logistics networks facilities such as warehouses or distribution centers operate at several throughput levels presented in Table 6.7. Multistate network reliability is then used to estimate the probability that a required volume of traffic or goods can be moved within a time window, or that at least a given fraction of demand is satisfied under congestion, incidents, and disruptions. This supports evaluating alternative routes and facility layouts, identifying critical links and hubs whose capacity reductions most harm performance, and informing risk management and business continuity plans for supply chains.

6.7.4 Emerging cyber-physical and multi-layer networks

In cyber-physical and multi-layer infrastructures, different layers (e.g., power, communication, information, transport) are modeled as interacting multistate networks. Table 6.8 captures physical power transfer capability,

Table 6.6 Road link capacity distribution (vehicles per hour)

State	*Description*	*Capacity (veh/h)*	*Probability*	*Comment*
0	Blocked	0	0.05	Accident, road works, or flooding
1	Severe congestion	400	0.15	Multiple incidents/peak demand
2	Moderate congestion	800	0.30	Typical peak-hour condition
3	Free flow/near free	1,200	0.50	Off-peak, good weather

Table 6.7 Warehouse/distribution center throughput distribution

State	*Description*	*Throughput (pallets/day)*	*Probability*	*Comment*
0	Closed	0	0.02	Strike, disaster, or total disruption
1	Strongly limited	200	0.13	Labor shortage/equipment failure
2	Reduced capacity	500	0.35	Staff partially available, minor issues
3	Normal operation	800	0.50	Full staffing and equipment

Table 6.8 Power layer: substation capacity distribution

State	*Description*	*Transfer capacity (MW)*	*Probability*	*Comment*
0	Substation offline	0	0.02	Major fault/loss of supply
1	Heavily derated	50	0.08	Transformer fault, temporary bypass
2	Partially derated	90	0.25	High loading, one transformer out
3	Full operating capacity	120	0.65	All equipment in service

while Table 6.9 captures communication capacity; the coupled table illustrates how the effective power capacity of a "substation + control" pair depends jointly on both layers. Multistate reliability analysis can then quantify (1) the probability that a critical service (e.g., delivering at least d MW with sufficient control bandwidth) is achievable, (2) joint performance levels across multiple infrastructures under shared hazards, and (3) the likelihood and severity of cascading degradations across layers. Building full multi-layer, dependency-aware models remains challenging, but the same multistate framework extends naturally to these interdependent systems.

Table 6.9 Communication/control layer: control link to substation

State	*Description*	*Usable bandwidth (kbps)*	*Probability*	*Comment*
0	No communication	0	0.03	Link failure, cyber-attack, or power loss
1	Severely degraded control	32	0.12	Only basic alarms, limited remote control
2	Partially degraded control	128	0.35	Reduced redundancy/ higher delays
3	Full control capability	512	0.50	Normal SCADA/PMU data and control

6.8 CONCLUSION

This chapter reviewed key methodologies for evaluating MFN reliability, highlighting their importance for modeling modern infrastructure systems whose components operate across multiple capacity levels. By synthesizing exact, approximate, and simulation-based approaches we showed how advances increasingly focus on reducing the combinatorial growth associated with d-MPs and their overlap. An illustrative example demonstrated the practical steps of applying MVI, while application cases across power, communication, transportation, and cyber-physical networks emphasized the realism and versatility of multistate modeling. Overall, MFN reliability remains computationally challenging, but ongoing progress in efficient disjointing, better minimal-path handling, and data-driven capacity modeling continue to expand its applicability to large, interdependent, and capacity variable systems.

REFERENCES

Abraham, J. A. (1979). An improved algorithm for network reliability. *IEEE Transactions on Reliability*, R-28(1), 58–61.

Aggarwal, K. K., Misra, K. B., & Gupta, J. S. (1975). A fast algorithm for reliability evaluation. *IEEE Transactions on Reliability*, 24(1), 83–85.

Bai, G., Tian, Z., & Zuo, M. J. (2018). Reliability evaluation of multistate networks: An improved algorithm using state-space decomposition and experimental comparison. *IISE Transactions*, 50(5), 407–418.

Bai, G., Zuo, M. J., & Tian, Z. (2015a). Ordering heuristics for reliability evaluation of multistate networks. *IEEE Transactions on Reliability*, 64(3), 1015–1023.

Bai, G., Zuo, M. J., & Tian, Z. (2015b). Search for all *d*-MPs for all *d* levels in multistate two-terminal networks. *Reliability Engineering & System Safety*, 142, 300–309.

Butler, D. A. (1982). Bounding the reliability of multistate systems. *Operations Research*, 30(3), 530–544.

Cancela, H., Robledo, F., Rubino, G., & Sartor, P. (2013). Monte Carlo estimation of diameter-constrained network reliability conditioned by pathsets and cutsets. *Computer Communications*, 36(6), 611–620.

Chakrabarti, P., & Goyal, N. K. (2024). A novel sum-of-disjoint-products technique for reliability evaluation of multistate consumer electronics communication flow network. *IEEE Transactions on Consumer Electronics*, 70(3), 5839–5847.

Chang, P.-C. (2019). Reliability estimation for a stochastic production system with finite buffer storage by a simulation approach. *Annals of Operations Research*, 277(1), 119–133.

Chang, P.-C., & Lin, Y.-K. (2015). Reliability analysis for an apparel manufacturing system applying fuzzy multistate network. *Computers & Industrial Engineering*, 88, 458–469.

Datta, E., & Goyal, N. (2023). An efficient sum of disjoint product method for reliability evaluation of stochastic flow networks using *d*-MPs. *International Journal of System Assurance Engineering and Management*, 14(4), 1228–1246.

Doulliez, P., & Jamoulle, E. (1972). Transportation networks with random arc capacities. *Revue Française d'automatique, Informatique, Recherche Opérationnelle. Recherche Opérationnelle*, 6(V3), 45–59.

Forghani-elahabad, M., Kagan, N., & Mahdavi-Amiri, N. (2019). An MP-based approximation algorithm on reliability evaluation of multistate flow networks. *Reliability Engineering & System Safety*, 191, 106566.

Fulkerson, D. R. (1966). Flow networks and combinatorial operations research. *The American Mathematical Monthly*, 73(2), 115–138.

Golpîra, H. (2020). Optimal integration of the facility location problem into the multi-project multi-supplier multi-resource construction supply chain network design under the vendor managed inventory strategy. *Expert Systems with Applications*, 139, 112841.

Hudson, J. C., & Kapur, K. C. (1985). Reliability bounds for multistate systems with multistate components. *Operations Research*, 33(1), 153–160.

Jia, H., Liu, D., Li, Y., Ding, Y., Liu, M., & Peng, R. (2020). Reliability evaluation of power systems with multi-state warm standby and multi-state performance sharing mechanism. *Reliability Engineering & System Safety*, 204, 107139.

Levitin, G. (2005). *The Universal Generating Function in Reliability Analysis and Optimization*. Springer.

Lin, Y.-K. (2001). A simple algorithm for reliability evaluation of a stochastic-flow network with node failure. *Computers & Operations Research*, 28(13), 1277–1285.

Lin, Y.-K., & Huang, C.-F. (2014). System reliability of assured accuracy rate for multi-state computer networks from service level agreements viewpoint. *Journal of Systems Science and Systems Engineering*, 23(2), 196–211.

Lisnianski, A., & Levitin, G. (2003). *Multi-state System Reliability: Assessment, Optimization and Applications*. World Scientific.

Locks, M. O. (2009). A minimizing algorithm for sum of disjoint products. *IEEE Transactions on Reliability*, 36(4), 445–453.

Niu, Y., & Shao, F.-M. (2011). A practical bounding algorithm for computing two-terminal reliability based on decomposition technique. *Computers & Mathematics with Applications*, 61(8), 2241–2246.

Ramirez-Marquez, J. E., & Coit, D. W. (2005). A Monte-Carlo simulation approach for approximating multi-state two-terminal reliability. *Reliability Engineering & System Safety*, 87(2), 253–264.

Shrestha, A., Xing, L., & Coit, D. W. (2010). An efficient multistate multivalued decision diagram-based approach for multistate system sensitivity analysis. *IEEE Transactions on Reliability*, 59(3), 581–592.

Ushakov, I. A. (1986). A universal generating function. *Soviet Journal of Computer System Science*, 24(5), 37.

Wu, W., Ning, A., & Ning, X. (2008). Evaluation of the reliability of transport networks based on the stochastic flow of moving objects. *Reliability Engineering & System Safety*, 93(6), 838–844.

Xing, L., & Dai, Y. S. (2008). A new decision-diagram-based method for efficient analysis on multistate systems. *IEEE Transactions on Dependable and Secure Computing*, 6(3), 161–174.

Yeh, W. C. (2015). An improved sum-of-disjoint-products technique for symbolic multi-state flow network reliability. *IEEE Transactions on Reliability*, 64(4), 1185–1193.

Zang, X., Wang, D., Sun, H., & Trivedi, K. S. (2003). A BDD-based algorithm for analysis of multistate systems with multistate components. *IEEE Transactions on Computers*, 52(12), 1608–1618.

Zuo, M. J., Tian, Z., & Huang, H. Z. (2007). An efficient method for reliability evaluation of multistate networks given all minimal path vectors. *IIE Transactions (Institute of Industrial Engineers)*, 39(8), 811–817.

Chapter 7

Maximum reliability travelling salesman tour in a stochastic network

Trust Tawanda, Adarsh Anand, and Santosh Kumar

7.1 INTRODUCTION

The Travelling Salesman Problem (TSP) is one of the combinatorial optimisation problems that is difficult to solve. Given a well-connected TSP network, the goal is to find the minimum cost tour that starts at a given node called the source and visits the remaining nodes in the network exactly once and returns to the source or home node while minimising the total travelled distance. Several methods have been developed to solve TSP, exact as well as heuristic algorithms. Exact algorithms (for example, see Tawanda et al. (2022, 2023, 2023a, Kumar et al., 2016, 2019) are computational procedures where one can compute the exact solution to small and medium TSP instances. On the other hand, the heuristic algorithms (Hong et al., 2025, Kumar and Munapo, 2025) are preferred when solving large TSP instances; this is because they compute near-optimal solutions within a reasonable computational time. In the Maximum Reliability TSP, the salesman needs to visit a specified group of cities only once and then return to the home city. The objective in this TSP variant is not minimising the distance or cost, but rather the salesman must choose a tour that is the most reliable. This variant is applicable in communication networks and supply chains with potential breakdowns. Yeh (2024) combined travel times and reliability in the constrained stochastic TSP. Mean and standard deviations of travel time, as well as reliability, were considered. The BAT algorithm was used to solve the multi-objective problem. The Steiner TSP, where network weights are correlated and stochastic, was considered by Letchford and Nasiri (2015). The mean and variance of the network were considered to make it a bi-objective optimisation problem. The problem was formulated and solved as an integer programming problem with up to 100 nodes. Wissink (2023) introduced a correlated probabilistic TSP to deal with the uncertainty of customer presence. Copula-based modelling was proposed to solve the problem. The research revealed that correlation impacts transportation routes' cost estimates and routing decisions. Some studies on reliable route include applications of reliable routing in mobile ad hoc

DOI: 10.1201/9781003787525-7

networks (Ye et al., 2003), wireless sensor networks (Mostafaei, 2018), vehicular ad hoc networks (Eiza et al., 2013), and wireless networks (Khandani et al., 2008). Zhang et al. (2017) solved the reliable shortest path problem using the Lagrangian relaxation based on a convex problem formulation. The effectiveness of the proposed method was validated using computational experiments on real road networks in terms of convergence time and optimality gap. The proposed algorithm performed better than the outer approximation algorithm. Some studies have been done on reliable shortest path (Chen et al., 2014, Chen et al., 2013, Zhang and Khani, 2019) and fuzzy reliable routes (Yang et al., 2017). Tu et al. (2020) investigated the constrained reliable shortest path problem. Other routing algorithms are documented in Tawanda (2013), Maposa et al. (2014), and Tawanda (2018). Mixed-integer programming was used to solve the problem, such as the Label setting algorithm with the aid of the A* technique to improve the convergence time. Exact solutions were computed within a reasonable computational time. Other studies were done on constrained reliable routing (Ruß et al., 2021, Chen et al., 2012). The k-reliable shortest path problem was investigated in Chen et al. (2016). This problem arises in stochastic networks with uncertain travel times. The deviation path algorithm, with the aid of the A* technique, was used to solve the problem. A case study was used to validate the algorithm, and the results revealed that good solutions were computed under various reliable network weights.

7.2 STOCHASTIC NETWORKS

Stochastic networks are useful in routing problems such as the TSP, Transportation problem, and shortest path problems. Weights of the links in these networks are associated with a probability p_{ij} or reliability $r_{ij} \in [0,1]$. These weights may represent accident probabilities, connection failure probabilities, and probability of successful message transmission, among others.

7.3 NODE LABELLING ALGORITHM

Node labelling algorithm was developed to solve network optimisation routing problems. The idea behind this method is to create unique node labels at each iteration (Kumar et al., 2025). The node labelling algorithm was used to solve TSP (Tawanda et al., 2022, 2023), and the shortest path problem (Tawanda et al., 2023a). In this chapter, a modification is proposed to solve the Maximum reliability TSP as presented in Algorithm 7.1.

Algorithm 7.1. Node Labelling Algorithm

Initialisation

Let node h be the home city. Define the set of unvisited cities as $L_0 = (h+1, h+2, \ldots, n)$.

For $j \in L_0$, initialise the label $R^*_{(h,j)^z} = r_{hj}$, where r_{hj} is the reliability of the edge between nodes h and j.

Set $z = 0$.

Step 1: Label Update

Set $j \rightarrow k$.

For all $j \in L_0$ not already visited in the path to k.

Compute new reliability labels for j as $R_{(h,j)^{z+1}} = R^*_{(h,k)^z} \times r_{kj}$

Step 2: Selection of Maximum Reliability Label

For each $j \in L_0$, select $R^*_{(h,j)^{z+1}}$ such that $R^*_{(h,j)^{z+1}} = Max\left\{R_{(h,j)^{z+1}}\right\}$
If multiple labels share the maximum value, consider all of them for further expansion in the next iteration.

Step 3: Iteration Check

If $z = n-2$, proceed to Step 4.

Otherwise, increment $z \leftarrow z+1$, and return to Step 1.

Set $j \rightarrow k$.

Step 4: Close Tours by Returning to the Home City

Compute $R_{(h,h)} = R^*_{(h,k)^{n-2}} \times r_{kh}$

Step 5: Selection of the Maximum Reliability Tour

Among all complete tours ending and returning to node 1, select the one with the highest reliability value $R^*_{(h,h)} = Max\left\{R_{(h,h)}\right\}$. The corresponding tour is the solution to the Maximum Reliability TSP.

7.4 NUMERICAL ILLUSTRATIONS

Example 7.1

Consider a four-node Maximum reliability TSP stochastic network in Figure 7.1. Node 1 is the home node, and the weights of the networks are probabilities as shown in Figure 7.1. The goal is to find the tour of maximum reliability.

Solution

The algorithm is initialised by linking the source node 1 with all the nodes that are directly connected to it, thus generating $0.99_{(1,2)}$, $0.95_{(1,3)}$, and $0.79_{(1,4)}$ node labels. Iteration 1 shows how these node labels are modified to get maximum reliability labels $R^*_{(1,j)^1}$ to each $j \in L_0$. Setting $j \to k$, node labels $R^*_{(1,k)^1}$ become the input for Iteration 2. Node labels $0.86_{(1,3,2)}$, $0.90_{(1,2,3)}$, and $0.88_{(1,2,4)}$ are modified in Iteration 2 to get new maximum reliability labels $R^*_{(1,j)^2}$. Setting $j \to k$, node labels $R^*_{(1,k)^2}$ become the input for Iteration 3. At this point all nodes have been visited and the salesman must return to the home city, node 1. Node labels $0.77_{(1,2,4,3)}$ and $0.79_{(1,2,3,4)}$ are modified by linking them back to the home city to give the maximum reliability label $R^*_{(1,1)^3}$ for the optimal tour given by $0.73_{(1,2,4,3,1)}$. Detailed iterations are shown in Table 7.1.

The best maximum reliability tour is given by $1-2-4-3-1$ as shown in Figure 7.2, with maximum reliability value of 0.73. In general, the tour is 73% reliable.

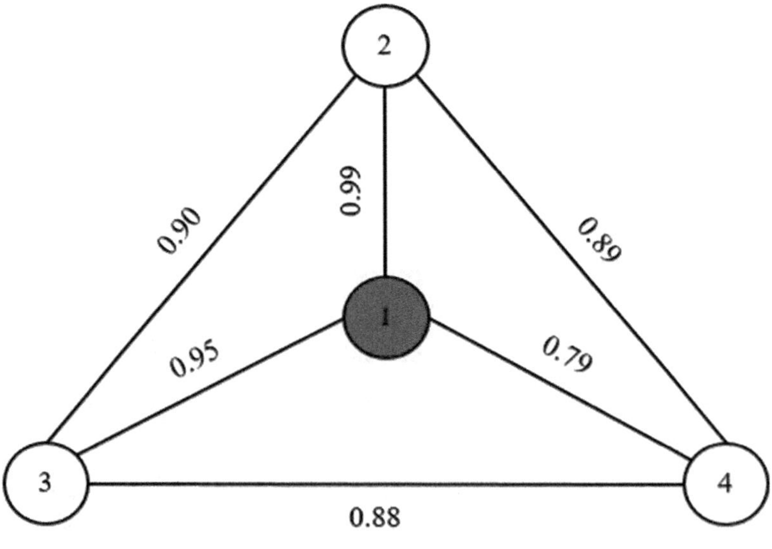

Figure 7.1 Four-node stochastic network.

Table 7.1 Iterations for the four-node stochastic network

Iteration 1				
$R^*_{(1,k)^0}$	1	2	3	4
$0.99_{(1,2)}$	-	-	$0.99 \times 0.90 = 0.90$	$0.99 \times 0.89 = 0.88$
$0.95_{(1,3)}$	-	$0.95 \times 0.90 = 0.86$	-	$0.95 \times 0.88 = 0.84$
$0.79_{(1,4)}$	-	$0.79 \times 0.89 = 0.70$	$0.79 \times 0.88 = 0.70$	-
R^*	-	0.86	0.90	0.88
$R^*_{(1,j)^1}$		$0.86_{(1,3,2)}$	$0.90_{(1,2,3)}$	$0.88_{(1,2,4)}$
Iteration 2				
$R^*_{(1,k)^1}$	1	2	3	4
$0.86_{(1,3,2)}$	-	-	-	$0.86 \times 0.89 = 0.77$
$0.90_{(1,2,3)}$	-	-	-	$0.90 \times 0.88 = 0.79$
$0.88_{(1,2,4)}$	-	-	$0.88 \times 0.88 = 0.77$	-
R^*	-	-	0.77	0.79
$R^*_{(1,j)^2}$	-	-	$0.77_{(1,2,4,3)}$	$0.79_{(1,2,3,4)}$
Iteration 3				
$R^*_{(1,k)^2}$	1	2	3	4
$0.77_{(1,2,4,3)}$	$0.77 \times 0.95 = 0.73$	-	-	-
$0.79_{(1,2,3,4)}$	$0.79 \times 0.79 = 0.62$	-	-	-
R^*	0.73	-	-	-
$R^*_{(1,1)^3}$	$0.73_{(1,2,4,3,1)}$	-	-	-

Example 7.2

The delivery driver for a critical logistics company is tasked with visiting four clients (nodes 2–5) and returning to the company (node 1). Each route between nodes has a reliability score, representing the likelihood of successful delivery as shown in Figure 7.3.

Solution

To initialise the algorithm, link the home node 1 with all the nodes that are directly connected to it, thus generating a set $R^*_{(1,k)^0}$ of node labels given by $0.25_{(1,2)}$, $0.77_{(1,3)}$, $0.10_{(1,4)}$, and $0.56_{(1,5)}$ node labels. Iteration 1 shows how these node labels are modified to get maximum reliability labels $R^*_{(1,j)^1}$ to each $j \in L_0$. These node labels are given by $0.76_{(1,3,2)}$, $0.50_{(1,5,3)}$, $0.74_{(1,3,4)}$, and $0.69_{(1,3,5)}$. Setting $j \rightarrow k$, node labels $R^*_{(1,k)^1}$ become the input for Iteration 2. Maximum reliability node labels

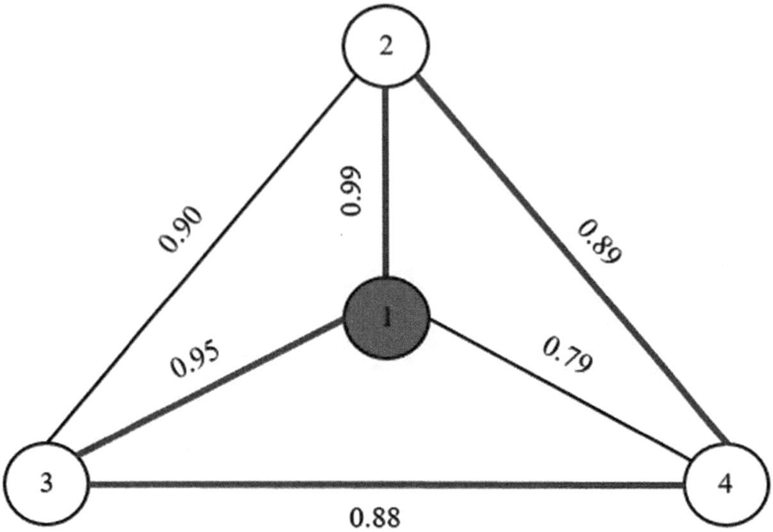

Figure 7.2 Maximum reliability optimal solution.

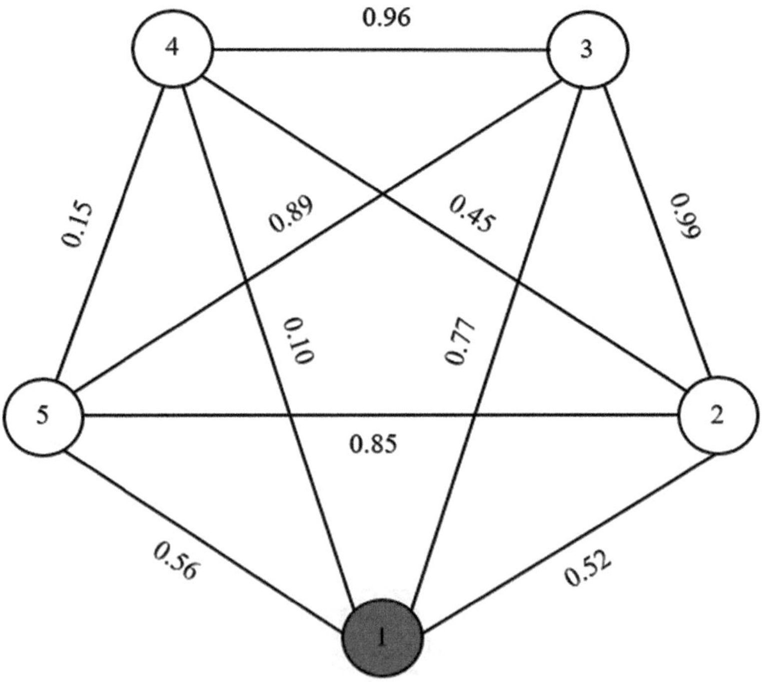

Figure 7.3 Five-node maximum reliability delivery problem.

$R^*_{(1,k)^1}$ are modified in Iteration 2 to get new maximum reliability labels $R^*_{(1,j)^2}$, thus $0.59_{(1,3,5,2)}$, $0.48_{(1,5,3,4)}$, and $0.65_{(1,3,2,5)}$ as shown in Iteration 2. Setting $j \to k$, node labels $R^*_{(1,k)^2}$ become the input for Iteration 3. Maximum reliability node labels $R^*_{(1,k)^2}$ are modified in Iteration 3 to get new maximum reliability labels $R^*_{(1,j)^3}$, thus $0.22_{(1,5,3,4,2)}$ and $0.10_{(1,3,5,2,4)}$. Setting $j \to k$, node labels $R^*_{(1,k)^3}$ are input for Iteration 4. At this point all nodes have been visited and the salesman must return to the home city, node 1. Node labels $R^*_{(1,k)^3}$ are modified by linking them back to the home city to give the maximum reliability node labels $R^*_{(1,1)^4}$ that give the optimal maximum reliability node label $0.06_{(1,5,3,4,2,1)}$. Detailed solution is shown in Table 7.2.

The best maximum reliability tour is given by $1-5-3-4-2-1$ as shown in Figure 7.4, with maximum reliability value of 0.06. The best tour is only 6% reliable.

7.5 CONCLUDING REMARKS

In this chapter, the Maximum reliability tour has been obtained by using the modified labelling method. Numerical illustrations have been used to demonstrate the computational power of the proposed algorithm. The algorithm is simple and reduces number of operations as it iterates. Future studies may consider extending the algorithm to maximum reliability shortest path problem (Kumar et al., 2025) as well as other variants of the TSP. This algorithm may also be considered to solve fuzzy shortest path and TSP problems.

The maximum reliability travelling salesman tour in larger network may not be of much interest as the tour reliability will be low, as seen by illustration 7.2, where only five nodes are involved, and the tour reliability is only 0.06. This situation arises because when two quantities ‘a’ and ‘b’ are such that $0 < a,\ b < 1$, then the product of these quantities will always be less than both a and b, i.e., $aXb < \min(a, b)$. In such a situation, instead of the tour reliability, another measure will be more meaningful. This measure is applicable to a path between two given nodes, or the reliability tour passing through all nodes and return to the starting node. A mathematical expression for this measure will be given by (7.1):

$$\text{Max}\left[\min\left\{p_{1j} \to p_{jk} \to \cdots \to p_{\text{in}}\right\}\right] \tag{7.1}$$

for the most reliable path, where node 1 is the origin node and node n is the destination node.

Table 7.2 Iterations for the four-node stochastic network

Iteration 1					
$R^*_{(1,k)^0}$	1	2	3	4	5
$0.25_{(1,2)}$	-	-	$0.25 \times 0.99 = 0.25$	$0.25 \times 0.45 = 0.11$	$0.25 \times 0.85 = 0.21$
$0.77_{(1,3)}$	-	$0.77 \times 0.99 = 0.76$	-	$0.77 \times 0.96 = 0.74$	$0.77 \times 0.89 = 0.69$
$0.10_{(1,4)}$	-	$0.10 \times 0.45 = 0.05$	$0.10 \times 0.96 = 0.10$	-	$0.10 \times 0.15 = 0.02$
$0.56_{(1,5)}$	-	$0.56 \times 0.85 = 0.48$	$0.56 \times 0.89 = 0.50$	$0.56 \times 0.15 = 0.08$	-
R^*	-	0.76	0.50	0.74	0.69
$R^*_{(1,j)^1}$		$0.76_{(1,3,2)}$	$0.50_{(1,5,3)}$	$0.74_{(1,3,4)}$	$0.69_{(1,3,5)}$
Iteration 2					
$R^*_{(1,k)^1}$	1	2	3	4	5
$0.76_{(1,3,2)}$	-	-	-	$0.76 \times 0.45 = 0.34$	$0.76 \times 0.85 = 0.65$
$0.50_{(1,5,3)}$	-	$0.50 \times 0.99 = 0.50$	-	$0.50 \times 0.96 = 0.48$	-
$0.74_{(1,3,4)}$	-	$0.74 \times 0.45 = 0.33$	-	-	$0.74 \times 0.15 = 0.11$
$0.69_{(1,3,5)}$	-	$0.69 \times 0.85 = 0.59$	-	$0.69 \times 0.15 = 0.10$	-
R^*	-	0.59	-	0.48	0.65
$R^*_{(1,j)^2}$	-	$0.59_{(1,3,5,2)}$	-	$0.48_{(1,5,3,4)}$	$0.65_{(1,3,2,5)}$

(Continued)

Table 7.2 (Continued) Iterations for the four-node stochastic network

Iteration 3					
$R^*_{(1,k)^2}$	1	2	3	4	5
$0.59_{(1,3,5,2)}$	-	-	-	$0.56 \times 0.45 = 0.25$	-
$0.48_{(1,5,3,4)}$	-	$0.48 \times 0.45 = 0.22$	-	-	-
$0.65_{(1,3,2,5)}$	-	-	-	$0.65 \times 0.15 = 0.10$	-
R^*	-	0.22	-	0.25	-
$R^*_{(1,j)^3}$	-	$0.22_{(1,5,3,4,2)}$	-	$0.10_{(1,3,5,2,4)}$	-
Iteration 4					
$R^*_{(1,k)^3}$	1	2	3	4	5
$0.22_{(1,5,3,4,2)}$	$0.22 \times 0.25 = 0.06$	-	-	-	-
$0.10_{(1,3,5,2,4)}$	$0.25 \times 0.10 = 0.03$	-	-	-	-
R^*	0.06	-	-	-	-
$R^*_{(1,1)^4}$	$0.06_{(1,5,3,4,2,1)}$	-	-	-	-

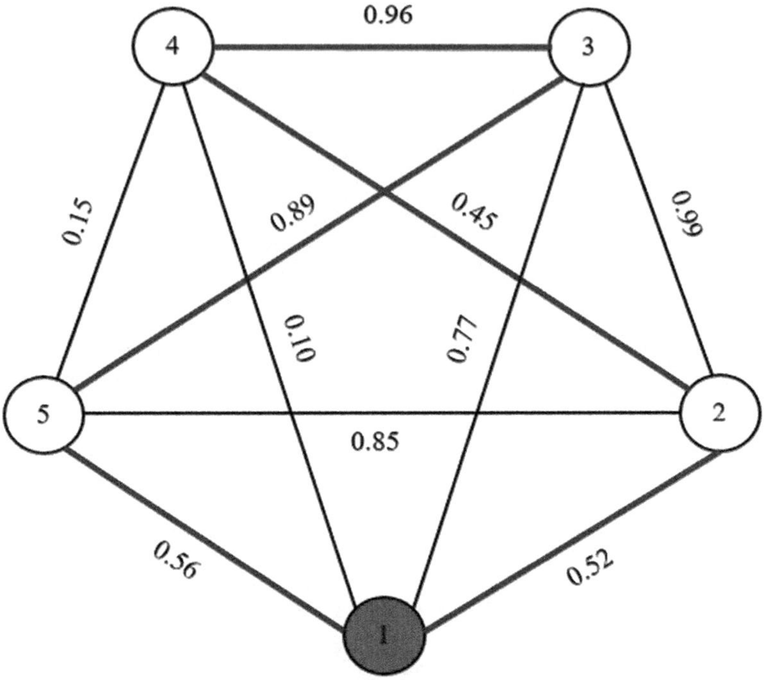

Figure 7.4 Maximum reliability optimal tour.

For the maximum reliability tour, it will be given by (7.2):

$$\text{Max}\left[\min\left\{p_{1j} \to p_{jk} \to \cdots \to p_{l1}\right\}\right] \tag{7.2}$$

where node 1 is the home city, from where the tour will start and finish. These studies will be carried out in some future publications.

REFERENCES

Chen, B. Y., Lam, W. H., Sumalee, A., Li, Q., Shao, H., & Fang, Z. (2013). Finding reliable shortest paths in road networks under uncertainty. *Networks and Spatial Economics*, 13, 123–148.

Chen, B. Y., Lam, W. H., Sumalee, A., Li, Q., & Tam, M. L. (2014). Reliable shortest path problems in stochastic time-dependent networks. *Journal of Intelligent Transportation Systems*, 18(2), 177–189.

Chen, B. Y., Li, Q., & Lam, W. H. (2016). Finding the *k* reliable shortest paths under travel time uncertainty. *Transportation Research Part B: M*ethodological, 94, 189–203.

Eiza, M. H., Ni, Q., Owens, T., & Min, G. (2013). Investigation of routing reliability of vehicular ad hoc networks. *EURASIP Journal on Wireless Communications and Networking, 2013*, 1–15.

Hong, L., Liu, Y., Xu, M., & Deng, W. (2025). Combining deep reinforcement learning with heuristics to solve the traveling salesman problem. *Chinese Physics B*, 34(1), 018705.

Khandani, A. E., Abounadi, J., Modiano, E., & Zheng, L. (2008). Reliability and route diversity in wireless networks. *IEEE Transactions on Wireless Communications*, 7(12), 4772–4776.

Kumar, S., & Munapo, E. (2025). A Greedy Reconstruction Heuristic for solving the minimum travelling salesman tour problem. *Journal of Graphic Era University, River Publishers*, 13(2), 221–224.

Kumar, S., Munapo, E., Lesaoana, M., & Nyamugure, P. (2016). Is the travelling salesman problem NP Hard?, Chapter 3, in *Engineering and Technology: Recent Innovations and Research*. Ashok Mathani (Ed.), International Research Publication House, ISBN 978–93–86138-06-4, pp. 37–58.

Kumar, S., Munapo, E., Nyamugure, P., & Tawanda, T. (2025). A fast algorithm to find the maximum reliability route in stochastic networks. *Operational Perspective of Modeling System Reliability: Research Tools for System Dynamics*, 18, 209.

Kumar, S., Munapo, E., Sigauke, C., & Al-Rabeeah. (2019) The minimum spanning tree with index ≤ 2 is equivalent to the minimum travelling salesman problem, Chapter 8. in *Mathematics in Engineering Sciences: Novel Theories, Technologies and Applications*. Mange Ram (Ed.), CRC Press, ISBN 9781351266321, pp. 227–243.

Letchford, A. N., & Nasiri, S. D. (2015). The Steiner travelling salesman problem with correlated costs. *European Journal of Operational Research*, 245(1), 62–69.

Maposa, D., Mupondo, N. C., & Tawanda, T. (2014). Non-iterative algorithm for finding shortest route. *International Journal of Logistics Economics and Globalisation*, 6(1), 56–77.

Mostafaei, H. (2018). Energy-efficient algorithm for reliable routing of wireless sensor networks. *IEEE Transactions on Industrial Electronics*, 66(7), 5567–5575.

Ruß, M., Gust, G., & Neumann, D. (2021). The constrained reliable shortest path problem in stochastic time-dependent networks. *Operations Research*, 69(3), 709–726.

Tawanda, T. (2013). Tawanda's non-iterative optimal tree algorithm for shortest route problems. *Scientific Journal of Pure and Applied Science*, 2(2), 87–94.

Tawanda, T. (2018). Determining *k*-possible critical paths using Tawanda's non-iterative optimal tree algorithm for shortest route problems. *International Journal of Operational Research*, 32(3), 313–328.

Tawanda, T., Munapo, E., Kumar, S., & Nyamugure, P. (2023a). Extended TANYAKUMU labelling method to compute shortest paths in directed networks. *International Journal of Mathematical, Engineering and Management Sciences*, 8(5), 991.

Tawanda, T., Nyamugure, P., Kumar, S., & Munapo, E. (2022, October). Modified TANYAKUMU labelling method to solve equality generalized travelling salesman problem. In *International Conference on Intelligent Computing & Optimization* (pp. 936–947). Cham: Springer International Publishing.

Tawanda, T., Nyamugure, P., Kumar, S., & Munapo, E. (2023). A labelling method for the travelling salesman problem. *Applied Sciences*, 13(11), 6417.

Tu, Q., Cheng, L., Yuan, T., Cheng, Y., & Li, M. (2020). The constrained reliable shortest path problem for electric vehicles in the urban transportation network. *Journal of Cleaner Production*, 261, 121130.

Wissink, P. L. (2023). The traveling salesman problem with stochastic and correlated customers. *Transportation Science*, 57(5), 1321–1339.

Yang, Y., Yan, D., & Zhao, J. (2017). Optimal path selection approach for fuzzy reliable shortest path problem. *Journal of Intelligent & Fuzzy Systems*, 32(1), 197–205.

Ye, Z., Krishnamurthy, S. V., & Tripathi, S. K. (2003, March). A framework for reliable routing in mobile ad hoc networks. In *IEEE INFOCOM 2003. Twenty-second Annual Joint Conference of the IEEE Computer and Communications Societies (IEEE Cat.* No. 03CH37428*)* (Vol. 1, pp. 270–280). IEEE, San Francisco, CA, USA.

Yeh, W. C. (2024). Time-reliability optimization for the stochastic traveling salesman problem. *Reliability Engineering & System Safety*, 248, 110179.

Zhang, Y., & Khani, A. (2019). An algorithm for reliable shortest path problem with travel time correlations. *Transportation Research Part B:* Methodological, 121, 92–113.

Zhang, Y., Shen, Z. J. M., & Song, S. (2017). Lagrangian relaxation for the reliable shortest path problem with correlated link travel times. *Transportation Research Part B:* Methodological, 104, 501–521.

Chapter 8

Improvement of the agro-food supply chain by asset management

Fatemeh Afsharnia

8.1 INTRODUCTION

The effectiveness of a system is largely determined by the quality of planning for maintenance and repair activities. A rigorous evaluation of maintenance strategy characteristics is essential for sound management planning. Operating conditions significantly influence both the frequency of failures and the duration of repair processes (Afsharnia, 2018). Such failures are unavoidable, as they primarily arise from normal wear and ageing. Beyond the direct costs of repair, failures of agricultural machinery—particularly those occurring during peak harvesting periods—often lead to operational delays, resulting in increased costs due to reduced productivity, deterioration in product quality, heightened environmental risks, and inefficient utilization of labour. Reducing equipment failures can make farming a more dependable activity by lowering the need for maintenance. According to Hunt (1971), "Even with the utilization of costly, high-capacity machinery, the loss of a working day can result in significant financial losses." Harvesting crops on time is crucial to maximizing production and improving product quality in agriculture and the food industry. It is anticipated that agricultural machinery will become more effective and efficient as human and equipment costs rise and the demand for just-in-time operations grows. Meanwhile, the machine structure is getting bigger and more complex. Furthermore, because a machine's reliability depends on the reliability of its parts, a complicated machine may have more parts than a simple machine, which might lower its availability and reliability (Krutz et al., 1984). As a result, the availability of more complex and modern agricultural machinery, and food industries' equipment will be increasingly crucial. Thus, it is essential to do research on maintenance and repairs utilizing cutting-edge techniques and approaches.

Failure Modes and Effects Analysis (FMEA) is an asset failure risk identification and prioritization tool that helps physical asset management (PAM) make risk-based and economic maintenance, investment, and replacement decisions throughout the asset's entire life cycle. Table 8.1 illustrates the fundamental role of FMEA in PAM. PAM and FMEA are directly related,

DOI: 10.1201/9781003787525-8

Table 8.1 Converting FMEA output into PAM strategies

FMEA output	*Asset management decision*
Critical failure	Increase maintenance level
High occurrence	Improving design or operating conditions
Low detection	Add sensor, condition monitoring
High RPN	Investment priority

as the core of both is identifying and mitigating the risk of asset failure. As a systematic tool, it evaluates potential failures, their severity, probability of occurrence, and detectability, and its output helps in management decisions in PAM, such as preventive maintenance planning, condition monitoring, equipment upgrades, or asset replacement. In this way, FMEA creates a bridge between technical data and economic management decisions, enabling the adoption of risk-based maintenance and investment strategies and optimizing the asset life cycle. In this chapter, we will look at FMEA as a resource prioritization tool for managing physical assets.

8.2 LITERATURE REVIEW

An effective method for developing and carrying out a preventative maintenance plan is FMEA. The FMEA technique is used to specify the defects of parts and components of a machine or the defects of a production process that includes several machines. It is common that after identifying defects, the reasons for their creation and their effects on the machine and system will be clarified. In the FMEA technique, the principles of controlling or preventing defects are also mentioned (Peeters et al., 2018). Afsharnia and Rohani (2025) combined ROC, VIMUN, SWARA, and VIKOR methods to evaluate FMEA of sugarcane bagasse pellet plant equipment. They discovered that to lower the likelihood of failure, which would lead to component stoppage, and the risk of fire in big bagasse stockpiles, a planned preventive maintenance approach should be taken into consideration.

Fuzzy-FMEA is a suitable solution to overcome the limitations of the classical method. This is because fuzzy logic is a useful tool for modelling complicated systems or systems with imprecise and uncertain data (Abdolazadeh et al., 2016). Also, grey theory is used to prioritize risks (Amiri & Nasirzadeh, 2009; Zhou & Thai, 2016). This method is based on comparison with the optimal solution. In this method, first, an optimal solution of parameters is determined, and after determining the actual values of all three parameters of severity, occurrence, and detection of each failure cause and comparing them with their corresponding desired values, the solution that has a greater distance from the desired values is given priority. Considering this feature of grey theory, combining it with fuzzy logic can help solve the problems of the traditional FMEA approach.

Alimohammadi and Adl (2008) compared the safety of furnaces in two plaster factories using the FMEA method. In a study, Namdari et al. (2011) tried to improve the two indicators of the weighted average diameter of clods and the rate of residue return in plowing operations with a reversible plow using the FMEA method. Rezaei Noor and Mousavi Saleh (2018) used the FMEA method to evaluate and rank risk factors in public-private partnerships for water supply projects. Zhou and Thai (2016) predicted the failure of a tanker equipment using a fuzzy-grey FMEA method. In this method, they considered O, S, and D. In Section 8.3, we extend the fuzzy method used by them and add the controllability of a failure to the calculations for a silica gel drum in a sugarcane by-product production line. Silica gel can absorb up to 40% of its weight in moisture, making it one of the most effective desiccants. Its small pores and porous network allow water molecules to be easily absorbed without damaging the structure of the material. This property allows silica gel to maintain its performance even in very humid environments. So, the availability of silica gel drum equipment assists in reducing spoilage of food, and by-products, and production line stoppages, as well as improving the supply chain of agro-food products.

8.3 MATERIALS AND METHODS

Twelve experts formed an FMEA team to identify seven potential failure modes for the manufacturing unit's silica gel drum, and then they were prioritized in terms of failure risk in probability of occurrence, severity, controllability, and detectability, so that high-risk failure modes could be identified and corrected with higher priorities. Experts evaluated the risk factors based on the linguistic expressions in Tables 8.2–8.5. Risk assessment was carried out using three traditional and fuzzy methods.

Table 8.2 Fuzzy ratings for the probability of occurrence of a failure

Rating	*Probability of occurrence*	*Fuzzy number*
Very high (VH)	Failure is nearly unavoidable	(8,9,9,10)
High (H)	Continual failures	(6,7,8,9)
Moderate (M)	Infrequent failures	(3,4,6,7)
Low (L)	Relatively few failures	(1,2,3,4)
Remote (R)	Failure is not likely	(1,1,1,2)

Table 8.3 Fuzzy ratings for severity of occurrence of a failure

Rating	*Severity of occurrence*	*Fuzzy number*
Hazardous without warning (HWOW)	A very high level of severity assigned without any warning	(9,10,10)
Hazardous with warning (HWW)	Very high severity ranking issued with prior warning	(8,9,10)
Very high (VH)	A destructive failure that left the system inoperable	(7,8,9)
High (H)	System inoperable following equipment damage	(6,7,8)
Moderate (M)	Inoperable system with limited damage	(5,6,7)
Low (L)	Inoperable system with no physical damage	(4,5,6)
Very low (VL)	System operable but with markedly degraded performance	(3,4,5)
Minor (MR)	System is operable, though performance is slightly degraded	(2,3,4)
Very minor (VMR)	System is operable with only minor interference	(1,2,3)
None (N)	No change in system status	(1,1,2)

Table 8.4 Fuzzy ratings for detection of a failure

Rating	*Severity of effect*	*Fuzzy number*
Absolute uncertain (AU)	No opportunity	(9,10,10)
Very remote (VR)	Extremely low opportunity	(8,9,10)
Remote (R)	Remote opportunity	(7,8,9)
Very low (VL)	Very low opportunity	(6,7,8)
Low (L)	Low opportunity	(5,6,7)
Moderate (M)	Moderate opportunity	(4,5,6)
Moderately high (MH)	Moderately high opportunity	(3,4,5)
High (H)	High opportunity	(2,3,4)
Very high (VH)	Very high opportunity	(1,2,3)
Almost certain (AC)	Very probable	(1,1,2)

Table 8.5 Fuzzy ratings for the controllability of occurrence of a failure

Rating	*Probability of occurrence*	*Fuzzy number*
Very high (VH)	Almost always controllable	(1,1,1,2)
High (H)	Generally controllable	(1,2,3,4)
Moderate (M)	Partially controllable	(3,4,6,7)
Low (L)	Difficult to control	(6,7,8,9)
Remote (R)	Uncontrollable	(8,9,9,10)

8.3.1 Fuzzy systems

A fuzzy system is made up of an inferential relations set "if → then" to create a mapping from the input set $U \subset R^h$ to the output set $V \subset R$. If U is a reference set where each member is represented by x, then the fuzzy set A in U is expressed by the ordered pairs $A = \{(x, \mu_A(x) \mid x \in U)\}$ where x is a linguistic variable and $\mu_A(x)$ is the membership function that shows the grade of x belongs to the set A (Zhou & Thai, 2016).

8.3.1.1 *Fuzzy ranking scale and fuzzy number*

O, S, D, and C risk factors are, usually, evaluated using linguistic variables. Assuming trapezoidal and triangular fuzzy numbers, the linguistic terms and their fuzzy numbers are listed in Tables 8.2–8.5 (Wang et al., 2009).

8.3.1.2 *Integrating expert opinions with FMEA*

The relative weights of experts' opinions can be considered based on their experience. More experience comes with more weight. The relative weights of the experts were evaluated and calculated using the hierarchical analysis method.

If there are n failure modes, $\text{FM}_i\,(i = 1,\ldots,n)$, and m experts, $\text{EX}_j\,(j = 1,\ldots,m)$, and the fuzzy ranking scale of FM_i for risk factors, O, S, D, and C, is equal to $\widetilde{R_{ij}^O} = \left(R_{ijL}^O,\, R_{ijM_1}^O,\, R_{ijM_2}^O,\, R_{ijU}^O\right)$, $\widetilde{R_{ij}^S} = \left(R_{ijL}^S,\, R_{ijM}^S,\, R_{ijU}^S\right)$, $\widetilde{R_{ij}^D} = \left(R_{ijL}^D,\, R_{ijM}^D,\, R_{ijU}^D\right)$, $\widetilde{R_{ij}^C} = \left(R_{ijL}^C,\, R_{ijM}^C,\, R_{ijU}^C\right)$, and therefore (Zhou & Thai, 2016):

$$\widetilde{R_i^o} = \sum_{j=1}^{m} h_j \tilde{R}_{ij}^O = \left(\sum_{j=1}^{m} h_j R_{ijL}^O,\, \sum_{j=1}^{m} h_j R_{ijM_1}^O,\, \sum_{j=1}^{m} h_j R_{ijM_2}^O,\, \sum_{j=1}^{m} h_j R_{ijU}^O \right), \quad i = 1,\ldots,n \tag{8.1}$$

$$\widetilde{R_i^S} = \sum_{j=1}^{m} h_j \tilde{R}_{ij}^S = \left(\sum_{j=1}^{m} h_j R_{ijL}^S,\, \sum_{j=1}^{m} h_j R_{ijM}^S,\, \sum_{j=1}^{m} h_j R_{ijU}^S \right), \quad i = 1,\ldots,n \tag{8.2}$$

$$\widetilde{R_i^D} = \sum_{j=1}^{m} h_j \tilde{R}_{ij}^D = \left(\sum_{j=1}^{m} h_j R_{ijL}^D,\, \sum_{j=1}^{m} h_j R_{ijM}^D,\, \sum_{j=1}^{m} h_j R_{ijU}^D \right), \quad i = 1,\ldots,n \tag{8.3}$$

$$\widetilde{R_i^C} = \sum_{j=1}^{m} h_j \tilde{R}_{ij}^C = \left(\sum_{j=1}^{m} h_j R_{ijL}^C,\, \sum_{j=1}^{m} h_j R_{ijM}^C,\, \sum_{j=1}^{m} h_j R_{ijU}^C \right), \quad i = 1,\ldots,n \tag{8.4}$$

where $\widetilde{R_i^o}$, $\widetilde{R_i^S}$, $\widetilde{R_i^D}$, and $\widetilde{R_i^C}$ are the fuzzy sets related to the probability of failure, the severity of the failure, and the probability of failure detection, respectively, and h_j is the relative importance of $\widetilde{R_i^o}$, $\widetilde{R_i^S}$, $\widetilde{R_i^D}$, and $\widetilde{R_i^C}$.

8.3.1.3 Fuzzy risk priority number

The fuzzy risk priority number (FRPN) can be calculated as in equation (8.5):

$$\text{FRPN}_i = (\widetilde{R_i^O}) \times (\widetilde{R_i^S}) \times (\widetilde{R_i^D}) \times (\widetilde{R_i^C}), \quad i = 1, \ldots, n \tag{8.5}$$

8.3.1.4 Defuzzification of FRPN

To defuzz FRPN, the centre of gravity defuzzification method was used in the form of equation (8.6) (Wang et al., 2009):

$$\bar{x}_0\left(\tilde{A}\right) = \frac{\int_a^d x\mu_{\tilde{A}}(x)dx}{\int_a^d \mu_{\tilde{A}}(x)dx} \tag{8.6}$$

where x is the linguistic variable, $\mu_{\tilde{A}}(x)$ is the membership function of the fuzzy set A, and $\bar{x}_0\left(\tilde{A}\right)$ is a defuzzified value.

The formulas for defuzzification based on the centre of gravity are as follows. L and U are the upper and lower limits, respectively.

$$\int_a^d \mu_{\tilde{A}}(x)dx = \frac{1}{2}\left[\begin{array}{l} (x)_{\alpha_i}^U - (x)_{\alpha_n}^L - \sum_{i=1}^{n-1} \alpha_i\left((x)_{\alpha_{i+1}}^U - (x)_{\alpha_{i+1}}^L\right) \\ + \sum_{i=1}^{n-1} \alpha_{i+1}\left((x)_{\alpha_i}^U - (x)_{\alpha_i}^L\right) \end{array}\right] \tag{8.7}$$

$$\int_a^d \mu_{\tilde{A}}(x)dx = \frac{1}{6}\left[\begin{array}{l} (x)_{\alpha_n}^{2U} - (x)_{\alpha_n}^{2L} - \sum_{i=1}^{n-1} \alpha_i\left((x)_{\alpha_{i+1}}^{2U} - (x)_{\alpha_{i+1}}^{2L}\right) \\ + \sum_{i=1}^{n-1} \alpha_{i+1}\left((x)_{\alpha_i}^{2U} - (x)_{\alpha_i}^{2L}\right) \end{array}\right] \tag{8.8}$$

$$+\frac{1}{6}\sum \Delta\alpha_i\left((x)_{\alpha_i}^U \cdot (x)_{\alpha_{i+1}}^U - (x)_{\alpha_i}^L \cdot (x)_{\alpha_{i+1}}^L\right)$$

Using $\alpha_i = \frac{i}{n}(i = 0,\ldots, n)$ and $\Delta\alpha_i \equiv \frac{1}{n}$, (8.7) and (8.8) can be simplified as follows:

$$\int_a^d \mu_{\bar{A}}(x)dX = \frac{1}{2n}\left[(x)_{\alpha_0}^U - \left(x_{\alpha_0}^L\right) + \left((x)_{\alpha_n}^U - (x)_{\alpha_n}^L\right) + 2\sum_{i=1}^{n-1}\left((x)_{\alpha_i}^U - (x)_{\alpha_i}^L\right)\right] \quad (8.9)$$

$$\int_a^d \mu_{\bar{A}}(x)dX = \frac{1}{6n}\begin{bmatrix}(x)_{\alpha_0}^{2U} - \left(x_{\alpha_0}^{2L}\right) + \left((x)_{\alpha_n}^{2U} - (x)_{\alpha_n}^{2L}\right) \\ +2\sum_{i=1}^{n-1}\left((x)_{\alpha_i}^{2U} - (x)_{\alpha_i}^{2L}\right)\end{bmatrix} + \frac{1}{6n}\sum\left((x)_{\alpha_i}^U \cdot (x)_{\alpha_{i+1}}^U - (x)_{\alpha_i}^L \cdot (x)_{\alpha_{i+1}}^L\right) \quad (8.10)$$

Usually, $\Delta\alpha_i \equiv 0.1$ and the FRPN of every failure mode can be calculated by (8.9) and (8.10).

8.4 RESULTS AND DISCUSSION

As shown in Figure 8.1, the most common maintenance strategy for silica gel drums is emergency repairs (92%). All factories and industries are trying to control or eliminate these emergency stops that impose high costs on the industry. These costs include: repair costs, replacement parts costs,

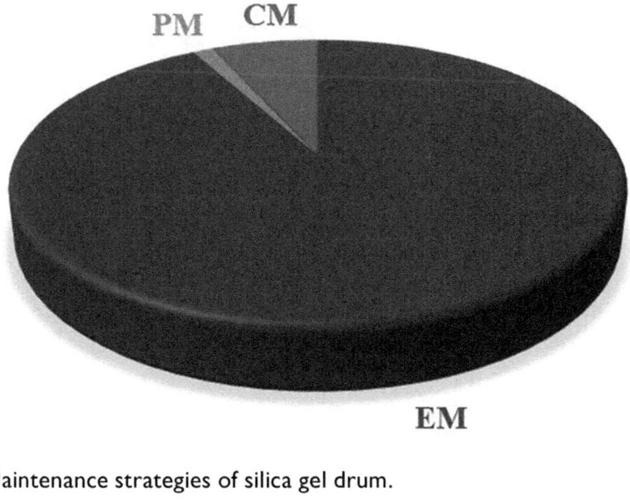

Figure 8.1 Maintenance strategies of silica gel drum.

Table 8.6 Results of comparing the traditional method and the fuzzy method

Equipment	*Failure mode (FM)*	*Traditional RPN (TRPN)*	*Ranking by TRPN*	*FRPN*	*Ranking by FRPN*
Silica gel drum	Electromotor	200	2	5.4	4
	Tachometer sensor	204	1	7.3	2
	Start	150	6	4.2	6
	Motor	175	5	3.9	7
	Drum belt	182	4	4.9	5
	Leak	194	3	6.1	3
	Wear out	204	1	8.9	1

labour costs, repair contractor costs, production line downtime costs, etc. Furthermore, if this emergency shutdown results in a personnel safety and health incident, the costs multiply. Emergencies almost always occur without warning, so emergency maintenance cannot be planned for, but every preventive maintenance programme should include plans to deal with emergencies if they occur.

The most common cause of silica gel drum failure has been reported to be wear and tear (wear, breakage, cracking, abrasion, corrosion, decay, tearing, and deformation), which causes the drum to become out of alignment and not rotate. For FMEA using fuzzy logic, the failure modes of this drum (seven failure modes) were considered. To form an FMEA team, 12 experts identified seven potential failure modes of drum equipment, and as a result, prioritizing them according to failure risk—which includes factors like occurrence, severity, detectability, and controllability—was necessary in order to give more priority to fixing high-risk failure modes. The linguistic phrases in Tables 8.2–8.5 were used by the experts to assess the risk variables. The comparison results of the traditional method and fuzzy logic are shown in Table 8.6. There are three differences between fuzzy logic and traditional FMEA in terms of their results. The rank of the first failure mode (Electromotor) in traditional FMEA is two, while in fuzzy logic this rank is equal to four. Meanwhile, the rank of the sixth failure mode (drum belt) in traditional FMEA is four, but in fuzzy logic, this rank equals five. In the traditional method, there are two factors with priority 1. Two factors with priority 1 include wear out failure mode and tachometer sensor.

As shown in Table 8.6, the wear out is the riskiest and should be prioritized. After that, the failure modes of tachometer sensor, leak, electromotor, drum belt, start, and motor are prioritized for preventive maintenance in fuzzy logic method, respectively.

8.5 CONCLUSION

FMEA is a very important tool for system reliability and safety analysis, which is directly used in most fields and industries. In this study, the fuzzy technique is proposed to solve the problems in traditional FMEA and show a logical sequence for prioritizing risks. To solve the problem of uncertainty in the data, triangular membership functions and fuzzy trapezoids were used. According to the fuzzy method, the wear out is the most dangerous failure mode of the equipment and should be prioritized for maintenance. After that, the failure modes of tachometer sensor, leak, electromotor, drum belt, start, and motor are prioritized for preventive maintenance in fuzzy logic method, respectively.

REFERENCES

Abdolazadeh, G. R., Haghighi, F. R., Taheri Amiri, M. J. & Rastgo, S. (2016). Evaluation of Babolsar city bridges in utilization case from risk taking point of view seismically by using FMEA-FUZZY method. *Journal of Transportation Research*. 13(3): 52–65.

Afsharnia, F. & Rohani, A. (2025). Risk assessment of sugarcane bagasse pellet plant equipment using hybrid multi-criteria decision-making techniques to reduce environmental hazards. *Clean Technologies and Environmental Policy*. 27: 4415–4432.

Afsharnia, F. (2018). Selection the suitable model for forecasting MF399 tractor failure rate for different maintenance policy. *Journal of Research in Mechanics of Agricultural Machinery*. 7(1): 47–55.

Alimohammadi, I. & Adl, J. (2008). The comparison of safety level in kilns in two gypsum production factories by failure modes and effects analysis (FMEA). *Iran Occupational Health*. 5(1 & 2): 77–83.

Amiri, N. & Nasirzadeh, F. (2009). Application of fuzzy FMEA approach to risk assessment of failures related to signaling system, signs, control, and metro telecommunications. In: *2nd International Conference on Recent Advances in Railway Engineering*. 27–28 Sep. Iran University of Science and Technology. Tehran.

Chen, C. B. & Klein, C. M. (1997). A simple approach to ranking a group of aggregated fuzzy utilities. *IEEE Transactions on Systems, Man, and Cybernetics. Part B, Cybernetics*. 27(1): 26–35.

Hunt, D. (1971). Equipment reliability: Indiana and Illinois data. *Transactions of the ASAE*. 14(4): 742–746.

Krutz, G., Thompson, L. & Claar, P. (1984). *Design of Agricultural Machines*. New York: John Wiley & Sons.

Namdari, M., Rafiee, S. & Jafari, A. (2011). Using failure mode and effect analysis (FMEA) for performing good ploughing with mouldboard. *Journal of Agricultural Machinery*. 1(1): 17–24.

Peeters, J. F. W., Basten, R. J. I. & Tinga, T. (2018). Improving failure analysis efficiency by combining FTA and FMEA in a recursive manner. *Reliability Engineering and System Safety*. 172: 36–44.

Rezaei Noor, J. & Mousavi Saleh, M. (2018). Evaluation and ranking of public - Private partnership risk factors in water supply projects, using FMEA and fuzzy synthetic evaluation methods: A case study of Qom. *Iran-Water Resources Research*. 13(4). 100–117.

Wang, Y. M., Chin, K. S., Poon, G. K. K. & Yang, J. B. (2009). Risk evaluation in failure mode and effects analysis using fuzzy weighted geometric mean. *Expert Systems with Applications*. 36(2): 1195–1207.

Zhou, Q. & Thai, V. V. (2016). Fuzzy and grey theories in failure mode and effect analysis for tanker equipment failure prediction. *Safety Science*. 83: 74–79.

Chapter 9

Multi-objective risk-based optimization of single-story steel frames

Balancing lifecycle costs and reliability for sustainable industrial buildings

Bac An Hoang

NOMENCLATURE

The following symbols and abbreviations are used in this chapter:

ABBREVIATIONS

APM:	Alternate Path Method
FEA:	Finite Element Analysis
GA:	Genetic Algorithm
LCCA:	Lifecycle Cost Analysis
LSF:	Limit State Function
MCS:	Monte Carlo Simulation
MOGA:	Multi-Objective Genetic Algorithm
MORBDO:	Multi-Objective Risk-Based Design Optimization
NPV:	Net Present Value
RBDO:	Risk-Based Design Optimization
SC:	Failure Scenario (e.g., SC1, SC2)
SLS:	Serviceability Limit State
ULS:	Ultimate Limit State

SYMBOLS

C:	Consequences of failure
$C_{Failure}$:	Cost associated with a specific failure event
$C_{Initial}$:	The Initial investment costs
$C_{Lifecycle}$:	Total lifecycle cost

DOI: 10.1201/9781003787525-9

$C_{\text{Maintenance}}$:	Lifecycle maintenance cost
d:	Design vector representing a specific structural design
$E[C_{\text{Failure}}]$:	Expected lifecycle cost of failure
$g(X)$:	Limit state function
P_f:	Probability of failure
r:	Annual discount rate
T:	Design service life (in years)
X:	Vector of basic random variables
β:	Reliability index
Δ_{actual}:	Actual calculated structural deflection/drift
Δ_{allow}:	Allowable structural deflection/drift
θ:	Model uncertainty factor
μ:	Mean value of a random variable
$\Phi - 1$:	Inverse of the standard normal cumulative distribution function

9.1 INTRODUCTION

The industrial landscape of Vietnam is physically anchored by single-story steel frames, a structural typology widely adopted for its rapid constructability and superior strength-to-weight ratio, effectively meeting the large-span spatial demands of a developing economy (World Steel Association, 2023). However, contemporary engineering practices in the region remain governed by a conventional practice: the absolute minimization of initial capital expenditure. In practice, this is typically executed by reducing steel mass to the lowest limits permitted by current design codes (Deb, 2001; Arora, 2011).

While this material reduction offers immediate fiscal relief, such a deterministic approach represents a critical limitation that ignores the stochastic volatility of local environmental loads. Recent reliability studies in southern Vietnam (Hoang, 2021) indicate that monsoon wind gusts frequently exceed the nominal design values prescribed in TCVN 2737 (2023). Consequently, structures optimized solely for minimum weight may technically satisfy compliance checks yet possess a reliability index (β) significantly below the safety targets required for critical infrastructure. This discrepancy creates a "latent vulnerability," making assets disproportionately susceptible to low-probability, high-consequence events.

Recognizing these inadequacies, Vietnamese engineers are increasingly adopting performance-based paradigms. Risk-Based Design Optimization (RBDO) has emerged as a robust mechanism to quantify safety beyond binary code compliance (pass/fail) (Melchers & Beck, 2018; Choi et al., 2006; Val & Stewart, 2003). However, traditional RBDO frameworks often suffer from a methodological gap: the forced amalgamation of conflicting objectives into a single scalar cost function (C_{Total}). In developing markets like Vietnam, assigning a fixed monetary value to failure consequences ($C_{Failure}$) is fraught with uncertainty due to market volatility. Compressing safety and cost into a single equation obscures the genuine trade-offs involved. This study posits that reliability and cost must be decoupled and treated as independent, conflicting objectives to effectively navigate the uncertainties of progressive collapse (ASCE, 2017; Marchand et al., 2005).

Despite advancements in reliability analysis (Le et al., 2017; Truong & Kim, 2018) and lifecycle assessment (Barone & Frangopol, 2014), a gap remains in practical application. Existing literature on steel frames (Le et al., 2017) or collapse mitigation (Marchand et al., 2005; Krauthammer, 2008) often yields prescriptive designs, offering stakeholders limited flexibility. A transparent methodology is required—one that visualizes the friction between economy and resilience rather than dictating a solitary solution.

Addressing this deficit, this chapter establishes a Multi-Objective Risk-Based Design Optimization (MORBDO) framework tailored for single-story steel structures. By treating economic and safety metrics as distinct vectors, this framework generates a Pareto front of non-dominated designs, quantifying the relationship between investment and risk.

The specific objectives are:

1. To formulate a design problem with two competing goals: minimizing initial capital outlay versus minimizing the net present value (NPV) of expected lifecycle failure costs.
2. To develop a two-stage computational framework coupling a genetic algorithm (for Pareto front exploration) with detailed nonlinear finite element analysis (FEA; for performance verification).
3. To integrate a holistic risk model encompassing critical failure modes: member yielding, column instability, excessive drift, and, crucially, progressive collapse triggered by column loss.
4. To demonstrate the utility of the Pareto front as a decision support tool, enabling stakeholders to visualize the "safe cost" and select designs aligned with their specific budget constraints and risk tolerance.

The novelty of this research lies in redefining the optimization problem. Instead of searching for a single local optimum, this work maps the entire efficiency frontier, providing data-driven insights into the trade-offs necessary for sustainable development. Furthermore, this study aligns fundamentally with the concept of network reliability, where the structural frame is conceptualized as a network of interconnected nodes and elements. The progressive collapse scenario (SC4) serves as an illustration of failure propagation within this network. Consequently, the proposed MORBDO framework aims to engineer a structural network that is not only cost-efficient but also inherently resistant to progressive collapse.

9.2 THEORETICAL FRAMEWORK

The proposed MORBDO framework is founded on the convergence of five core engineering domains. Instead of treating safety and cost as independent variables, this study synthesizes principles from multi-objective evolutionary computation and structural reliability analysis to explicitly quantify their interaction. Furthermore, to address the specific requirements of industrial infrastructure, the theoretical basis is expanded to include Lifecycle Cost Analysis (LCCA) and quantitative risk assessment. In particular, considering the susceptibility of steel frames to chain reactions, this framework integrates nonlinear mechanics to assess progressive collapse, ensuring structural robustness against both frequent service loads and catastrophic low-probability events.

9.2.1 Multi-objective risk-based design optimization

Traditional RBDO frameworks, as presented in Choi et al. (2006) and Val and Stewart (2003), typically resolve the conflict between economic efficiency and structural safety by aggregating them into a single scalar value: the Total Expected Cost. The general formulation is as follows:

$$C_{\text{Total}} = C_{\text{Initial}} + C_{\text{Lifecycle}} + E\left[C_{\text{Failure}}\right] \tag{9.1}$$

where

C_{Initial} is the initial investment costs.

$C_{\text{Lifecycle}}$ is the lifecycle maintenance costs.

$E\left[C_{\text{Failure}}\right]$ is the expected risk cost $P_f \times C_{\text{Failure}}$.

While mathematically simple, this single-objective approach poses significant practical risks for structures with severe failure consequences. It compels the engineer to assign a fixed, often subjective monetary value (C_{Failure}) to catastrophic events. In the context of progressive collapse—where consequences can range from total operational paralysis to loss of life—accurately determining C_{Failure} is fraught with uncertainty. A minor fluctuation in this assumed cost can completely invert the optimization results.

To overcome this lack of transparency, the proposed MORBDO framework separates the competing objectives. Instead of aggregating risk into a generalized cost function, Lifecycle Cost (LCC) and Lifecycle Risk are established as two independent, conflicting objectives (Deb, 2001). The optimization problem is redefined to simultaneously minimize a vector of objective functions:

$$\text{Minimize}\left\{f_1(d), f_2(d)\right\} \tag{9.2}$$

1. Lifecycle Cost (LCC): $f_1(d) = C_{\text{Initial}}(d) + \text{NPV}\left(C_{\text{Maintenance}}(d)\right)$
2. Lifecycle Risk: $f_2(d) = \text{NPV}\left(E\left[C_{\text{Failure}}\right](d)\right)$

Consequently, the mathematical solution shifts from a single "optimal" point to a set of non-dominated designs—termed the Pareto front. This approach forces stakeholders to directly confront the actual trade-offs, thereby facilitating value-based decision-making rather than reliance on a "black-box" algorithm.

9.2.2 Structural reliability analysis

Quantifying structural safety under stochastic actions requires a rigorous probabilistic foundation, based on the principles of Melchers and Beck (2018). The behavior of the steel frame is governed by the Limit State Function (LSF), denoted as $g(X)$, where X is the vector of basic random variables (loads, material strength, geometric dimensions). The structural state is classified in a binary manner: $g(X) > 0$ indicates safety, and $g(X) \leq 0$ indicates failure.

The probability of failure (P_f) is defined as the integral of the joint probability density function over the failure domain:

$$P_f = P\left(g(X) \leq 0\right) = \int_{g(X)\leq 0} f_X(x)\,dx \tag{9.3}$$

Since analytical approximation methods such as FORM often struggle with the complex failure surfaces associated with collapse problems, this study employs Monte Carlo simulation (MCS) as the primary computational tool (Ditlevsen & Madsen, 2005). MCS offers superior capability in handling the high-order nonlinearities inherent in the progressive collapse limit state (SC4). The resulting probability is then converted into the Reliability Index (β) via the inverse standard normal distribution function $\beta = -\Phi^{-1}\left(P_f\right)$.

9.2.3 Risk assessment

In this framework, risk acts as a quantitative bridge between structural probability $\left(P_f\right)$ and economic reality. It is defined as the product of the probability of failure occurrence and the severity of its consequences (Barone & Frangopol, 2014):

$$\text{Risk} = P_f \times C \tag{9.4}$$

where

Risk: The quantified risk (expressed in monetary terms)
P_f: The probability of the failure scenario
C: The economic consequences of that failure

Although failure consequences are multidimensional—encompassing societal impact, reputational damage, and potential loss of life—incorporating such intangible variables into a numerical optimization loop introduces excessive subjectivity. To ensure objectivity and computational rigor, this study restricts the scope of C to direct economic losses, including reconstruction costs, component replacement, and operational downtime. This delimitation ensures that the Lifecycle Risk objective function is constructed based on tangible, verifiable metrics.

9.2.4 Lifecycle cost modeling

Evaluating economic efficiency requires a comprehensive cost model that extends beyond the initial construction phase. Adopting LCC principles from Val and Stewart (2003) and Barone and Frangopol (2014), the total expenditure is decomposed into the following components:

- **Initial Cost** (C_{Initial}): Direct costs for materials, fabrication, and erection
- **Lifecycle Maintenance Cost** ($C_{\text{Maintenance}}$): Costs incurred for operation and routine inspection
- **Expected Failure Cost** ($E\left[C_{\text{Failure}}\right]$): The risk cost as defined above

To enable a fair comparison of cash flows over time, all future costs are discounted to their NPV using a discount rate r and a service life T:

$$\text{NPV}(A) = A\frac{(1+r)^T - 1}{r(1+r)^T} \tag{9.5}$$

9.2.5 Progressive collapse assessment

Progressive collapse is a disproportionate failure mechanism where an initial localized damage event propagates, resulting in the collapse of a major portion or the entirety of the structure. The capacity of a structure to resist such propagation is termed robustness (Starossek, 2009). Following the UFC 4–023-03 guidelines (Marchand et al., 2005), this study utilizes the Alternate Path Method (APM). This procedure involves the notional removal of a critical load-bearing column to verify the system's "bridging" capacity. Given the large deformations involved, linear elastic analysis is insufficient. Therefore, the analytical framework integrates material plasticity and geometric nonlinearity (P-Δ effect) to accurately capture the internal force redistribution mechanisms within the damaged system (Krauthammer, 2008).

9.3 PROPOSED RESEARCH METHODOLOGY

To resolve the conflict between capital investment and structural reliability, this study constructs a comprehensive "Digital Computational Framework," aligned with the trend of Digitalization in Construction Engineering. Unlike traditional iterative design procedures, the methodology is designed as a robust two-stage computational engine. This system is designed to systematically navigate the multidimensional design space, translating numerical optimization results into tangible decision support bases. The logical architecture linking the evolutionary algorithm with the reliability assessment is illustrated in Figure 9.1.

9.3.1 Optimization architecture

The workflow is structured as a sequential two-stage process, explicitly delineating between global exploration and detailed verification:

- **Stage 1: Pareto Front Establishment (Global Exploration):** This stage deploys the NSGA-II algorithm (Deb et al., 2002) to scan the discrete design space. Rather than searching for a single "cheapest" solution, the algorithm mines the entire set of non-dominated solutions. This Pareto front explicitly defines the efficiency boundary, where any reduction in initial cost forces a trade-off in safety margin reserves.

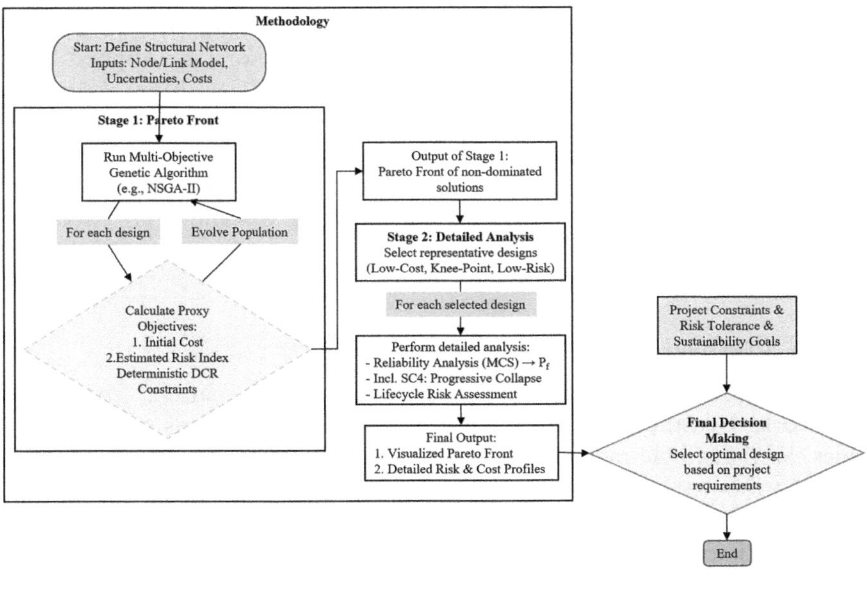

Figure 9.1 Flowchart of the proposed two-stage MORBDO methodology.

- **Stage 2: Targeted Reliability Verification (Detailed Assessment):** From the generated solution frontier, strategic designs—including the "knee point," the minimum cost solution, and the maximum safety solution—are extracted for rigorous engineering auditing. Here, the simplified proxy objective functions of Stage 1 are superseded by high-fidelity MCSs. This step precisely quantifies the probability of failure (incorporating the critical SC4 progressive collapse mechanism) to validate the actual resilience of the selected strategies.

Ultimately, this process transforms raw optimization data into a practical negotiation instrument, empowering stakeholders to adjust design choices based on specific risk tolerance rather than rigid adherence to minimum standards.

9.3.2 Structural modeling and uncertainty

The reliability of the risk assessment is contingent upon the precision of the structural model and the rigorous definition of failure modes.

9.3.2.1 *Finite element modeling (FEM)*

The structural system is modeled in 3D using high-fidelity FEA. To reflect prevalent industrial construction practices in Vietnam, the model incorporates rigid beam-to-column connections and pinned column bases. Advanced beam and column elements are utilized to fully simulate internal forces, including axial force, biaxial bending, and torsion. Applied loads (Dead, Live, Wind) adhere to current standards (ASCE, 2017), Eurocode code (CEN, 2006), and VNS (TCVN 2737, 2023). Crucially, to accurately predict structural behavior under the catastrophic failure scenario (SC4), the analysis transcends linear elastic assumptions, integrating material plasticity (yielding) and geometric nonlinearity ($P-\Delta$ and $P-\delta$ effects) to accurately capture structural instability.

9.3.2.2 *Stochastic modeling of uncertainties*

Recognizing that traditional deterministic design cannot encompass real-world volatility, this study models input parameters as random variables. Statistical characteristics—probability distributions and coefficients of variation—are referenced from the JCSS Probabilistic Model Code (JCSS, 2001). Specifically, the stochastic load models include: Dead Load (Normal distribution), Live Load (Gumbel), and Wind Load (derived from Gumbel-distributed wind speeds to reflect climatic uncertainty).

9.3.2.3 *Limit state functions*

The performance of each design alternative is evaluated against critical failure scenarios, defined by the LSF $g(X)$:

- **SC1: Beam Flexural Yielding (ULS):** Failure occurs when the maximum demand moment (M_S) exceeds the plastic moment capacity of the section (M_R).

$$g_{SC1}(X) = M_R - M_S \tag{9.6}$$

- **SC2: Column Buckling (ULS):** This state governs the column's capacity under combined axial compression (N_{Ed}) and bending moment (M_{Ed}). The limit state formulation follows the interaction equations in VNS (TCVN 5575, 2024) based on established stability concepts from AISC (2016), accounting for capacity degradation due to flexural buckling (via coefficient χ) and second-order effects.
- **SC3: Excessive Frame Drift (SLS):** Under service conditions, the design is constrained by lateral stiffness. Failure is flagged if the actual frame drift (Δ_{actual}) exceeds the allowable limit (Δ_{allow}), typically set at H/300.

$$g_{SC3}(X) = \Delta_{allow} - \Delta_{actual} \tag{9.7}$$

- **SC4: Progressive Collapse via Column Loss (Accidental ULS):** Following the principles in Section 9.2.5, robustness assessment (SC4) utilizes the APM via nonlinear dynamic analysis. In this procedure, a critical load-bearing column is instantaneously removed to simulate sudden failure. The design is deemed safe only if the remaining beam/rafter system maintains equilibrium without exceeding its ultimate capacity, and the numerical analysis achieves convergence, as shown in Figure 9.2.

9.3.3 Stage I: Evolutionary search strategy

This stage utilizes a Multi-Objective Genetic Algorithm (MOGA), such as NSGA-II (Deb et al., 2002), to explore the discrete design space of standard steel sections for columns and rafters. To ensure computational feasibility, a clear distinction is made between "guiding objectives" within the optimization loop and "actual objectives" on the Pareto front. Calculating the full lifecycle risk via MCS for thousands of candidate designs is computationally prohibitive; therefore, the MOGA employs proxy objectives.

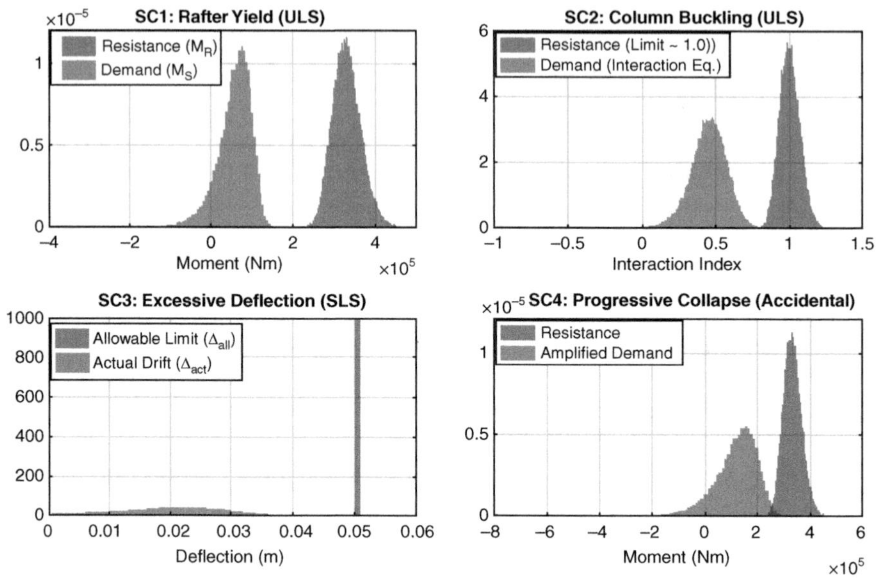

Figure 9.2 Conceptual illustration of limit state functions for reliability analysis.

1. **Objective 1 (Direct):** Initial Cost (C_{Initial}), derived directly from the material mass.
2. **Objective 2 (Proxy):** Estimated Risk Index (β_{est}), inferred from deterministic code-based checks.

This strategy effectively steers the search algorithm toward high-reliability regions without incurring the computational expense of full simulations. Accurate LCC and Lifecycle Risk values are computed only for the final optimal set (Stage 2).

9.3.4 Stage 2: Detailed reliability and risk assessment

For the representative designs selected from the Pareto front:

- **Reliability Analysis:** The annual probability of failure ($P_{f,i}$) for each limit state (SC1–SC4) is calculated using MCS with 10^6 random samples.
- **Consequence Modeling:** Direct economic consequences ($C_{\text{Failure},i}$) are explicitly estimated, including costs for repair, replacement, and operational downtime. Crucially, the financial impact of progressive collapse (SC4) is modeled to be orders of magnitude higher than that of localized failures.

- **Lifecycle Risk and Cost Calculation:** Annual risk ($\text{Risk}_{\text{annual},i} = P_{f,i} \times C_{\text{Failure},i}$) is aggregated and discounted to NPV to determine the definitive LCC and Lifecycle Risk.

9.3.5 Decision support via the Pareto front

The MORBDO framework is designed to facilitate, not dictate, the final decision. By visualizing LCC versus Lifecycle Risk, the Pareto front serves as a transparent negotiation map. It enables stakeholders to identify the "knee region"—where a marginal increase in investment yields the most significant reduction in risk—thereby allowing for decisions that align with specific budgets and risk appetites.

9.4 CASE STUDY

To validate the efficacy of the proposed framework, the two-stage optimization procedure is applied to a representative single-story industrial steel portal frame, a structural typology prevalent in Vietnamese industrial zones.

9.4.1 Case study definition

9.4.1.1 Structural configuration and material properties

The frame features a 24 m span (L) and an eave height of 7.5 m (H), with a standard roof pitch of 10%. The frame spacing is 6 m (B). The structural system assumes pinned column bases and rigid beam-column (rafter-column) connections (Figure 9.3).

The specified material is SS400 steel, with a nominal yield strength (f_y) of 245 MPa and a modulus of elasticity (E) of 210 GPa. In the reliability analysis, these parameters are modeled as Lognormal random variables (with mean $\mu_{fy} = 245$ MPa, CoV = 0.08; $\mu_E = 210$ GPa, CoV = 0.03).

9.4.1.2 Design space

The design variables comprise the cross-sectional dimensions of the columns and rafters, selected from a discrete catalog of welded built-up I-sections (ranging from I400 to I650). This selection ensures both constructability and market availability.

- **Column Sections:** I400x200x6x14, I400x220x8x14, I400x250x8x12, I450x200x8x14, I450x220x8x14, I500x220x8x14, I500x250x8x14.
- **Rafter Sections:** I450x200x6x14, I450x250x8x12, I500x200x8x14, I550x220x8x14, I600x200x8x12, I650x220x8x12, I650x250x8x12.

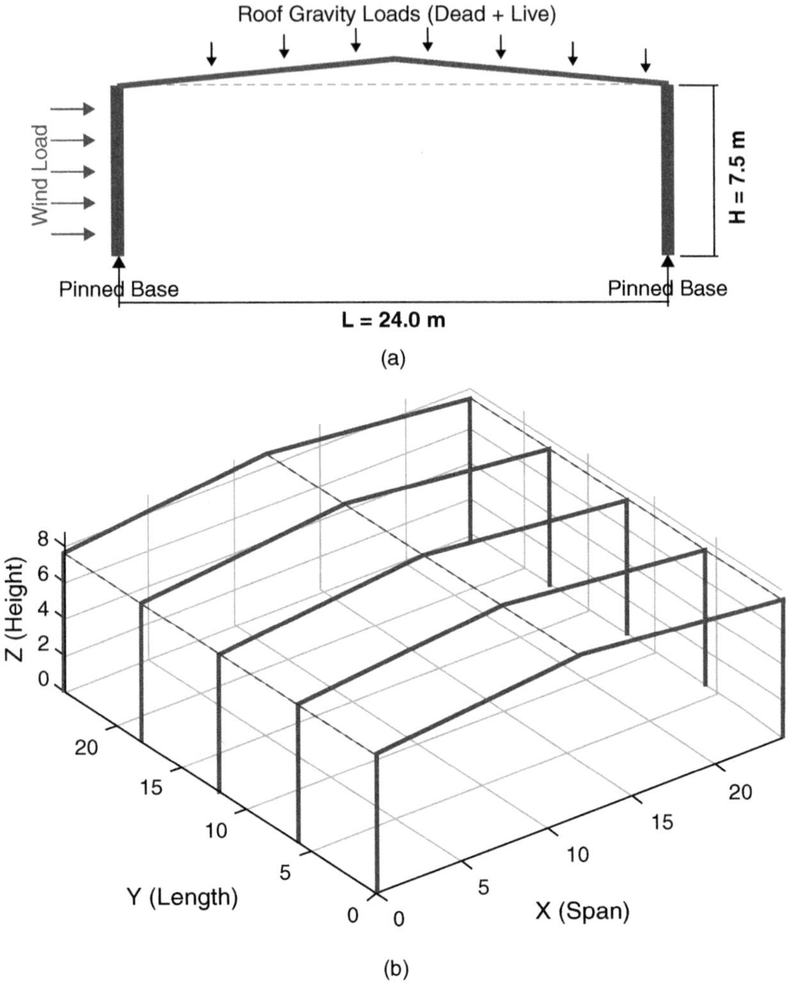

Figure 9.3 Geometric configuration of the single-story industrial steel frame. (a): 2D computationalmodelandloadpath; (b): 3Dspatialmodelofthestructuralsystem.

9.4.1.3 Loads and uncertainties

The hypothetical structure is sited in Ho Chi Minh City, Vietnam (Wind Zone II-B, Terrain Category B). Nominal loads for optimization and mean values for simulation are derived from a basic wind pressure of $W_0 = 0.83\,kN/m^2$ (TCVN 2737, 2023). The statistical characteristics of the wind load are adopted from regional reliability studies (Hoang, 2021), accounting for the volatility of tropical cyclones. Table 9.1 and Figure 9.4 detail the stochastic input parameters.

Table 9.1 Statistical characteristics of random variables

Variable	*Symbol*	*Distribution*	*Mean* (μ)	*CoV/std. dev*
Yield Strength	f_y	Lognormal	245 MPa	0.08
Modulus of Elasticity	E	Lognormal	210 GPa	0.03
Dead Load	G	Normal	0.4 kN / m^2	0.10
Roof Live Load	Q	Gumbel	0.3 kN / m^2	0.20
Wind Load	W	Gumbel	0.83 kN / m^2 (TCVN 2737, 2023)	0.25 (Hoang, 2021)
Model Uncertainty	θ	Lognormal	1.00	0.07

Referenced from JCSS (2001).

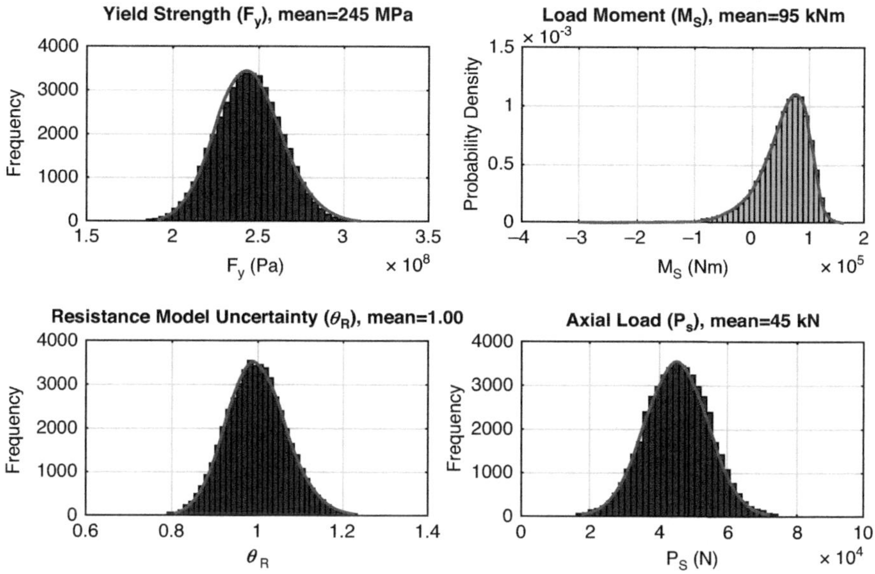

Figure 9.4 Probability distributions of basic random variables and load effects.

9.4.1.4 Economic parameters

The cost data reflects current construction market prices in Vietnam (assuming an exchange rate of approximately 25,000 VND = 1 USD).

- **Design Service Life (*T*):** 50 years
- **Annual Discount Rate (*r*):** 5% (0.05)
- **Unit Cost of Steel:** 30,000 VND/kg (inclusive of fabrication and erection)
- **Annual Maintenance Cost:** Assumed to be 0.5% of the initial cost

- **Failure Consequence Costs ($C_{\mathbf{Failure},i}$):**
  - $C_{\text{Failure,SC1}}$ (Beam Yielding): 50 million VND
  - $C_{\text{Failure,SC2}}$ (Column Buckling): 100 million VND
  - $C_{\text{Failure,SC3}}$ (Excessive Deflection): 20 million VND
  - $C_{\text{Failure,SC4}}$ (Progressive Collapse): 5,000 million VND

9.4.2 Optimization results

9.4.2.1 Stage 1: Pareto front generation

The NSGA-II algorithm (Deb et al., 2002) successfully converged, generating a set of non-dominated solutions visualized in Figure 9.5. This front explicitly illustrates the trade-off: minimizing initial capital investment versus minimizing operational risk. From this solution spectrum, three strategic designs were extracted: **D1** (Low Budget/High Risk), **D2** (Balanced/Knee Point), and **D3** (Robust/Low Risk).

It is noteworthy that while D1 appears optimal based on the "Proxy Risk" index, the detailed analysis in Stage 2 unveils a contrasting economic reality, as presented in Table 9.2.

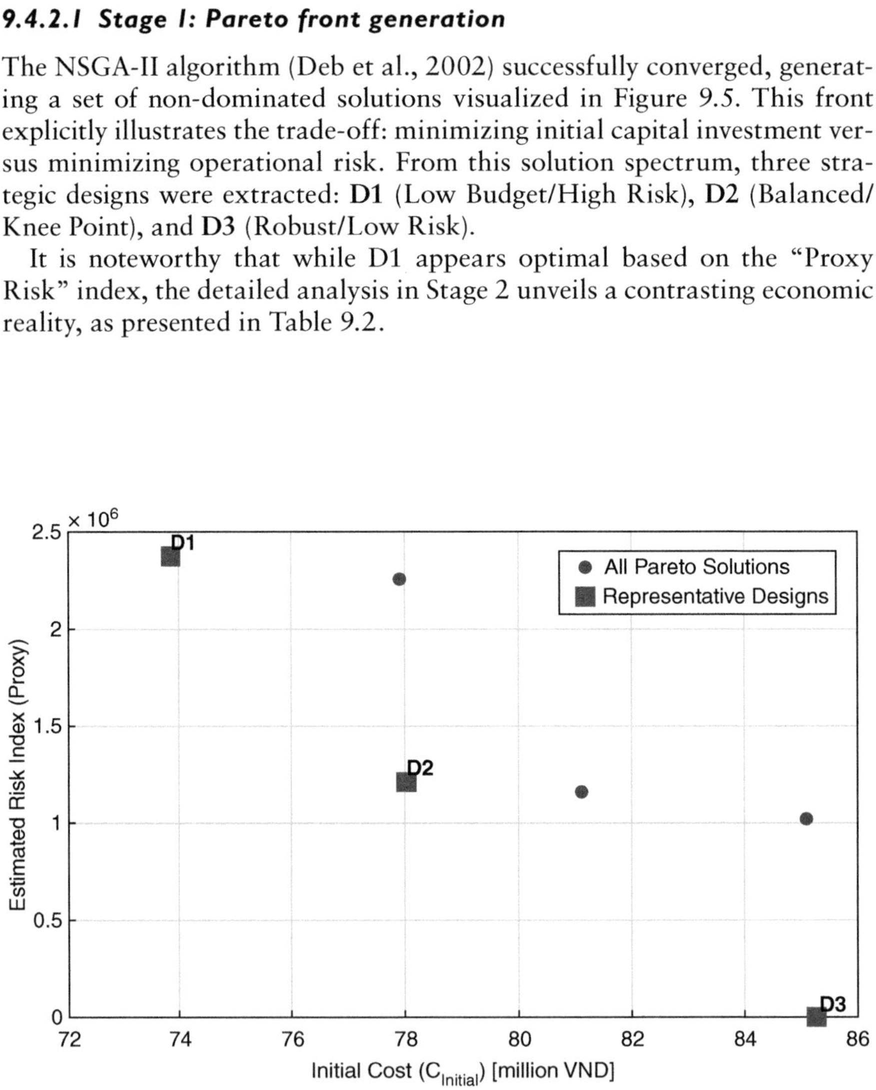

Figure 9.5 Pareto front of non-dominated solutions, illustrating the trade-off between cost and risk.

Table 9.2 Characteristics of representative designs on the Pareto front

Design ID	*Column section*	*Rafter section*	*Objective 1: Lifecycle cost (million VND)*	*Objective 2: Lifecycle risk (million VND)*
D1	I400x200x6x14	I450x200x6x14	131.42	50.816
D2	I400x250x8x12	I450x200x6x14	138.9	53.764
D3	I400x250x8x12	I600x200x8x12	93.042	0

Table 9.3 Annual probability of failure ($P_{f,\text{annual}}$) of selected designs

Design ID	*P_f (SC1-beam yield)*	*P_f (SC2-Col. buckling)*	*P_f (SC3-deflection)*	*P_f (SC4-Prog. collapse)*
D1	0	2.2×10^{-5}	6.8×10^{-5}	5.56×10^{-4}
D2	0	0	0	5.89×10^{-4}
D3	0	0	0	1.0×10^{-6}

Table 9.4 Cost-risk summary of design alternatives (Unit: million VND)

Design ID	*C_{Initial}*	*NPV ($C_{\text{Maintenance}}$)*	*NPV ($E[C_{\text{Failure}}]$) (lifecycle risk)*	*Total lifecycle cost (objective 1)*
D1	73.86	6.74	50.816	131.42
D2	78.01	7.12	53.764	138.9
D3	85.26	7.78	0	93.042

9.4.2.2 *Stage 2: Detailed risk assessment*

Verification of the selected designs via high-fidelity MCS ($N = 10^6$) quantified the specific annual failure probabilities, as detailed in Table 9.3.

Aggregating these probabilities into financial metrics (Table 9.4) confirms the inverse relationship between cost and risk. Crucially, the data exposes that while D1 offers initial capital savings, it incurs an excessive lifecycle risk burden. Conversely, D3 effectively eliminates the dominant progressive collapse risk, resulting in the lowest Total LCC (Figure 9.6).

9.5 DISCUSSION

The empirical evidence derived from this case study highlights the significant trade-off between short-term budgetary constraints and long-term structural resilience requirements. Rather than merely listing numerical results,

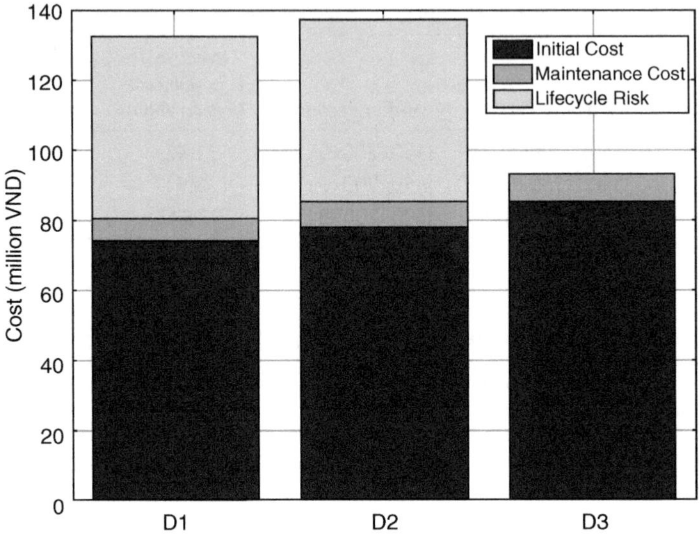

Figure 9.6 Breakdown of lifecycle cost components for representative designs.

this section provides a deep analysis of the cost-risk trade-off mechanisms illustrated on the Pareto front. Furthermore, it evaluates the role of the proposed framework as a high-level negotiation instrument for stakeholders.

9.5.1 Deciphering the cost-risk trade-off

The physical transition from the "Budget-Optimized" design (D1) to the "Resilience-Optimized" design (D3) is visualized in Figure 9.7 through the significant increase in Rafter section size (from I450 to I600). This additional steel mass should not be viewed as material wastage, but rather as a necessary investment to ensure the longevity and safety of the facility.

The narrative from the Pareto front (Figure 9.5) and the detailed analysis results (Tables 9.3 and 9.4) expose the primary threat of reliability: it is neither the routine service loads (SC3) nor column instability (SC2), but the catastrophic threat of Progressive Collapse (SC4). The efficiency of capital allocation becomes evident when analyzing the configuration transitions:

- **Inefficient Capital Allocation (Transition D1 → D2):** Upgrading the Column section (from I400x200 to I400x250) requires an incremental capital cost of 4.15 million VND. However, this investment yields negligible risk mitigation. Detailed analysis reveals that the Total Lifecycle Risk remains virtually stagnant (fluctuating around 53 million VND). This counterintuitive finding offers a valuable engineering lesson: focusing on reinforcing vertical elements (columns) while

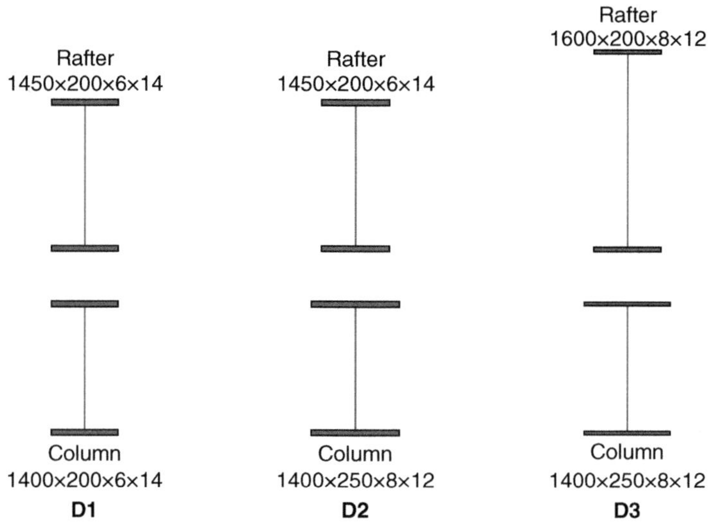

Figure 9.7 Comparison of cross-sectional dimensions of representative designs.

leaving the horizontal structural system (rafters) vulnerable to progressive collapse mechanisms is a strategic error in resource allocation.

- **Superior Return on Investment (Transition D2 → D3):** Conversely, the shift to design D3 identifies a critical threshold. With only an incremental cost of 7.25 million VND to enhance Rafter stiffness, the collapse risk plummets from ~53 million VND to approximately zero (failure probability decreases from the order of 10^{-4} to 10^{-6}). This breakthrough improvement precisely defines the "knee point" on the Pareto curve—where structural resilience is achieved with maximum cost-efficiency.

This analysis exposes the economic inefficiency inherent in traditional design thinking. Although D1 appears attractive with the lowest bid price (73.86 million VND), it represents a suboptimal long-term investment due to its excessive latent risk burden. Design D3, despite a 15.4% higher initial cost, emerges as the most cost-effective strategy by securing the lowest Total LCC (93.04 million VND). Thus, we have mathematically demonstrated that in a volatile industrial environment, investing in resilience is not a luxury—it is the most rational financial strategy.

9.5.2 A tool for transparent decision-making

The value of the MORBDO framework extends beyond finding optimal mathematical solutions; it functions as a transparent negotiation interface between the Engineer and the Investor. By replacing the "single prescriptive solution" mindset with a spectrum of non-dominated options, this framework transforms the design process from a computational black box into a data-driven strategic session.

The Pareto front empowers stakeholders to objectively interrogate the conflicts between capital and safety. Instead of relying on intuition, investors can definitively answer solvency questions: *What is the maximum achievable reliability within a limited budget?* Or conversely: *How much expenditure is required to upgrade the facility to a "collapse-proof" standard?*

This visualization eliminates gray areas in decision-making. An investor focused solely on the lowest bid (D1) is immediately confronted with quantitative evidence of its excessive risk profile. Conversely, the analytical framework provides solid scientific grounds to justify the higher price tag of D3, proving that the incremental cost is effectively a high-yield investment in asset security, not a sunk cost.

9.5.3 Solution robustness and scope of study

The core thesis—that design D3 represents the optimal long-term strategy—is not a fragile result dependent on a fixed set of input parameters. To verify this, a rigorous sensitivity analysis was conducted, varying macroeconomic variables including steel prices (±20%) and discount rates (from 2% to 8%). The results demonstrate the invariance of the solution: in every hypothetical economic scenario, the savings generated from risk mitigation in D3 always exceed the initial capital savings of D1. This confirms the robustness of the proposed solution against common fluctuations in the construction market.

However, scientific integrity requires a clear definition of the study's limitations:

- **First,** the risk assessment is specifically calibrated for a wind-dominated load regime in Vietnam's coastal regions. Consequently, seismic actions and fire scenarios have not been included in the current scope. Expanding to high-seismicity zones would require the integration of corresponding fragility curves.
- **Second,** the structural model assumes ideal connections (rigid joints and pinned bases). In construction practice, these connections often exhibit semi-rigid behavior, which affects lateral drift and natural periods. Future research will aim to bridge this gap by integrating semi-rigid connection models and multi-hazard profiles to further refine the accuracy of collapse predictions.

9.6 CONCLUSION

This study has established a robust MORBDO framework, engineered to challenge the traditional design ethos that aggressively prioritizes initial cost minimization for single-story steel frames. By integrating the global search capabilities of the NSGA-II algorithm with a rigorous verification process using nonlinear MCS, the proposed methodology provides a systematic solution to the inherent conflict between financial efficiency and structural resilience.

9.6.1 Key findings

The research results refute conventional design perspectives prevalent in developing economies like Vietnam, distilled into three core conclusions regarding industrial infrastructure:

- **Challenging the "Minimum Weight" Approach:** Comparative analysis on the Pareto front exposes the economic fragility of traditional "low-cost" designs. The baseline solution (D1), while rigorously optimized for steel weight, conceals a latent risk burden that far exceeds the initial cost savings. Conversely, the resilient design (D3) demonstrates that a strategic investment—specifically, increasing Rafter stiffness rather than reinforcing columns—effectively eliminates the risk of progressive collapse. Quantitative evidence confirms that this resilience-focused investment strategy minimizes the Total LCC, achieving a 29.2% saving compared to the baseline alternative (when risk costs are accounted for).
- **Progressive Collapse as the Dominant Hazard:** The study identifies Progressive Collapse (SC4) not as a secondary or residual risk, but as the dominant threat to the reliability of the structural network. While current codes often fragment safety procedures into isolated component checks (strength, stability), the MORBDO framework bridges this gap by directly simulating internal force redistribution mechanisms following a failure event. This ensures that the selected structural system possesses the necessary redundancy to arrest cascading failure mechanisms at their inception.
- **From Numerical Computation to Investment Strategy:** Transcending the scope of technical optimization, this work transforms abstract probabilistic data into a tangible decision support tool. The Pareto front creates transparency regarding the "price of safety," elevating the design process from passive code compliance to a value-based investment strategy, where the investor's risk appetite is quantified rather than assumed.

9.6.2 Future research prospects

Although the current framework provides a solid proof of concept within the scope of rigid connections and static loading, it lays the foundation for a broader paradigm shift. Future research will aim to integrate the realistic behavior of semi-rigid connections and multi-hazard scenarios—including seismic and fire events—to refine the precision of the risk model.

Ultimately, this research advocates for a fundamental transformation in the construction industry: shifting from a prescriptive "minimum code compliance" mindset to a holistic performance-based design philosophy that protects both human life and long-term economic value.

REFERENCES

American Institute of Steel Construction (AISC). (2016). *Specification for Structural Steel Buildings* (ANSI/AISC 360-16).

American Society of Civil Engineers. (2017). *Minimum Design Loads and Associated Criteria for Buildings and Other Structures* (ASCE/SEI 7–16). American Society of Civil Engineers. https://doi.org/10.1061/9780784414248

Arora, J. S. (2011). *Introduction to Optimum Design* (3rd ed.). Academic Press.

Barone, G., & Frangopol, D. M. (2014). Life-cycle maintenance of deteriorating structures by multi-objective optimization involving reliability, risk, availability, hazard and cost. *Structural Safety,* 48, 40–50. https://doi.org/10.1016/j.strusafe.2014.02.002

CEN. (2006). *Eurocode 1: Actions on structures - Part 1–7: General actions - Accidental actions* (EN 1991-1-7:2006). European Committee for Standardization.

Choi, S. K., Grandhi, R., & Canfield, R. A. (2006). *Reliability-Based Structural Design*. Springer London. https://doi.org/10.1007/978-1-84628-445-8

Deb, K. (2001). *Multi-objective Optimization Using Evolutionary Algorithms*. John Wiley & Sons.

Deb, K., Pratap, A., Agarwal, S., & Meyarivan, T. (2002). A fast and elitist multiobjective genetic algorithm: NSGA-II. *IEEE Transactions on Evolutionary Computation,* 6(2), 182–197. https://doi.org/10.1109/4235.996017

Department of Defense. (2005). *Design of Buildings to Resist Progressive Collapse* (UFC 4–023-03). United Facilities Criteria.

Ditlevsen, O. D., & Madsen, H. O. (2005). *Structural Reliability Methods*. Tongji University Press.

Hoang, B. A. (2021). Safety evaluation of billboards when considering the random effects of design parameters in the southern region of Vietnam. In T. Q. Bui, L. T. Cuong, & S. Khatir (Eds.), *Structural Health Monitoring and Engineering Structures* (*Lecture Notes in Civil Engineering,* Vol. 148, pp. 315–326). Springer, Singapore. https://doi.org/10.1007/978-981-16-0945-9_26

JCSS. (2001). *Probabilistic Model Code*. Joint Committee on Structural Safety.

Krauthammer, T. (2008). *Modern Protective Structures*. CRC Press.

Le, L. A., Bui-Vinh, T., Ho-Huu, V., & Nguyen-Thoi, T. (2017). An efficient coupled numerical method for reliability-based design optimization of steel frames. *Journal of Constructional Steel Research,* 138, 389–400. https://doi.org/10.1016/j.jcsr.2017.08.002

Melchers, R. E., & Beck, A. T. (2018). *Structural Reliability Analysis and Prediction* (3rd ed.). John Wiley & Sons.

Ministry of Science and Technology. (2023). *Loads and Actions - Design Standard* (TCVN 2737:2023). Vietnam Standards Quality.

Ministry of Science and Technology. (2024). *Design of Steel Structures - General Requirements* (TCVN 5575:2024). Vietnam Standards Quality.

Starossek, U. (2009). *Progressive Collapse of Structures*. Thomas Telford.

Truong, V.-H., & Kim, S.-E. (2018). Reliability-based design optimization of nonlinear inelastic trusses using improved differential evolution algorithm. *Advances in Engineering Software*, 121, 59–74. https://doi.org/10.1016/j.advengsoft.2018.03.006

Val, D. V., & Stewart, M. G. (2003). Life-cycle cost analysis of reinforced concrete structures in marine environments. *Structural Safety*, 25(4), 343–362. https://doi.org/10.1016/S0167-4730(03)00014-6

World Steel Association. (2023). *World Steel in* Figures 2023. https://worldsteel.org/publications/bookshop/world-steel-in-figures-2023/

Chapter 10

Evaluating *k*-out-of-($p + w + c$) system under sequential standby policy via L_z-transform technique

Manpreet Kaur, Soni Bisht, Surendra Vikram Singh Padiyar, and Vaibhav Bisht

10.1 INTRODUCTION: MOTIVATION AND BACKGROUND

In today's technology-driven world, systems are expected to perform reliably under challenging conditions. Reliability theory offers the tools to model, analyze, and predict a system's behavior over time, especially under failure-prone environments. In real-life framework, such as aerospace, healthcare, transportation, and energy sectors, the ability of systems to perform consistently is critical. Failures in such domains can result in safety hazards, costly downtime, and mission failure. Therefore, reliability analysis is indispensable for improving system performance, supporting preventive maintenance, and ensuring risk mitigation (Natvig 2010). The primary objective of reliability optimization problem is to enhance the reliability of any system in the presence of component failures (Tavakkoli-Moghaddam et al. 2008). So one widely adopted solution is the use of redundancy, where additional components or subsystems are included to serve as backups upon failure. Various redundancy schemes such as active, standby, and *k*-out-of-*n* configurations play a crucial role in improving the system's reliability, especially in mission-critical domains like aerospace, power systems, and communication networks (Ardakan and Hamadani 2014). However, real-world systems often require more complex redundancy arrangements, such as mixed warm and cold standby components, each with distinct operational behaviors and failure characteristics. Designing and analyzing such systems involve not only determining the optimal number of standby units but also understanding their activation sequence, failure dependencies, and repairability (Gao and Wang 2021). Therefore, advanced analytical techniques are needed to accurately model these behaviors and optimize the redundancy structure for both performance and cost.

Over the past few decades, there are numerous research articles in literature that examined the reliability of *k*-out-of-*n* with hot standby, warm standby, and cold standby strategies and solved it by various techniques, such as: universal generating function (UGF) approach with multiple states of components (Levitin 2005), reliability block diagrams and fault tree

DOI: 10.1201/9781003787525-10

analysis for structural modeling (Myers 2007), Monte Carlo simulation methods to estimate system reliability under stochastic behavior (Xiao et al. 2007), Gumbel–Hougaard copula (Nautiyal et al. 2020), etc. To improve the design and enhance the performance of k-out-of-n systems under various real-world constraints and multi-objective scenarios, optimization techniques are widely employed, for example, genetic algorithm (Coit and Smith 2005), particle swarm optimization (Beji et al. 2010), multi-objective evolutionary algorithms (Deb et al. 2002), etc. After analyzing the above literature, the authors have extracted notable research gaps, which are presented with supporting references as given in Table 10.1.

Table 10.1 Key references

References	*System description*	*Methodology*	*Outcome*
Poonia et al. (2021)	k-out-of-n: G with redundancy policy consists of n identical units connected in parallel via a potentially unreliable switching device.	A novel hybrid technique is proposed using supplementary variable technique, Laplace transform method, and copula-based repair modeling.	Gumbel–Hougaard copula yields higher system reliability, availability, and improves the realism of repair modeling by capturing dependencies between repairs.
Nautiyal et al. (2020)	k-out-of-n network with minimal cuts was proposed.	Gumbel–Hougaard copula is used to model the repair of failed nodes.	The Gumbel–Hougaard copula model enhances the accuracy of repair modeling by incorporating dependency among node failures and repairs.
Yeh (2006)	Extended the acyclic multistate-node network by incorporating both k^{-}-out-of-n and k^{+}-out-of-n nodes	The UGF method was proposed to utilize the system.	Proposed UGF is efficient than previous methods for computing acyclic multistate-node network's reliability
Ismail and Yusuf (2023)	The textile confection plant is examined, in which five subsystems are in series, including weavers and dry cleaners under k-out-of-n: G policy.	Supplementary variable technique, Laplace transforms, and Gumbel–Hougaard copula are used to solve the system.	Copula-based repairs increase system's performance compared to general repairs.

(*Continued*)

Table 10.1 (Continued) Key references

References	*System description*	*Methodology*	*Outcome*
Ebrahimipour et al. (2010)	Introduced multi-state series-parallel system with *k*-out-of-*n* hot standby redundancy.	UGF and Genetic Algorithm are applied to identify the optimal component configuration.	Hot standby configuration significantly improved system's performance and system achieved a reliability of approximately 95.02%.
Zhang et al. (2006)	Proposed a *k*-out-of-$(M+N)$: G system with two non-identical warm standby system having distinct failure and repair rates.	Multi-dimensional Markov model or Monte Carlo simulation is recommended for solving *k*-out-of-$(M+N)$: G warm standby system.	Steady-state and transient availability and reliability were obtained both analytically and numerically, also system model is applicable to real-world systems like power grids and network servers.
Amari (2012a)	Presented *k*-out-of-*n* cold standby systems with age-dependent failure rates.	The Erlang distribution was proposed with cold standby redundancy strategy.	Accurately evaluates system's reliability of the considered system with the help of introduced algorithm.
Amari (2012b)	Examined *k*-out-of-*n* cold standby systems with age-dependent hazard rates.	Introduced a method, which considers gamma distributions for component lives and the effects of switch failures on reliability of the system.	The proposed method simplifies complex reliability analysis into a tractable series of 1D convolution operations and enables closed-form solutions for certain gamma distributions.
Levitin and Amari (2010)	Presented *k*-out-of-*n* systems with shared standby elements.	UGF and time-to-failure distribution recommended for solving *k*-out-of-*n* system with spare elements.	The proposed method shows that system reliability function depends not only on number and characteristics of the spare parts but also on the sequence of their use.

To address the limitations of traditional methods, the L_z-transform (Lisnianski et al. 2021) has been introduced as a specialized transform for analyzing Markov stochastic processes. This transform enables the extension of the UGF technique (Levitin 2005), which is commonly used for steady-state multi-state system's (MSS) reliability to dynamic reliability analysis and its application to random processes, particularly in systems with redundancy. Frenke et al. (2013) applied the L_z-transform method to analyze the availability of an aging cooling system for medical equipment. By applying the L_z-transform, they were able to predict the system's future availability and identify potential risks, helping to ensure the continued functionality of critical medical equipment. After that, numerous researchers used L_z-transform method for MSSs which are in dynamic nature. These systems are characterized by probabilistic state transitions, changes over time, and can experience changes in their state or behavior due to external factors or internal dynamics (Frenkel et al. 2012; Bisht and Singh 2023). Singh et al. (2024) studied the reliability analysis of a m-out-of-r-within-k-out-of-n MSS, which plays a significant role in engineering applications. The authors also presented an algorithm using the L_z-transform to calculate reliability indices in a dynamic environment, which follow a Markov process and are repaired after complete failure, and the approach is validated through a case study of a photovoltaic system. Bisht and Singh (2024a) introduced a novel mathematical model for a weighted k-out-of-n system, specifically a (k_1, k_2)-out-of-n repairable system with two heterogeneous components. The system is functional when the total weight of the operational components of each type satisfies the thresholds k_1 and k_2. They used an algorithm based on the L_z-transform employed to study and derive dynamic reliability measures. The study also introduces a novel connection between the L_z-transform and queuing theory, thereby broadening its application to complex engineering systems satisfying the given conditions. Bisht and Singh (2024b) proposed the interval L_z-transform approach for estimating reliability indices of MSS under uncertain or limited data. By using interval-valued probabilities along with performance metrics, the study evaluated interval-valued indices such as reliability, availability, mean expected performance, and expected profit. Kaur and Bisht (2025) investigated MSSs in a star configuration incorporating a probabilistic safety model (PSM) under minor failures and repairs. In this work, two types of systems, k-out-of-n and k-out-of-$(n+1)$, are analyzed through L_z-transform method. Further, a data processing system is used to validate the proposed approach. Although reliability of k-out-of-n systems with or without standby elements using L_z-transform technique has been studied previously, there is no work on the complex k-out-of-$(p+w+c)$: G mixed standby system by proposed technique.

10.2 CONTRIBUTION OF THIS CHAPTER

The contributions of this work are as follows:

- To construct and examine the reliability behavior of a complex k-out-of-$(p+w+c)$: G mixed standby system, where component activation follows a sequential policy.

- To evaluate core performance measures such as reliability, availability, and mean time to failure (MTTF), and sensitivity under different initial conditions and complex standby configuration.
- To examine how changes in failure rates affect system performance through comprehensive sensitivity analysis, thereby providing critical insights for effective maintenance scheduling.
- The effectiveness of the proposed model is illustrated through case studies that compare system performance under various operating scenarios, such as full functionality of primary components and partial system failures triggering standby activation.
- To implement an innovation L_z-transform-based algorithm that effectively models complex standby system behavior, highlighting its potential as a powerful tool in advanced reliability computation.

10.3 MODEL FORMULATION

This study models a complex k-out-of-$(p+w+c)$: G mixed standby system, in which 'p' primary components operate in parallel, with 'w' warm and 'c' cold standby units available. At $t=0$, all units are assumed new and fully operational. The system requires a minimum $k(1 \leq k \leq p)$ components are in working condition for its effective operation. When there are $(k-1)$ operating components, the system fails. The assumptions of the model are given as follows:

- Initially, all components are considered to be in fully working condition.
- The warm standby components have a failure rate that is lower than primary components, whereas cold standby components experience no failures until activation.
- The replacement policy operates as follows: upon failure of the primary component, it is immediately replaced by an available warm spare component. If a warm component fails, it is subsequently replaced by an available cold standby component. The failed component is sent to the repair facility without any delay.

The following notations are used throughout this chapter:

p Number of the primary components
w Number of warm standby components
c Number of cold standby components
k Minimum number of operating components
λ^p Failure rate of primary components
λ^w Failure rate of warm standby components
λ^c Failure rate of cold standby components
μ^p Repair rate of primary components
μ^w Repair rate of warm standby components

α^p Probability of primary component
α^w Probability of warm spare component
α^c Probability of cold standby component

10.4 L_z-TRANSFORM OF MULTI-STATE COMPONENT

Obtaining the L_z-transform of each element is the first step in computing the dynamic reliability indices of a multi-state repairable k-out-of-$(p+w+c)$ mixed standby system. In order to calculate it, firstly, obtain the states' set represented by $g = g_{1,1}, g_{1,2}, g_{1,3}, \ldots, g_{1,L}$, presented in Figure 10.1 where $g_{i,j}$ is the performance levels of component i, which is in state j, $j = 1,2,3,\ldots,L_i$. Now, probability distribution of component i in state j, i.e., $\alpha_{1,j}$, can be obtained by solving simultaneously the system of differential equation by the help of initial conditions $\alpha_{i,1} = 1$, $\alpha_{i,2} = 0$, $\alpha_{i,3} = 0, \ldots, \alpha_{i,L} = 0$.

$$\left.\begin{aligned}
\frac{d\alpha_{i,L_i}(t)}{dt} &= -\sum_{e=1}^{L_i-1} \lambda_{e=1}^{(i)} \alpha_{i,e}(t) + \alpha_{e,L_i}^{(i)}(t) \sum_{e=1}^{L_i-1} \mu_{e,L_i}^{(i)} \\
\frac{d\alpha_{i,j}(t)}{dt} &= \sum_{e=j+1}^{L_i} \lambda_{e,j}^{(i)} \alpha_{i,e}(t) + \sum_{e=1}^{j-1} \mu_{e,j}^{(i)} \alpha_{i,e}(t) - \alpha_{i,j}(t) \left(\sum_{e=1}^{j-1} \lambda_{j,e}^{(i)} + \sum_{e=1}^{L_i} \mu_{j,e}^{(i)} \right) \\
\frac{d\alpha_{i,1}(t)}{dt} &= \sum_{e=2}^{L_i} \lambda_{e,1}^{(i)} \alpha_{i,e}(t) - \alpha_{i,1}(t) \sum_{e=2}^{L_i} \mu_{e=2}^{L_i}
\end{aligned}\right\} \tag{10.1}$$

The L_z-transform of one element i can be obtained subsequently as:

$$L_z\{G_i(t)\} = \sum_{j=1}^{L_i} \alpha_{i,j}(t) z^{g_{i,j}} \tag{10.2}$$

Once one element's L_z-transform has been determined, a similar procedure is used to determine the L_z-transform for each of the k-out-of-$(p+w+c)$: G mixed standby system's multi-state component i, $1 \le i \le (p+w+c)$.

10.5 L_z-TRANSFORM OF MULTI-STATE k-OUT-OF-$(p+w+c)$ MIXED STANDBY SYSTEM

The L_z-transform of the entire multi-state k-out-of-$(p+w+c)$ mixed standby system can be obtained by operator $\otimes$. Let $L_z\{G_1(t), G_2(t), \ldots, G_M(t)\}$ be

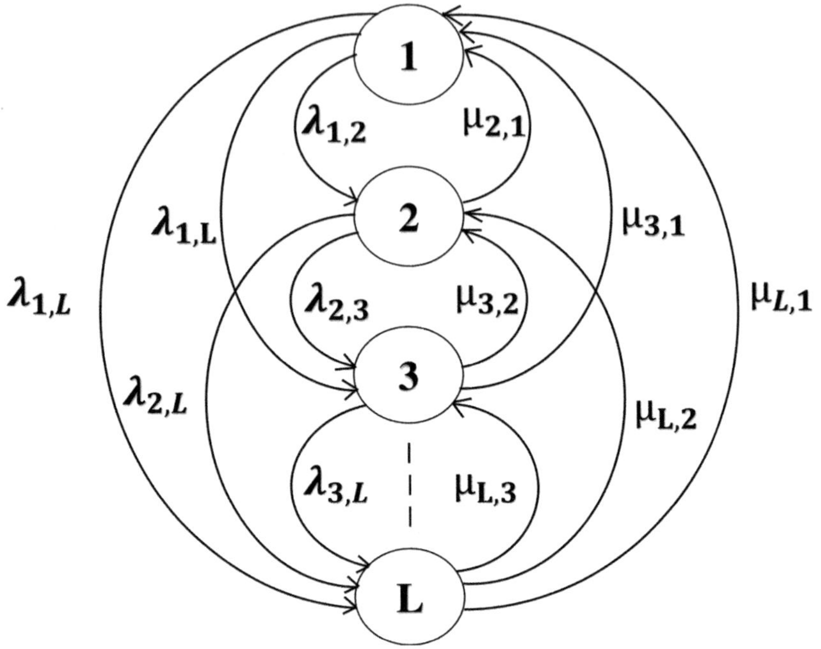

Figure 10.1 Transition diagram of multi-state component.

the L_z-transform of 1st, 2nd, …, Mth component of considered system. The following properties of the L_z-transform will be used for series and parallel configuration:

- **Property-1**: If components are connected in series configuration, then L_z-transform of system is derived as:

$$L_z\{G_1(t),G_2(t),\ldots,G_M(t)\}$$

$$=\underset{\min}{\otimes}\{L_z\{G_1(t),G_2(t),\ldots,G_M(t)\}\}$$

$$=\underset{\min}{\otimes}\left\{\sum_{e=1}^{L_1}\alpha_{1,e}(t)z^{g_{1,j}},\sum_{e=1}^{L_2}\alpha_{2,e}(t)z^{g_{2,j}},\ldots,\sum_{e=1}^{L_M}\alpha_{M,e}(t)z^{g_{M,j}}\right\}$$

$$=\sum_{e=1}^{L_1}\sum_{e=1}^{L_2}\cdots\sum_{e=1}^{L_M}\left(\alpha_{1,e}(t)\alpha_{2,e}(t)\ldots\alpha_{M,e}(t)\right)z^{\min\left(g_{1,j},g_{2,j},\ldots,g_{M,j}\right)} \tag{10.3}$$

- **Property-2:** If components are arranged in parallel configuration, then L_z-transform of system is derived as:

$$L_z\{G_1(t),G_2(t),\dots,G_M(t)\}$$
$$=\underset{+}{\otimes}\{L_z\{G_1(t),G_2(t),\dots,G_M(t)\}\}$$
$$=\underset{+}{\otimes}\left\{\sum_{e=1}^{L_1}\alpha_{1,e}(t)z^{g_{1,j}},\sum_{e=1}^{L_2}\alpha_{2,e}(t)z^{g_{2,j}},\dots,\sum_{e=1}^{L_M}\alpha_{M,e}(t)z^{g_{M,j}}\right\}$$
$$=\sum_{e=1}^{L_1}\sum_{e=1}^{L_2}\cdots\sum_{e=1}^{L_M}(\alpha_{1,e}(t)\alpha_{2,e}(t)\dots\alpha_{M,e}(t))z^{+(g_{1,j},g_{2,j},\dots,g_{M,j})} \tag{10.4}$$

10.6 SEQUENTIAL POLICY

In the proposed k-out-of-$(p+w+c)$ mixed standby system, a sequential activation policy is adopted. According to this policy, whenever a primary component fails, the next warm standby component is immediately activated. If no warm standby remains, then the cold standby is activated. This ensure that cold standby is preserved until all primary and warm components are exhausted.

Mathematically, sequential rule is given as:

$$w^W=\min(p^F,w-w^F) \tag{10.5}$$

$$c^W=\max(0,p^F-(w-w^F)) \tag{10.6}$$

where
w^W number of working warm standby components
c^W number of working cold standby components
w^F number of failed warm standby components
c^F number of failed cold standby components

10.7 RELIABILITY $R(t)$ AND AVAILABILITY $A(t)$ OF PROPOSED SYSTEM

The reliability $R(t)$ and availability $A(t)$ of the complex k-out-of-$(p+w+C)$ mixed standby system at any time t for minimum number of required component X can be obtained as

$$R(t) = A(t) = \sum_{j=1}^{L} \alpha_{i,j}(t) z^{(g_{1,j}, g_{2,j}, \ldots, g_{M,j}) - X} \tag{10.7}$$

The formulas for system availability and reliability may appear identical, but they are not the same. This difference arises because availability accounts for failure rate.

10.8 MTTF OF SYSTEM

MTTF of complex k-out-of-$(p+w+C)$ mixed standby system, which represents the average operational time until the failures of primary components of system, is mathematically written as:

$$\text{MTTF} = \int_{0}^{\infty} \sum_{j=1}^{L} \alpha_{i,j}(t) z^{(g_{1,j}, g_{2,j}, \ldots, g_{M,j}) - X} \tag{10.8}$$

10.9 SENSITIVITY $S(t)$ OF CONSIDERED SYSTEM

Sensitivity $S(t)$ of complex k-out-of-$(p+w+C)$ mixed standby system can be written as:

$$S(t) = \frac{\partial \sum_{j=1}^{L} \alpha_{i,j}(t) z^{(g_{1,j}, g_{2,j}, \ldots, g_{M,j}) - X}}{\partial \lambda_{i,j}^{p}} \tag{10.9}$$

It shows the effect of different failures in primary components to reliability.

10.10 PRACTICAL EXAMPLE OF THE CONSIDERED SYSTEM

In this chapter, our model is designed to represent some inverters in solar power plant. In a solar power plant, inverters play a crucial role in converting and managing electrical energy generated by solar panels. Their function is not just conversion, but also includes protection, synchronization, monitoring, and system control. Assume that such a system comprises p primary, w warm standby, and c cold standby inverters. For the system to operate effectively, at least k inverters must be operational. The system fails, if $(p+w+c)-(k-1) = p+w+c-k+1$ inverters fail. All inverters (primary, warm spare, and cold standby) are assumed to be initially in a brand-new state. All primary inverters fail independently of one another and independently of the warm inverters, and vice versa. Upon failure, both primary and warm inverters are immediately sent to the repair facility. A failed primary inverter is replaced by an available warm inverter, while a failed warm inverter is replaced by an available cold inverter.

10.11 NUMERICAL EXAMPLE

In this section, we analyze the effect of mixed standby components on system performance measures, including reliability, availability, MTTF, and sensitivity. We take $p = 3$, $w = 2$, $c = 2$, and $k = 3$ as fixed values. Each component contains three operational states: fully working, partially working, completely failure, and connected in series-parallel configuration as shown in Figure 10.2. The transition intensities for considered system are given in Table 10.2 and steady-state diagram for each component is presented in Figure 10.3.

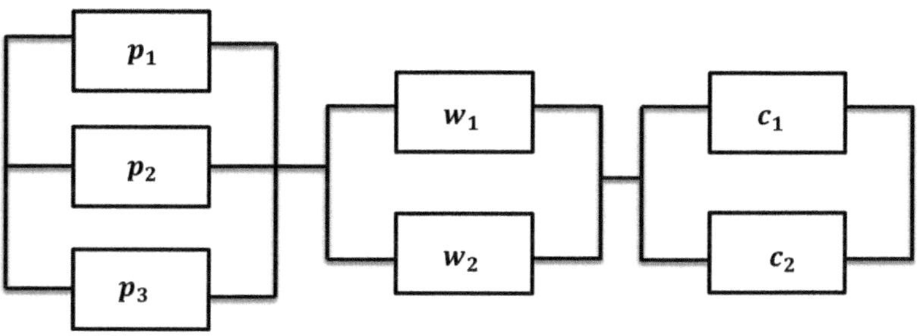

Figure 10.2 Complex 3-out-of-(3 + 2 + 2) mixed standby system.

Table 10.2 Transition intensities of primary and warm standby components

Primary components		*Warm standby components*	
$\lambda_{1,2}^{p_i}$	0.32	$\lambda_{1,2}^{w_j}$	0.29
$\lambda_{1,3}^{p_i}$	0.38	$\lambda_{1,3}^{w_j}$	0.32
$\lambda_{2,3}^{p_i}$	0.34	$\lambda_{2,3}^{w_j}$	0.30
$\mu_{2,1}^{p_i}$	0.38	$\mu_{2,1}^{w_j}$	0.38
$\mu_{3,1}^{p_i}$	0.42	$\mu_{3,1}^{w_j}$	0.42
$\mu_{3,2}^{p_i}$	0.40	$\mu_{3,2}^{w_j}$	0.40

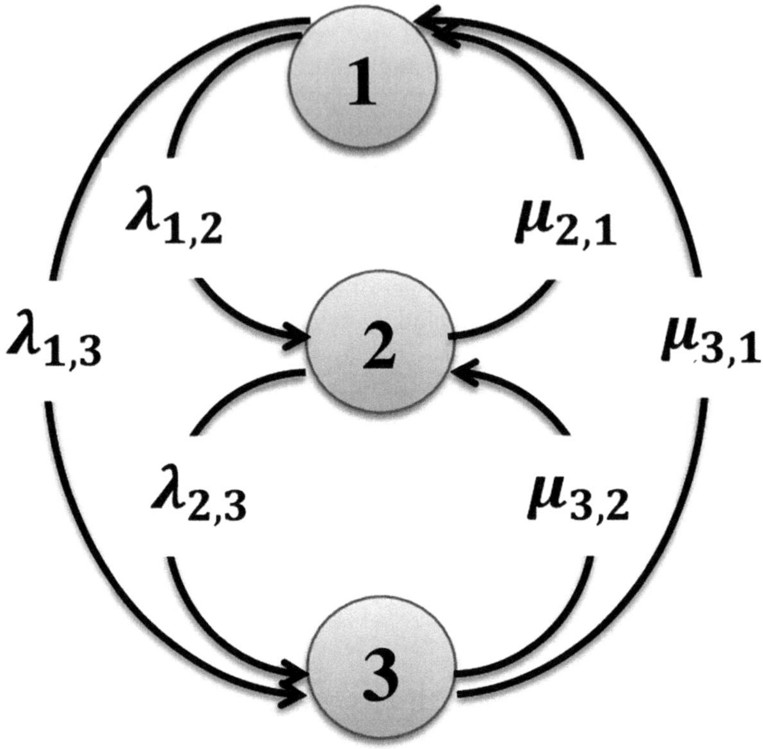

Figure 10.3 Steady-state diagram for each element of system.

The next step is to develop a system of differential equations using equation (10.10). Firstly, the system of differential equations is constructed to determine the transient state probabilities related to the reliability of the system, which are written as:

$$\left.\begin{aligned} \frac{d\alpha_{1,l}(t)}{dt} &= -(\lambda_{1,2}+\lambda_{1,3})\alpha_{1,l}(t) \\ \frac{d\alpha_{1,2}(t)}{dt} &= \lambda_{1,2}\alpha_{1,l}(t)-\lambda_{2,3}\alpha_{1,2}(t) \\ \frac{d\alpha_{1,3}(t)}{dt} &= \lambda_{1,3}\alpha_{1,l}(t)+\lambda_{2,3}\alpha_{1,2}(t) \end{aligned}\right\} \tag{10.10}$$

Numerically, the system of differential equations for primary and warm standby components is expressed as:

$$\left.\begin{aligned}
\frac{d\alpha_{1,1}^{p}(t)}{dt} &= -0.70\alpha_{1,1}^{p}(t) \\
\frac{d\alpha_{1,2}^{p}(t)}{dt} &= 0.32\alpha_{1,1}^{p}(t) - 0.34\alpha_{1,2}^{p}(t) \\
\frac{d\alpha_{1,3}^{p}(t)}{dt} &= 0.38\alpha_{1,1}^{p}(t) + 0.34\alpha_{1,2}^{p}(t)
\end{aligned}\right\} \tag{10.11}$$

$$\left.\begin{aligned}
\frac{d\alpha_{1,1}^{w}(t)}{dt} &= -0.61\alpha_{1,1}^{w}(t) \\
\frac{d\alpha_{1,2}^{w}(t)}{dt} &= 0.29\alpha_{1,1}^{w}(t) - 0.30\alpha_{1,2}^{w}(t) \\
\frac{d\alpha_{1,3}^{w}(t)}{dt} &= 0.32\alpha_{1,1}^{w}(t) + 0.30\alpha_{1,2}^{w}(t)
\end{aligned}\right\} \tag{10.12}$$

By solving these system of differential equations with initial conditions $\alpha_{1,1} = 1,0,0$, the transient state probabilities $\{\alpha_{1,1}, \alpha_{1,2}, \alpha_{1,3}\}$ are obtained as:

$$\left.\begin{aligned}
\alpha_{1,1}^{p}(t) &= \exp\left(-\frac{7t}{10}\right) \\
\alpha_{1,2}^{p}(t) &= \frac{8\exp(-17t/50)}{9} - \frac{8\exp(-7t/10)}{9} \\
\alpha_{1,3}^{p}(t) &= 1 - \frac{8\exp(-17t/50)}{9} - \frac{\exp(-7t/10)}{9}
\end{aligned}\right\} \tag{10.13}$$

$$\left.\begin{aligned}
\alpha_{1,1}^{w}(t) &= \exp\left(\frac{-61t}{100}\right) \\
\alpha_{1,2}^{w}(t) &= \frac{29\exp(-3t/10)}{31} - \frac{29\exp(-61t/100)}{31} \\
\alpha_{1,3}^{w}(t) &= 1 - \frac{2\exp(-61t/100)}{31} - \frac{29\exp(-3t/10)}{31}
\end{aligned}\right\} \tag{10.14}$$

Similarly, we formulate the system of differential equations to calculate the transient probability for system's availability, which are given as follows:

$$\left.\begin{aligned}
\frac{d\alpha_{1,l}(t)}{dt} &= -\left(\lambda_{1,2}+\lambda_{1,3}\right)\alpha_{1,l}(t)+\mu_{2,1}\alpha_{1,2}(t)+\mu_{3,1}\alpha_{1,3}(t) \\
\frac{d\alpha_{1,2}(t)}{dt} &= \lambda_{1,2}\alpha_{1,l}(t)-\left(\mu_{2,1}+\lambda_{2,3}\right)\alpha_{1,2}(t)+\mu_{3,2}\alpha_{1,3}(t) \\
\frac{d\alpha_{1,3}(t)}{dt} &= \lambda_{1,3}\alpha_{1,l}(t)+\lambda_{2,3}\alpha_{1,2}(t)-\left(\mu_{3,1}+\mu_{3,2}\right)\alpha_{1,3}(t)
\end{aligned}\right\} \qquad (10.15)$$

$$\left.\begin{aligned}
\frac{d\alpha_{1,1}(t)}{dt} &= -0.70\alpha_{1,1}^{p}(t)+0.38\alpha_{1,2}^{p}(t)+0.42\alpha_{1,3}^{p}(t) \\
\frac{d\alpha_{1,2}^{p}(t)}{dt} &= 0.32\alpha_{1,1}^{p}(t)-0.72\alpha_{1,2}^{p}(t)+0.40\alpha_{1,3}^{p}(t) \\
\frac{d\alpha_{1,3}^{p}(t)}{dt} &= 0.38\alpha_{1,1}^{p}(t)+0.34\alpha_{1,2}^{p}(t)-0.82\alpha_{1,3}^{p}(t)
\end{aligned}\right\} \qquad (10.16)$$

$$\left.\begin{aligned}
\frac{d\alpha_{1,1}^{w}(t)}{dt} &= -0.61\alpha_{1,1}^{w}(t)+0.38\alpha_{1,2}^{w}(t)+0.42\alpha_{1,3}^{w}(t) \\
\frac{d\alpha_{1,2}^{w}(t)}{dt} &= 0.29\alpha_{1,1}^{w}(t)-0.68\alpha_{1,2}^{w}(t)+0.40\alpha_{1,3}^{w}(t) \\
\frac{d\alpha_{1,3}^{w}(t)}{dt} &= 0.32\alpha_{1,1}^{w}(t)+0.30\alpha_{1,2}^{w}(t)-0.82\alpha_{1,3}^{w}(t)
\end{aligned}\right\} \qquad (10.17)$$

Now, L_z-transform for each primary component can be written as:

$$L_z\left\{\alpha_1^{p}(t)\right\}=\alpha_{1,1}^{p}(t)z^{1}+\alpha_{1,2}^{p}(t)z^{0.7}+\alpha_{1,3}^{p}(t)z^{0}$$

$$L_z\left\{\alpha_2^{p}(t)\right\}=\alpha_{2,1}^{p}(t)z^{1}+\alpha_{2,2}^{p}(t)z^{0.7}+\alpha_{2,3}^{p}(t)z^{0}$$

$$L_z\left\{\alpha_3^{p}(t)\right\}=\alpha_{3,1}^{p}(t)z^{1}+\alpha_{3,2}^{p}(t)z^{0.7}+\alpha_{3,3}^{p}(t)z^{0}$$

Similarly, L_z-transform for warm and cold standby components can be obtained as:

$$L_z\left\{\alpha_1^w(t)\right\}=\alpha_{1,1}^w(t)z^1+\alpha_{1,2}^w(t)z^{0.7}+\alpha_{1,3}^w(t)z^0$$

$$L_z\left\{\alpha_2^w(t)\right\}=\alpha_{2,1}^w(t)z^1+\alpha_{2,2}^w(t)z^{0.7}+\alpha_{2,3}^w(t)z^0$$

$$L_z\left\{\alpha_1^c(t)\right\}=\alpha_1^c(t)z^1$$

$$L_z\left\{\alpha_2^c(t)\right\}=\alpha_2^c(t)z^1$$

While cold standby components have no failures, these components are fully operational until activation, i.e., $\alpha^{c_m}=1$, where $m=1,2$. The reliability indices of complex k-out-of-$(p+w+c)$: G mixed standby system for five cases are derived in the following subsection.

10.12 PARTICULAR CASES

Case-1: This case considers a system where the primary components are in working condition and arranged in parallel, as shown in Figure 10.4. The associated warm and cold spare components are connected in series with the primary components, enabling sequential activation in case of primary component failure.

$$L_z\left\{\alpha_1(t)\right\}=L_z\left\{\alpha_1^p(t)\right\}\underset{+}{\otimes}L_z\left\{\alpha_2^p(t)\right\}\underset{+}{\otimes}L_z\left\{\alpha_3^p(t)\right\}$$

$$=\left[\alpha_{1,1}^p(t)z^1+\alpha_{1,2}^p(t)z^{0.7}+\alpha_{1,3}^p(t)z^0\right]\underset{+}{\otimes}\left[\alpha_{2,1}^p(t)z^1+\alpha_{2,2}^p(t)z^{0.7}\right.$$

$$\left.+\alpha_{2,3}^p(t)z^0\right]\underset{+}{\otimes}\left[\alpha_{3,1}^p(t)z^1+\alpha_{3,2}^p(t)z^{0.7}+\alpha_{3,3}^p(t)z^0\right]$$

$$=\left[\alpha_{1,1}^p(t)\alpha_{2,1}^p(t)\alpha_{3,1}^p(t)\right]z^3+\left[\alpha_{1,1}^p(t)\alpha_{2,1}^p(t)\alpha_{3,2}^p(t)+\alpha_{1,1}^p(t)\alpha_{2,2}^p(t)\right.$$

$$\left.\alpha_{3,1}^p(t)+\alpha_{1,2}^p(t)\alpha_{2,1}^p(t)\alpha_{3,1}^p(t)\right]z^{2.7}+\left[\alpha_{1,1}^p(t)\alpha_{2,2}^p(t)\alpha_{3,2}^p(t)+\right.$$

$$\left.\alpha_{1,2}^p(t)\alpha_{2,1}^p(t)\alpha_{3,2}^p(t)+\alpha_{1,2}^p(t)\alpha_{2,2}^p(t)\alpha_{3,1}^p(t)\right]z^{2.4}$$

$$+\left[\alpha_{1,2}^p(t)\alpha_{2,2}^p(t)\alpha_{3,2}^p(t)\right]z^{2.1}+\left[\alpha_{1,1}^p(t)\alpha_{2,1}^p(t)\alpha_{3,3}^p(t)\right.$$

$$\left.+\alpha_{1,1}^p(t)\alpha_{2,3}^p(t)\alpha_{3,1}^p(t)+\alpha_{1,3}^p(t)\alpha_{2,1}^p(t)\alpha_{3,1}^p(t)\right]z^2$$

$$+\left[\alpha_{1,1}^{p}(t)\alpha_{2,2}^{p}(t)\alpha_{3,3}^{p}(t)+\alpha_{1,2}^{p}(t)\alpha_{2,1}^{p}(t)\alpha_{3,3}^{p}(t)\right.$$

$$+\alpha_{1,1}^{p}(t)\alpha_{2,3}^{p}(t)\alpha_{3,2}^{p}(t)+\alpha_{1,3}^{p}(t)\alpha_{2,1}^{p}(t)\alpha_{3,2}^{p}(t)+\alpha_{1,2}^{p}(t)\alpha_{2,3}^{p}(t)\alpha_{3,1}^{p}(t)$$

$$\left.+\alpha_{1,3}^{p}(t)\alpha_{2,2}^{p}(t)\alpha_{3,1}^{p}(t)\right]z^{1.7}+\left[\alpha_{1,2}^{p}(t)\alpha_{2,2}^{p}(t)\alpha_{3,3}^{p}(t)\right.$$

$$\left.+\alpha_{1,2}^{p}(t)\alpha_{2,3}^{p}(t)\alpha_{3,2}^{p}(t)+\alpha_{1,3}^{p}(t)\alpha_{2,2}^{p}(t)\alpha_{3,2}^{p}(t)\right]z^{1.4}+\left[\alpha_{1,1}^{p}(t)\alpha_{2,3}^{p}(t)\right.$$

$$\left.\alpha_{3,3}^{p}(t)+\alpha_{1,3}^{p}(t)\alpha_{2,1}^{p}(t)\alpha_{3,3}^{p}(t)+\alpha_{1,3}^{p}(t)\alpha_{2,3}^{p}(t)\alpha_{3,1}^{p}(t)\right]z^{1}+\left[\alpha_{1,2}^{p}(t)\right.$$

$$\left.\alpha_{2,3}^{p}(t)\alpha_{3,3}^{p}(t)+\alpha_{1,3}^{p}(t)\alpha_{2,2}^{p}(t)\alpha_{3,3}^{p}(t)+\alpha_{1,3}^{p}(t)\alpha_{2,3}^{p}(t)\alpha_{3,2}^{p}(t)\right]z^{0.7}$$

$$+\left[\alpha_{1,3}^{p}(t)\alpha_{2,3}^{p}(t)\alpha_{3,3}^{p}(t)\right]z^{0}$$

Case-2: This scenario involves the failure of one primary component, while the remaining two continue to operate, as presented in Figure 10.5. Upon failure, a warm spare component is activated to replace the failed component and sustain the system's performance.

$$L_z\left\{\alpha_2(t)\right\}=\left\{L_z\left\{\alpha_1^{p}(t)\right\}\underset{+}{\otimes}L_z\left\{\alpha_2^{p}(t)\right\}\right\}\underset{\min}{\otimes}\left\{L_z\left\{\alpha_1^{w}(t)\right\}\right\}$$

$$=\left\{\left[\alpha_{1,1}^{p}(t)z^{1}+\alpha_{1,2}^{p}(t)z^{0.7}+\alpha_{1,3}^{p}(t)z^{0}\right]\underset{+}{\otimes}\left[\alpha_{2,1}^{p}(t)z^{1}+\alpha_{2,2}^{p}(t)z^{0.7}+\alpha_{2,3}^{p}(t)z^{0}\right]\right\}$$

$$\underset{\min}{\otimes}\left\{\alpha_{1,1}^{w}(t)z^{1}+\alpha_{1,2}^{w}(t)z^{0.7}+\alpha_{1,3}^{w}(t)z^{0}\right\}$$

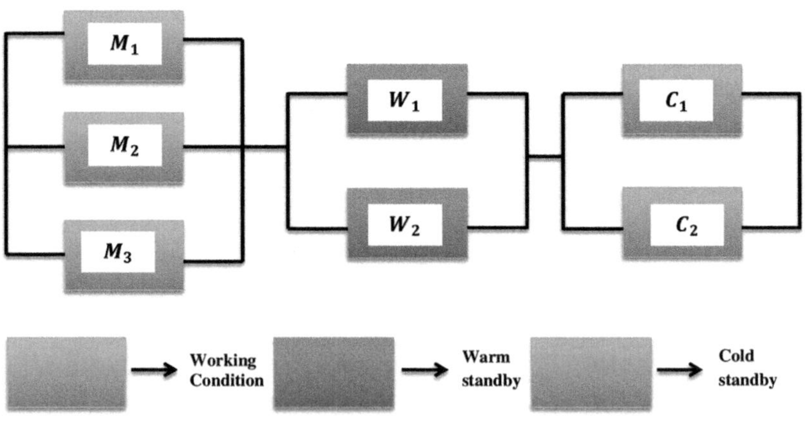

Figure 10.4 Complex 3-out-of-(3 + 2 + 2) mixed standby system for case-1.

$$
\begin{aligned}
&= \Big[\alpha_{1,1}^{p}(t)\alpha_{2,1}^{p}(t)\alpha_{1,1}^{w}(t) + \alpha_{1,1}^{p}(t)\alpha_{2,2}^{p}(t)\alpha_{1,1}^{w}(t) + \alpha_{1,2}^{p}(t)\alpha_{2,1}^{p}(t)\alpha_{1,1}^{w}(t) \\
&+ \alpha_{1,2}^{p}(t)\alpha_{2,1}^{p}(t)\alpha_{1,1}^{w}(t) + \alpha_{1,1}^{p}(t)\alpha_{2,3}^{p}(t)\alpha_{1,1}^{w}(t) + \alpha_{1,3}^{p}(t)\alpha_{2,1}^{p}(t)\alpha_{1,1}^{w}(t) \Big] z^{1} \\
&+ \Big[\alpha_{1,1}^{p}(t)\alpha_{2,1}^{p}(t)\alpha_{1,2}^{w}(t) + \alpha_{1,1}^{p}(t)\alpha_{2,2}^{p}(t)\alpha_{1,2}^{w}(t) + \alpha_{1,2}^{p}(t)\alpha_{2,1}^{p}(t)\alpha_{1,2}^{w}(t) \\
&+ \alpha_{1,2}^{p}(t)\alpha_{2,2}^{p}(t)\alpha_{1,2}^{w}(t) + \alpha_{1,1}^{p}(t)\alpha_{2,3}^{p}(t)\alpha_{1,2}^{w}(t) + \alpha_{1,3}^{p}(t)\alpha_{2,1}^{p}(t)\alpha_{1,2}^{w}(t) \\
&+ \alpha_{1,2}^{p}(t)\alpha_{2,3}^{p}(t)\alpha_{1,1}^{w}(t) + \alpha_{1,3}^{p}(t)\alpha_{2,2}^{p}(t)\alpha_{1,1}^{w}(t) + \alpha_{1,2}^{p}(t)\alpha_{2,3}^{p}(t)\alpha_{1,2}^{w}(t) \\
&+ \alpha_{1,3}^{p}(t)\alpha_{2,2}^{p}(t)\alpha_{1,2}^{w}(t) \Big] z^{0.7} + \Big[\alpha_{1,1}^{p}(t)\alpha_{2,1}^{p}(t)\alpha_{1,3}^{w}(t) + \alpha_{1,1}^{p}(t)\alpha_{2,2}^{p}(t)\alpha_{1,3}^{w}(t) \\
&+ \alpha_{1,2}^{p}(t)\alpha_{2,1}^{p}(t)\alpha_{1,3}^{w}(t) + \alpha_{1,2}^{p}(t)\alpha_{2,2}^{p}(t)\alpha_{1,3}^{w}(t) + \alpha_{1,1}^{p}(t)\alpha_{2,3}^{p}(t)\alpha_{1,3}^{w}(t) \\
&+ \alpha_{1,3}^{p}(t)\alpha_{2,1}^{p}(t)\alpha_{1,3}^{w}(t) + \alpha_{1,2}^{p}(t)\alpha_{2,3}^{p}(t)\alpha_{1,3}^{w}(t) + \alpha_{1,3}^{p}(t)\alpha_{2,2}^{p}(t)\alpha_{1,3}^{w}(t) \\
&+ \alpha_{1,3}^{p}(t)\alpha_{2,3}^{p}(t)\alpha_{1,1}^{w}(t) + \alpha_{1,3}^{p}(t)\alpha_{2,3}^{p}(t)\alpha_{1,2}^{w}(t) + \alpha_{1,3}^{p}(t)\alpha_{2,3}^{p}(t)\alpha_{1,3}^{w}(t) \Big] z^{0}
\end{aligned}
$$

Case-3: In this configuration, only one primary component remains in working condition, while the other two have completely failed. Consequently, two warm standby components are activated to replace the failed primary units as shown in Figure 10.6 and keep the system running properly.

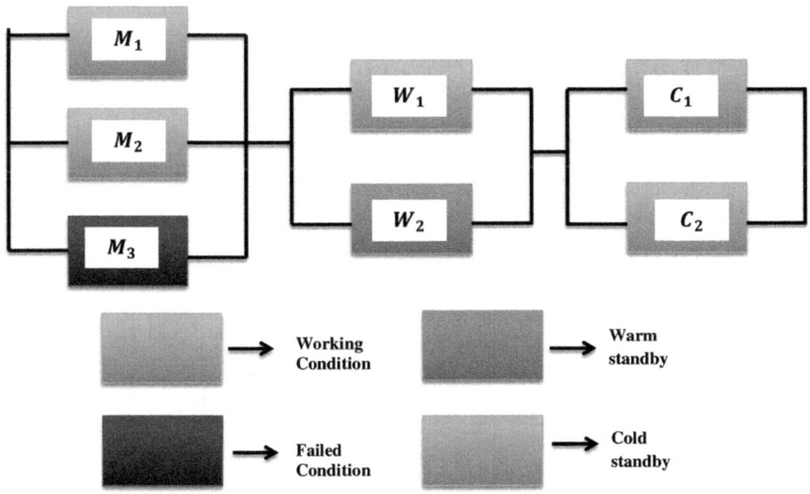

Figure 10.5 Complex 3-out-of-$(3+2+2)$ mixed standby system for case-2.

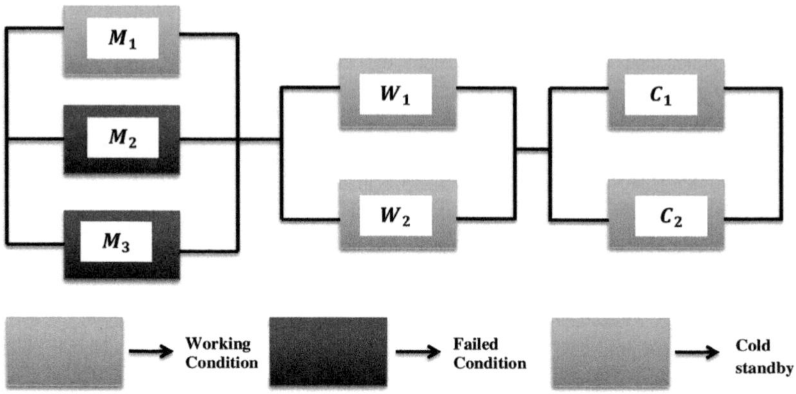

Figure 10.6 Complex 3-out-of-(3 + 2 + 2) mixed standby system for case-3.

$$
\begin{aligned}
L_z\{\alpha_3(t)\} &= \left\{L_z\{\alpha_1^p(t)\}\right\} \underset{\min}{\otimes} \left\{L_z\{\alpha_1^w(t)\} \underset{+}{\otimes} L_z\{\alpha_2^w(t)\}\right\} \\
&= \left\{\alpha_{1,1}^p(t)z^1 + \alpha_{1,2}^p(t)z^{0.7} + \alpha_{1,3}^p(t)z^0\right\} \underset{\min}{\otimes} \left\{\left[\alpha_{1,1}^w(t)z^1 + \alpha_{1,2}^w(t)z^{0.7} + \alpha_{1,3}^w(t)z^0\right]\right. \\
&\quad \left.\underset{+}{\otimes}\left[\alpha_{2,1}^w(t)z^1 + \alpha_{2,2}^w(t)z^{0.7} + \alpha_{2,3}^w(t)z^0\right]\right\} \\
&= \left[\alpha_{1,1}^p(t)\alpha_{1,1}^w(t)\alpha_{2,1}^w + \alpha_{1,1}^p(t)\alpha_{1,1}^w(t)\alpha_{2,2}^w + \alpha_{1,1}^p(t)\alpha_{1,2}^w(t)\alpha_{2,1}^w\right. \\
&\quad \left.+\alpha_{1,1}^p(t)\alpha_{1,2}^w(t)\alpha_{2,2}^w + \alpha_{1,1}^p(t)\alpha_{1,1}^w(t)\alpha_{2,3}^w + \alpha_{1,1}^p(t)\alpha_{1,3}^w(t)\alpha_{2,1}^w\right]z^1 \\
&\quad +\left[\alpha_{1,1}^p(t)\alpha_{1,2}^w(t)\alpha_{2,3}^w + \alpha_{1,1}^p(t)\alpha_{1,3}^w(t)\alpha_{2,2}^w + \alpha_{1,2}^p(t)\alpha_{1,1}^w(t)\alpha_{2,1}^w\right. \\
&\quad +\alpha_{1,2}^p(t)\alpha_{1,1}^w(t)\alpha_{2,2}^w + \alpha_{1,2}^p(t)\alpha_{1,2}^w(t)\alpha_{2,1}^w + \alpha_{1,2}^p(t)\alpha_{1,2}^w(t)\alpha_{2,2}^w \\
&\quad +\alpha_{1,2}^p(t)\alpha_{1,1}^w(t)\alpha_{2,3}^w + \alpha_{1,2}^p(t)\alpha_{1,3}^w(t)\alpha_{2,1}^w + \alpha_{1,2}^p(t)\alpha_{1,2}^w(t)\alpha_{2,3}^w \\
&\quad \left.+\alpha_{1,2}^p(t)\alpha_{1,3}^w(t)\alpha_{2,2}^w\right]z^{0.7} + \left[\alpha_{1,1}^p(t)\alpha_{1,3}^w(t)\alpha_{2,3}^w + \alpha_{1,2}^p(t)\alpha_{1,3}^w(t)\alpha_{2,3}^w\right. \\
&\quad +\alpha_{1,3}^p(t)\alpha_{1,1}^w(t)\alpha_{2,1}^w + \alpha_{1,3}^p(t)\alpha_{1,1}^w(t)\alpha_{2,2}^w + \alpha_{1,3}^p(t)\alpha_{1,2}^w(t)\alpha_{2,1}^w \\
&\quad +\alpha_{1,3}^p(t)\alpha_{1,2}^w(t)\alpha_{2,2}^w + \alpha_{1,3}^p(t)\alpha_{1,2}^w(t)\alpha_{2,3}^w + \alpha_{1,3}^p(t)\alpha_{1,3}^w(t)\alpha_{2,2}^w \\
&\quad \left.+\alpha_{1,3}^p(t)\alpha_{1,3}^w(t)\alpha_{2,3}^w\right]z^0
\end{aligned}
$$

Case-4: In this case, one primary component, one warm spare, and one cold standby component in working condition, which are connected in series, are presented in Figure 10.7, while the remaining have failed or are inactive.

$$L_z\left\{\alpha_4(t)\right\} = L_z\left\{\alpha_1^p(t)\right\} \underset{\min}{\otimes} L_z\left\{\alpha_1^w(t)\right\} \underset{\min}{\otimes} L_z\left\{\alpha_1^c(t)\right\}$$

$$= \left[\alpha_{1,1}^p(t)z^1 + \alpha_{1,2}^p(t)z^{0.7} + \alpha_{1,3}^p(t)z^0\right] \underset{\min}{\otimes} \left[\alpha_{1,1}^w(t)z^1 + \alpha_{1,2}^w(t)z^{0.7} + \alpha_{1,3}^w(t)z^0\right]$$

$$\underset{\min}{\otimes}\left[\alpha_1^c(t)z^1\right]$$

$$= \left[\alpha_{1,1}^p(t)\alpha_{1,1}^w(t)\alpha_1^c(t)\right]z^1 + \left[\alpha_{1,1}^p(t)\alpha_{1,2}^w(t)\alpha_1^c(t) + \alpha_{1,2}^p(t)\alpha_{1,1}^w(t)\alpha_1^c(t)\right.$$

$$\left. + \alpha_{1,2}^p(t)\alpha_{1,2}^w(t)\alpha_1^c(t)\right]z^{0.7} + \left[\alpha_{1,1}^p(t)\alpha_{1,3}^w(t)\alpha_1^c(t) + \alpha_{1,2}^p(t)\alpha_{1,3}^w(t)\alpha_1^c(t)\right.$$

$$\left. + \alpha_{1,3}^p(t)\alpha_{1,1}^w(t)\alpha_1^c(t) + \alpha_{1,3}^p(t)\alpha_{1,2}^w(t)\alpha_1^c(t) + \alpha_{1,3}^p(t)\alpha_{1,3}^w(t)\alpha_1^c(t)\right]z^0$$

Case-5: This scenario involves a state where only one primary component and two cold standby components remain operational, with all warm standby units and other primary components being non-functional as given in Figure 10.8.

$$L_z\left\{\alpha_5(t)\right\} = \left\{L_z\left\{\alpha_1^p(t)\right\} \underset{\min}{\otimes} \left\{\ \right\} L_z\left\{\alpha_1^c(t)\right\} \underset{+}{\otimes} L_z\left\{\alpha_2^c(t)\right\}\right\}$$

$$= \left\{\alpha_{1,1}^p(t)z^1 + \alpha_{1,2}^p(t)z^{0.7} + \alpha_{1,3}^p(t)z^0\right\} \underset{\min}{\otimes} \left\{\left[\alpha_1^c(t)z^1\right] \underset{+}{\otimes} \left[\alpha_2^c(t)z^1\right]\right\}$$

$$= \left[\alpha_{1,1}^p(t)\alpha_1^c(t)\alpha_2^c(t)\right]z^1 + \left[\alpha_{1,2}^p(t)\alpha_1^c(t)\alpha_2^c(t)\right]z^{0.7} + \left[\alpha_{1,3}^p(t)\alpha_1^c(t)\alpha_2^c(t)\right]z^0$$

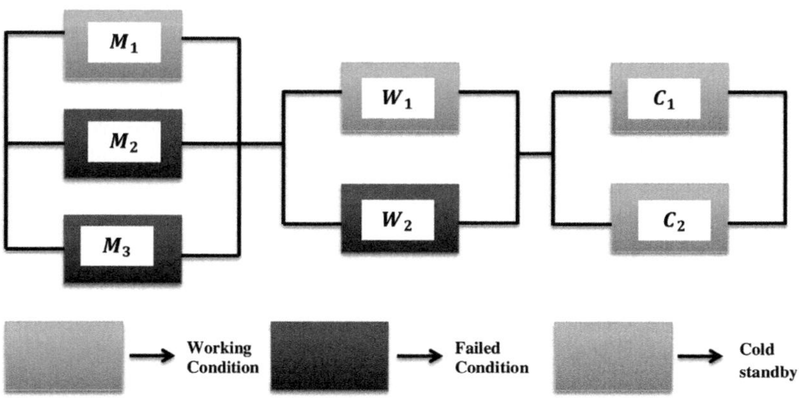

Figure 10.7 Complex 3-out-of-(3 + 2 + 2) mixed standby system for case-4.

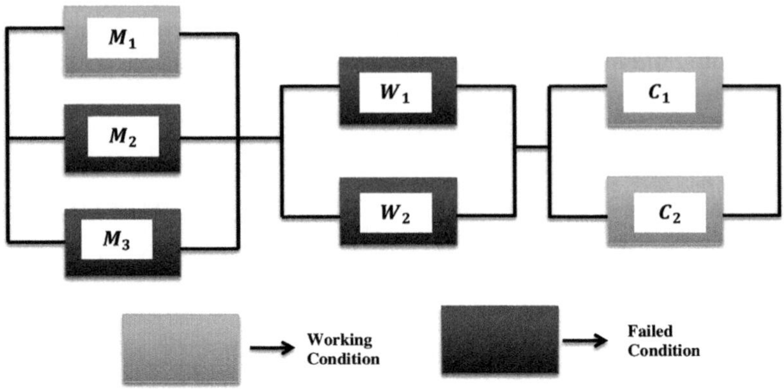

Figure 10.8 Complex 3-out-of-(3 + 2 + 2) mixed standby system for case-5.

Once the L_z-transform can be derived for five cases, then reliability attributes are derived and will be elaborated in the next section.

10.13 RESULTS AND DISCUSSION

In this section, we analyze the obtained reliability indices for considered cases of complex 3-out-of-(3+2+2) mixed standby system using L_z-transform.

10.14 RELIABILITY

The L_z-transform of each case is derived in Section 10.3. Reliability of system with minimum performance 0.7 is computed using equation (10.7). Five distinct operational cases were considered based on the working status of different component types, and the reliability of each case was evaluated over a 5-year operational period, initially the system is in full working condition, given in Table 10.3 as well as shown in Figure 10.9. Reliability for each case of considered system is discussed below:

- **Case-1: All primary components operational**
  In the first case, the system operates with all primary components in working condition. This scenario yields the highest system reliability among all five cases. Since primary components are connected in parallel, the system exhibits redundancy, and no standby components are required to be activated. The reliability decreases over time due to natural degradation.

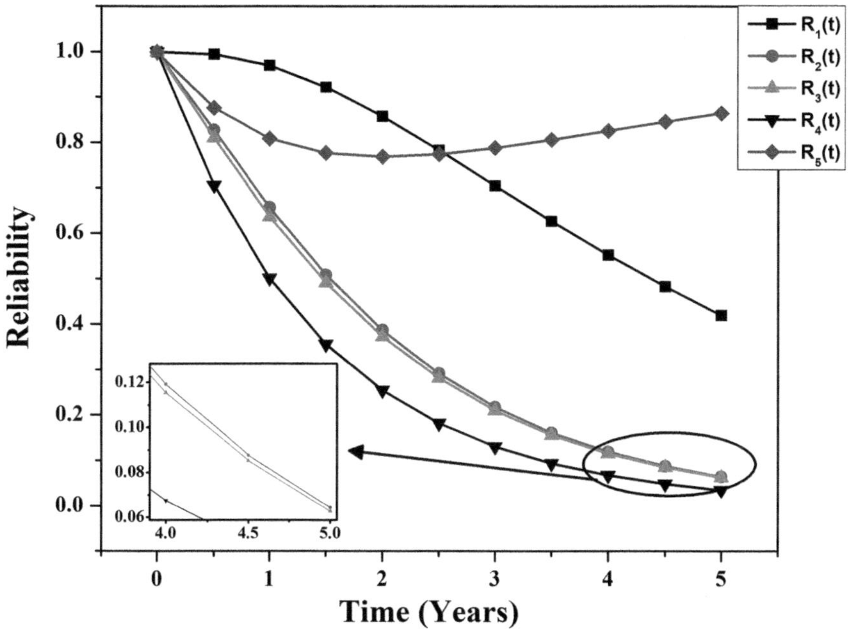

Figure 10.9 Reliability of complex 3-out-of-$(3+2+2)$ mixed standby system.

Table 10.3 Reliability of complex 3-out-of-$(3+2+2)$ mixed standby system

Time	$R_1(t)$	$R_2(t)$	$R_3(t)$	$R_4(t)$	$R_5(t)$
0.0	1.0000	1.0000	1.0000	1.0000	1.0000
0.5	0.9949	0.8276	0.8103	0.7063	0.8765
1.0	0.9696	0.6571	0.6370	0.5008	0.8087
1.5	0.9220	0.5087	0.4910	0.3564	0.7773
2.0	0.8575	0.3872	0.3733	0.2544	0.7689
2.5	0.7832	0.2914	0.2811	0.1820	0.7745
3.0	0.7048	0.2175	0.2101	0.1305	0.7883
3.5	0.6268	0.1613	0.1562	0.0938	0.8063
4.0	0.5521	0.1192	0.1156	0.0675	0.8259
4.5	0.4827	0.0877	0.0853	0.0487	0.8456
5.0	0.4194	0.0644	0.0628	0.0351	0.8645

- **Case-2: Two primary and one warm standby components operational**
  In the second case, the system still satisfies the minimum required number of functioning components due to the presence of redundant primary components and one warm standby. The reliability is slightly lower than case-1 but remains relatively high.

- **Case-3: One primary and two warm standby components operational**
  Here, the configuration satisfies the system's requirement but shows reliability close to that of case-2. However, the faster degradation rate of warm standby units limits the system's long-term performance.
- **Case-4: One primary, one warm, and cold standby components operational**
  In this case, due to sequential activation, the overall reliability is reduced. Among the five cases, this configuration results in the lowest system reliability.
- **Case-5: One primary and two cold standby components operational**
  Initially, this configuration exhibits a lower reliability due to some failures in the primary component. However, over time, as the cold standby units (which are fully functional when activated) come into operation, the system's reliability improves. Although the reliability remains lower than case-1, it surpasses the performance seen in cases 2–4.

10.15 AVAILABILITY

Availability was computed for each scenario using equation presented in Figure 10.10 and given in Table 10.4. Availability of system with minimum performance 0.7 over a 5-year operational period is discussed below:

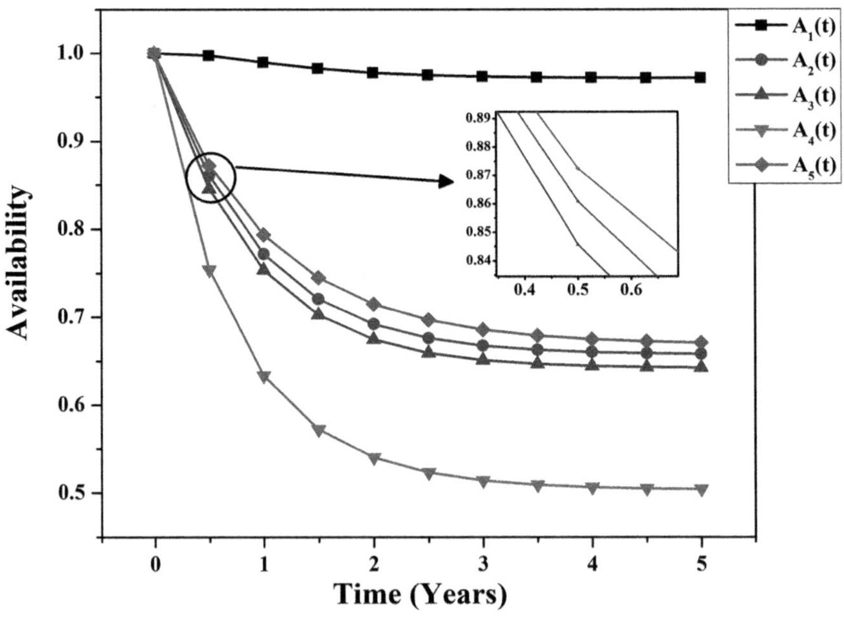

Figure 10.10 Availability of complex 3-out-of-(3 + 2 + 2) mixed standby system.

Table 10.4 Availability of complex 3-out-of-$(3+2+2)$ mixed standby system

Time	$A_1(t)$	$A_2(t)$	$A_3(t)$	$A_4(t)$	$A_5(t)$
0.0	1.0000	1.0000	1.0000	1.0000	1.0000
0.5	0.9972	0.8609	0.8456	0.7540	0.8726
1.0	0.9896	0.7721	0.7539	0.6339	0.7937
1.5	0.9826	0.7208	0.7027	0.5725	0.7450
2.0	0.9778	0.6922	0.6747	0.5404	0.7152
2.5	0.9749	0.6763	0.6595	0.5232	0.6969
3.0	0.9733	0.6677	0.6513	0.5141	0.6858
3.5	0.9724	0.6629	0.6469	0.5091	0.6791
4.0	0.9720	0.6604	0.6445	0.5065	0.6750
4.5	0.9717	0.6590	0.6433	0.5051	0.6725
5.0	0.9716	0.6582	0.6426	0.5043	0.6710

- **Case-1: All primary components operational**
  Due to parallel redundancy among the primary units, system downtime is minimal, and standby components are not required to operate. Availability remains close to 1 throughout the time span, only slightly decreasing due to some failures and natural wear.
- **Case-2: Two primary and one warm standby components operational**
  In this scenario, system remains functional, the activation and degradation behavior of the warm standby component lead to a noticeable reduction in availability over time. However, the overall availability remains high, closely comparable to case-3.
- **Case-3: One primary and two warm standby components operational**
  Here, the system experiences a decline in availability similar to case-2. The availability trend of this case is nearly identical to case-2, observation that warm standby components have a significant impact on long-term system performance.
- **Case-4: One primary, one warm, and one cold standby components operational**
  This case has the lowest availability among all cases. The mix of standby types and their staggered involvement in operation contributes to greatest downtime.
- **Case-5: One primary and two cold standby components operational**
  In the final scenario, the system maintains a moderate level of availability, higher than cases 2–4 but still lower than case-1.

10.16 MEAN TIME TO FAILURE

This section presents the analysis of MTTF for the proposed system, with a focus on the effect of primary component failures under various operational configurations given in Table 10.5 and also present in Figure 10.11. The system includes failure transition from fully working to partially working ($\lambda^p_{1,2}$), partially working to completely failed ($\lambda^p_{2,3}$), and fully working to completely failed ($\lambda^p_{1,3}$) states and varying failure rates from 0 to 0.1. Now, we discuss the MTTF with different failures:

- **MTTF with $\lambda^p_{1,2}$**

  The effect of failures transitioning from the fully working to partially working state has a significant influence on the system's MTTF as presented in Figure 10.11a. In case-1, where all primary components are operational, this gradual degradation allows the system to maintain functionality longer, thereby increasing MTTF. In case-2, the early involvement of a warm standby unit results in a quicker degradation path, leading to a decrease in MTTF. In case-3, there is a slight increase in MTTF compared to case-2. In case-4, the combination of warm and cold standby components introduces a staggered failure behavior, also leading to an increase in MTTF, although the overall value remains low due to transitional delays. Case-5 benefits from high reliability, resulting in an increased MTTF.

  Overall, the highest MTTF is observed in case-1, followed by case-5, with cases 2 and 3 showing moderate MTTF, and case-4 exhibiting the lowest MTTF due to standby configurations.
- **MTTF with $\lambda^p_{2,3}$**

  The transition from partially working to completely failed state adversely impacts the MTTF across all cases of the considered system as shown in Figure 10.11b. In case-1, the degradation of any component from a partially working to failed state results in a noticeable decrease in MTTF. In case-2, the involvement of a warm standby causes further reduction in MTTF. In cases 3, 4, and 5 the presence of warm or cold standby components does not prevent the system from being affected by partial-to-failure.

  In comparative terms, the highest MTTF is still achieved in case-1, followed by case-5, which benefits from the reliability of cold standby. Case-2 has a moderate MTTF, slightly better than case-3, while case-4 exhibits the lowest MTTF, primarily due to the compounding effects of warm and cold standby transitions and their delayed response.
- **MTTF with $\lambda^p_{1,3}$**

  The direct failure transition from the fully working to completely failed state leads to a decrease in MTTF across all five cases of the k-out-of-$(p+w+c)$ mixed standby system, which is given in Figure 10.11c. Overall, the highest MTTF is observed in case-1, followed by case-5, with cases 2 and 3 showing moderate MTTF, and case-4 exhibiting the lowest MTTF due to standby configurations.

Table 10.5 Mean time to failure of complex 3-out-of-$(3+2+2)$ mixed standby system

λ	*MTTF with* $\lambda_{1,2}^{P}$					*MTTF with* $\lambda_{2,3}^{P}$					*MTTF with* $\lambda_{1,3}^{P}$				
	$MTTF_1$	$MTTF_2$	$MTTF_3$	$MTTF_4$	$MTTF_5$	$MTTF_1$	$MTTF_2$	$MTTF_3$	$MTTF_4$	$MTTF_5$	$MTTF_1$	$MTTF_2$	$MTTF_3$	$MTTF_4$	$MTTF_5$
0.1	5.0826	2.3177	1.9114	1.4522	4.9642	6.2263	2.839	2.7982	2.0327	6.1079	7.9707	2.8980	2.4283	1.7553	7.8523
0.2	5.1127	2.1244	1.9258	1.4609	4.9943	5.6690	2.3406	2.4189	1.7876	5.5506	5.8929	2.2658	2.1641	1.6029	5.7745
0.3	5.1330	1.9908	1.9370	1.4678	5.0146	5.3253	2.0595	2.1274	1.5954	5.2069	5.2714	2.0141	1.9920	1.5010	5.1530
0.4	5.1473	1.8936	1.9461	1.4735	5.0289	5.0971	1.8778	1.8969	1.4406	4.9787	4.9913	1.8679	1.8715	1.4280	4.8729
0.5	5.1580	1.8201	1.9536	1.4782	5.0396	4.9373	1.7502	1.7102	1.3132	4.8189	4.8385	1.7690	1.7827	1.3731	4.7201
0.6	5.1662	1.7627	1.9599	1.4822	5.0478	4.8204	1.6553	1.5562	1.2066	4.7020	4.7451	1.6965	1.7146	1.3304	4.6267
0.7	5.1726	1.7167	1.9652	1.4856	5.0542	4.7321	1.5819	1.4270	1.1160	4.5844	4.7028	1.6521	1.7028	1.2986	4.5844
0.8	5.1778	1.6791	1.9698	1.4886	5.0594	4.6635	1.5232	1.3173	1.0380	4.5451	4.6400	1.5961	1.6175	1.2681	4.5216
0.9	5.1821	1.6479	1.9737	1.4912	5.0637	4.609	1.4754	1.2228	0.9703	4.4906	4.6085	1.5597	1.5818	1.2446	4.4901
1.0	5.1857	1.6215	1.9772	1.4934	5.0673	4.5648	1.4355	1.1408	0.9109	4.4464	4.5846	1.5292	1.5518	1.2248	4.4662

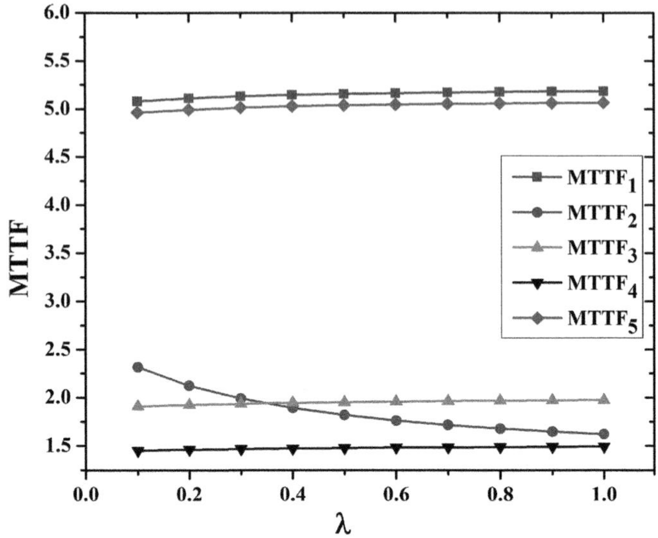

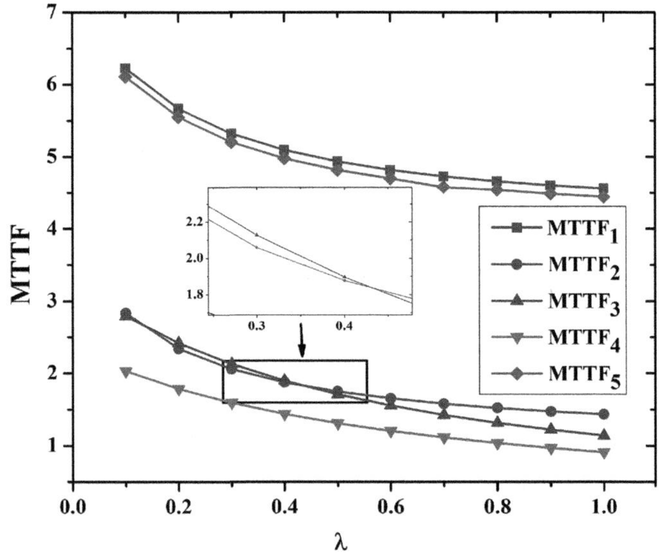

Figure 10.11 Mean time to failure of complex 3-out-of-$(3+2+2)$ mixed standby system. (a) MTTF with $\lambda_{1,2}^{p}$. (b) MTTF with $\lambda_{2,3}^{p}$. (c) MTTF with $\lambda_{1,3}^{p}$.

(Continued)

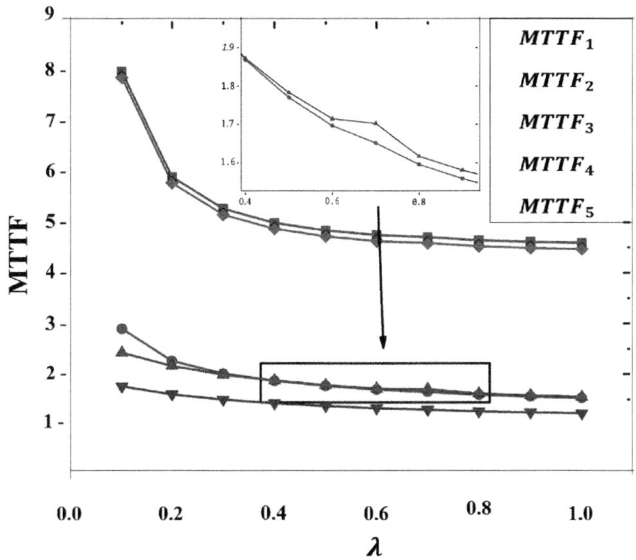

Figure 10.11 (Continued) Mean time to failure of complex 3-out-of-$(3+2+2)$ mixed standby system. (a) MTTF with $\lambda_{1,2}^{P}$. (b) MTTF with $\lambda_{2,3}^{P}$. (c) MTTF with $\lambda_{1,3}^{P}$.

10.17 SENSITIVITY

This section presents a sensitivity analysis conducted over a 0- to 5-year time span to evaluate the impact of primary component failures on system behavior under three distinct failure modes: from fully operational to partially working $\lambda_{1,2}^{p}$, from partially failed to completely failed $\lambda_{2,3}^{p}$, and from fully working to completely failed $\lambda_{1,3}^{p}$ given in Table 10.6 as well as shown in Figure 10.12.

- **Effect of $\lambda_{1,2}^{p}$ on sensitivity**

  The sensitivity analysis over time for the transition $\lambda_{1,2}^{p}$ reveals distinct behaviors across different system cases presented in Figure 10.12a. In case-1, sensitivity increases steadily over time, indicating that early-stage degradation has a growing impact on system reliability as the system ages. This trend suggests that even minimal initial failures gradually accumulate, reducing overall performance in systems primarily relying on fully functional components. In contrast, case-2 exhibits a sharp negative sensitivity, which decreases over time, reaching a peak negative value around 1.5 years and then gradually

Table 10.6 Sensitivity of complex 3-out-of-$(3+2+2)$ mixed standby system

Time (years)	Effect of $\lambda_{1,2}^{P}$					Effect of $\lambda_{2,3}^{P}$					Effect of $\lambda_{1,3}^{P}$				
	S_1	S_2	S_3	S_4	S_5	S_1	S_2	S_3	S_4	S_5	S_1	S_2	S_3	S_4	S_5
0.0	0.0000	0.0000	0.0000	0.0000	0.0000	0.0000	0.0000	0.0000	0.0000	0.0000	0.0000	0.0000	0.0000	0.0000	0.0000
0.5	0.0001	–0.0280	0.0037	0.0032	–0.3561	–0.0113	–0.0839	–0.3740	–0.3260	0.0300	–0.0009	–0.0330	–0.0311	–0.0271	0.0318
1.0	0.0011	–0.0839	0.0104	0.0082	–0.5078	–0.0571	–0.2149	–0.5431	–0.4270	0.0899	–0.0099	–0.1082	–0.0939	–0.0738	0.1014
1.5	0.0035	–0.1384	0.0163	0.0118	–0.5439	–0.1236	–0.3109	–0.5804	–0.4213	0.1521	–0.0332	–0.1884	–0.1560	–0.1133	0.1820
2.0	0.0069	–0.1780	0.0199	0.0136	–0.5186	–0.1901	–0.3568	–0.5445	–0.3710	0.2036	–0.0705	–0.2504	–0.2021	–0.1377	0.2586
2.5	0.0108	–0.1992	0.0211	0.0137	–0.4644	–0.2435	–0.3612	–0.4749	–0.3075	0.2401	–0.1168	–0.2862	–0.2278	–0.1475	0.3235
3.0	0.0145	–0.2040	0.0205	0.0128	–0.4001	–0.2788	–0.3383	–0.3954	–0.2457	0.2614	–0.1656	–0.2969	–0.2349	–0.1460	0.3736
3.5	0.0175	–0.1966	0.0188	0.0113	–0.3357	–0.2963	–0.3006	–0.3188	–0.1915	0.2696	–0.2117	–0.2880	–0.2278	–0.1368	0.4084
4.0	0.0196	–0.1811	0.0164	0.0096	–0.2767	–0.2989	–0.2572	–0.2513	–0.1467	0.2674	–0.2511	–0.2660	–0.2111	–0.1233	0.4290
4.5	0.0207	–0.1613	0.0139	0.0079	–0.2250	–0.2902	–0.2139	–0.1947	–0.1111	0.2574	–0.2818	–0.2367	–0.18917	–0.1079	0.4373
5.0	0.0211	–0.1398	0.0115	0.0064	–0.1813	–0.2737	–0.1741	–0.1489	–0.0833	0.2423	–0.3030	–0.2046	–0.1649	–0.0922	0.4354

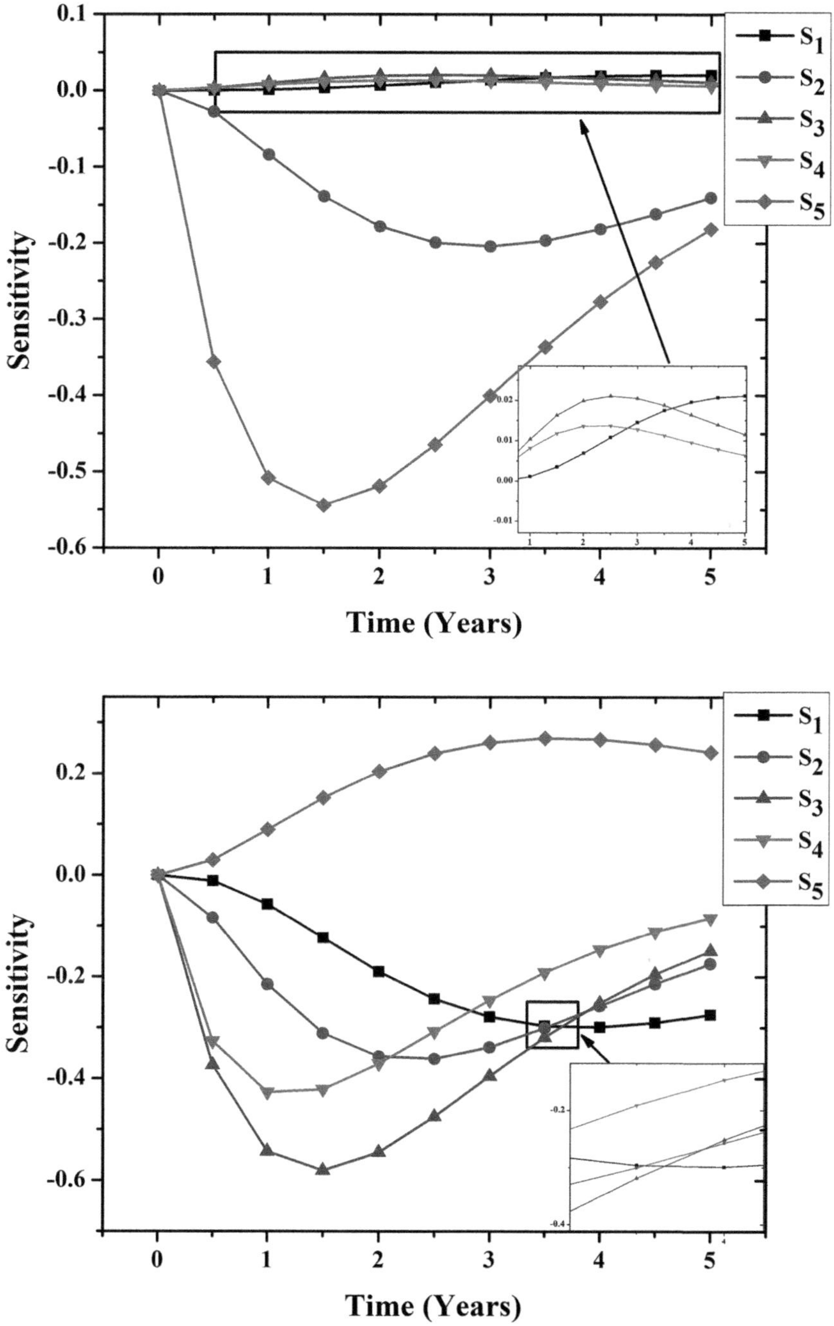

Figure 10.12 Sensitivity of complex 3-out-of-(3 + 2 + 2) mixed standby system. (a) Effect of $\lambda^{p}_{1,2}$. (b) Effect of $\lambda^{p}_{2,3}$. (c) Effect of $\lambda^{p}_{1,3}$.

(Continued)

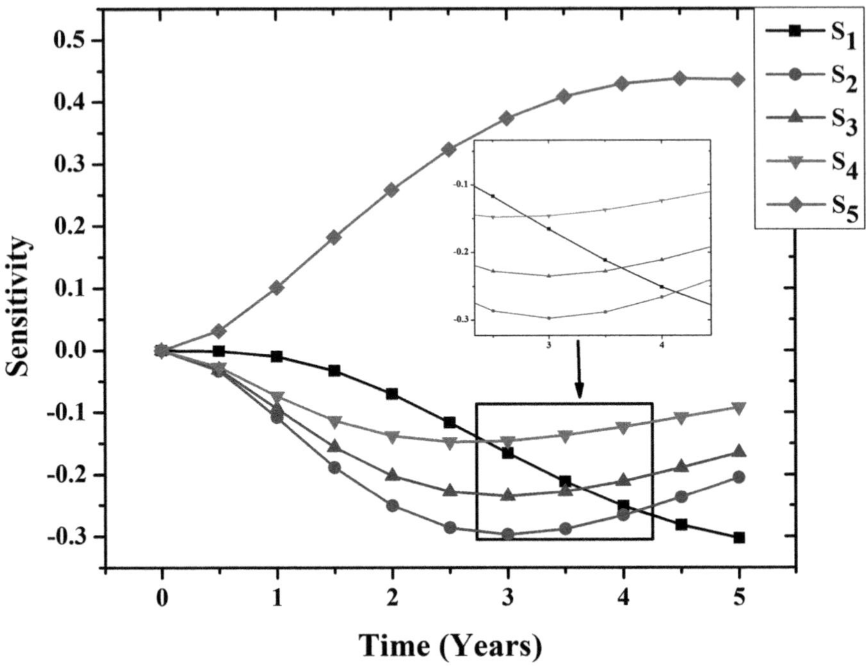

Figure 10.12 (Continued) Sensitivity of complex 3-out-of-(3 + 2 + 2) mixed standby system. (a) Effect of $\lambda_{1,2}^{p}$. (b) Effect of $\lambda_{2,3}^{p}$. (c) Effect of $\lambda_{1,3}^{p}$.

improving. This reflects a strong adverse impact from early partial failures, particularly when standby components (warm) are already involved. Cases 3 and 4 show moderate sensitivity, initially rising and then decreasing, indicating that their hybrid redundancy mitigates some early effects. Case-5, which uses cold standby units, initially shows significant negative sensitivity but improves steadily, implying that cold standby components begin to compensate as time progresses.

- **Effect of $\lambda_{2,3}^{p}$ on sensitivity**

  The sensitivity analysis for the failure $\lambda_{2,3}^{p}$ shows a consistently negative impact on system performance across all five cases as shown in Figure 10.12b. In case-1, the sensitivity starts near zero and becomes increasingly negative over time, reaching its peak negative value around year 4. This indicates that even with minimal degradation, fully functioning primary components become more vulnerable once partial failure begins. In case-2, the sensitivity drops more sharply in the early years due to the involvement of warm sharp components, which degrade faster once activated. The most significant sensitivity is observed in case-3, where the system relies heavily on multiple warm standby components. Here, the impact of transition to full failure is strongest between 1.5 and 2.5 years, showing the highest negative values, reflecting the system's vulnerability when all standby units are degrading simultaneously. Case-4 also experiences high sensitivity initially, but shows a relatively smoother curve as cold standby components begin to support the system. Case-5 is unique, although its sensitivity is slightly negative in the beginning, it turns positive after 0.5 years and steadily increases, indicating that cold standby units significantly enhance system robustness once activated, even in the presence of full failure transitions.

- **Effect of $\lambda_{1,3}^{p}$ on sensitivity**

  The sensitivity analysis for the failure $\lambda_{1,3}^{p}$ indicates a consistent increase in negative sensitivity across most cases, reflecting the growing impact of abrupt failures over time, which is given in Figure 10.12c. In case-1, the sensitivity starts very low and becomes progressively more negative, showing that even a sudden failure in highly redundant systems has a growing influence as the system ages and redundancy depletes. Case-2 shows a steeper decline in sensitivity in the early years due to reduced redundancy and the presence of degrading warm standby components. Case-3 heavily depends on warm standby units and shows a similar trend, but it is less sensitive. Case-4, involving both warm and cold standby units, displays moderate sensitivity due to mixed failure dynamics and delayed activation of cold standbys. In contrast, case-5 reveals a positive and increasing sensitivity throughout the 5-year period. This highlights the beneficial effect of cold standby components in absorbing the impact of direct failures. Once activated, cold standby units improve system performance, making it more resilient against sudden full failures. Overall, the analysis shows that cold standby components (case-5) provide strong compensation against sudden failures, while systems relying on warm or partially degraded components (cases 2–4) are increasingly sensitive to such failures over time.

10.18 CONCLUSION

This chapter presents an algorithm based on the L_z-transform to evaluate the reliability attributes of mixed k-out-of-$(p+w+c)$: G mixed standby system. The key findings of the research are summarized as follows:

- The reliability analysis demonstrates that system's reliability is highest when all primary components are operational and decreases as standby components are introduced, depending on their type and activation order. Cold standby components offer improved long-term reliability compared to warm standby units.

- The study reveals that system availability is highest when all primary components are operational and declines with increases in the use of standby units. Among standby configurations, cold standby components offer better long-term availability than warm ones.
- Among the three failure transitions of primary components, the system achieves the highest MTTF under $\lambda^p_{1,3}$. In contrast, early degradation $\lambda^p_{2,3}$ leads to the lowest MTTF due to continuous loss over time.
- From the three failure modes, the system shows highest sensitivity to failure $\lambda^p_{1,3}$, followed by $\lambda^p_{2,3}$, while $\lambda^p_{1,2}$ has the least negative impact.

Future Work: The developed models can be incorporated with the dependency of components, different types of failures, uncertainty, etc. Also, we will use this model in the environment of belief function, fuzzy set theory, and other versions of fuzzy sets to improve the reliability of the proposed model.

REFERENCES

Amari, S. V. (2012a). Reliability analysis of *k*-out-of-*n* cold standby systems with Erlang distributions. *International Journal of Performability Engineering*, 8(4), 417.

Amari, S. V. (2012b, January). Reliability of *k*-out-of-*n* standby systems with gamma distributions. In *2012 Proceedings Annual Reliability and Maintainability Symposium, Reno, Nevada, USA, IEEE*, 1–6.

Ardakan, M. A., & Hamadani, A. Z. (2014). Reliability–redundancy allocation problem with cold-standby redundancy strategy. *Simulation Modelling Practice and Theory*, 42, 107–118.

Beji, N., Jarboui, B., Eddaly, M., & Chabchoub, H. (2010). A hybrid particle swarm optimization algorithm for the redundancy allocation problem. *Journal of Computational Science*, 1(3), 159–167.

Bisht, V., & Singh, S. B. (2023). L_z-transform approach to evaluate reliability indices of multi-state repairable weighted *k*-out-of-*n* systems. *Quality and Reliability Engineering International*, 39(3), 1043–1057.

Bisht, V., & Singh, S. B. (2024a). Dynamic reliability measures of weighted *k*-out-of-*n* system with heterogeneous components and its application to solar power generating system. *Proceedings of the Institution of Mechanical Engineers, Part O: Journal of Risk and Reliability*, 239, 1178–1188.

Bisht, V., & Singh, S. B. (2024b). Interval valued reliability indices assessment of multi-state system using interval L_z-transform. *International Journal of System Assurance Engineering and Management*, 1–13.

Coit, D. W., & Smith, A. E. (2005). Reliability optimization of series-parallel systems using a genetic algorithm. *IEEE Transactions on Reliability*, 45(2), 254–260.

Deb, K., Pratap, A., Agarwal, S., & Meyarivan, T. A. M. T. (2002). A fast and elitist multiobjective genetic algorithm: NSGA-II. *IEEE Transactions on Evolutionary Computation*, 6(2), 182–197.

Ebrahimipour, V., Sheikhalishahi, M., Shoja, B. M., & Goldansaz, M. (2010, November). A universal generating function approach for redundancy optimization for hot-standby multi-state series parallel k-out-of-n systems. In *2010 Fourth UKSim European Symposium on Computer Modeling and Simulation, IEEE*, Pisa, Italy, 235–239.

Frenkel, I., Lisnianski, A., & Khvatskin, L. (2012). Availability assessment for aging refrigeration system by using L_z-transform. *Journal of Reliability and Statistical Studies*, 5(02), 33–43.

Frenkel, I., Lisnianski, A., & Khvatskin, L. (2013). On the L_z-transform application for availability assessment of an aging multi-state water cooling system for medical equipment. In Ilia B. Frenkel, Alex Karagrigoriou, Anatoly Lisnianski, Andre Kleyner (Eds.), *Applied Reliability Engineering and Risk Analysis: Probabilistic Models and Statistical Inference*, United Kingdom, Wiley, 59–77.

Gao, S., & Wang, J. (2021). Reliability and availability analysis of a retrial system with mixed standbys and an unreliable repair facility. *Reliability Engineering & System Safety*, 205, 107–240.

Ismail, A. L., & Yusuf, I. (2023). Reliability Metrics of Textile Confection Plant Using Copula Linguistic. In Garg, H. (ed.), *Advances in Reliability, Failure and Risk Analysis. Industrial and Applied Mathematics* (93–115). Singapore: Springer Nature Singapore.

Kaur, M., & Bisht, S. (2025). Reliability indices of multi-state star configuration with performance sharing using L_z-transform. *Life Cycle Reliability and Safety Engineering*, 14, 397–411.

Levitin, G. (2005). *The Universal Generating Function in Reliability Analysis and Optimization*. London: Springer.

Levitin, G., & Amari, S. V. (2010). Approximation algorithm for evaluating time-to-failure distribution of k-out-of-n system with shared standby elements. *Reliability Engineering & System Safety*, 95(4), 396–401.

Lisnianski, A., Frenkel, I., & Khvatskin, L. (2021). Modern Dynamic Reliability Analysis for Multi-state Systems. In Springer Series in Reliability Engineering. Springer, Switzerland.

Myers, A. F. (2007). k-out-of-n: G system reliability with imperfect fault coverage. *IEEE Transactions on Reliability*, 56(3), 464–473.

Natvig, B. (2010). *Multistate Systems Reliability Theory with Applications*. John Wiley & Sons, United Kingdom.

Nautiyal, N., Singh, S. B., & Bisht, S. (2020). Analysis of reliability and its characteristics of a k-out-of-n network incorporating copula. *International Journal of Quality & Reliability Management*, 37(4), 517–537.

Poonia, P. K., Sirohi, A., & Kumar, A. (2021). Cost analysis of a repairable warm standby k-out-of-n: G and 2-out-of-4: G systems in series configuration under catastrophic failure using copula repair. *Life Cycle Reliability and Safety Engineering*, 10(2), 121–133.

Singh, A., Bisht, V., & Singh, S. B. (2024). Reliability indices of multi-state repairable m-out-of-r-within-k-out-of-n system using L_z-transform method. *Communications in Statistics-Theory and Methods*, 54, 2741–2758.

Tavakkoli-Moghaddam, R., Safari, J., & Sassani, F. (2008). Reliability optimization of series-parallel systems with a choice of redundancy strategies using a genetic algorithm. *Reliability Engineering & System Safety*, 93(4), 550–556.

Xiao, G., Li, Z., & Li, T. (2007). Dependability estimation for non-Markov consecutive-k-out-of-n: F repairable systems by fast simulation. *Reliability Engineering & System Safety*, 92(3), 293–299.

Yeh, W. C. (2006). The k-out-of-n acyclic multistate-node networks reliability evaluation using the universal generating function method. *Reliability Engineering & System Safety*, 91(7), 800–808.

Zhang, T., Xie, M., & Horigome, M. (2006). Availability and reliability of k-out-of-$(M+N)$: G warm standby systems. *Reliability Engineering & System Safety*, 91(4), 381–387.

Chapter 11

Reliability and availability analysis of 4 × 4 Shuffle Exchange Network using L_z-transform

Vaibhav Bisht, S. Vishal Vignesh, G. Hemalatha, Priyanshu Arya, and Ayush Singh

11.1 INTRODUCTION

High-performance and parallel computing architectures depend entirely on the efficiency of communication between the processors and memory modules. Multistage interconnection networks (MINs) are the architecture that helps with such communications in multiprocessor systems, proving to be the most vital part of the data path. Different MIN architectures are available, but the most notable among them, owing to features such as logarithmic depth, simplicity in layout, and self-routing, are Shuffle Exchange Network (SEN), SEN + 1, and SEN + 2, which are a basic part of the design of reliable computing systems. A major determinant of the performance of such systems is network reliability (NR), defined as the probability that communication paths are successfully established between every specified input-output pair. Generally, network systems with n components have been traditionally analysed in terms of their reliability, specifically in the context of the 4 × 4 SEN, by employing static probabilistic approximations or methods such as the Universal Generating Function (UGF) technique. Although successful at analysing binary-state components in a steady-state regime, these traditional methods have major drawbacks in systems with multiple component states or in conditions where the time-dependent failure/recovery behaviour of components makes the system reliability profile complex.

MINs form the backbone of high-performance computing systems and manage data transmission between processors and memory units. The SEN, described by Wu and Feng (1980), is a prominent topology characterized by its simplicity and self-routing property. However, owing to the single-path connection between any pair of inputs and outputs, it is susceptible to single-point failures; thus, reliability becomes a significant performance factor (Blake & Trivedi, 1989). In a MIN, reliability is the probability of maintaining successful connectivity among all required input-output pairs, even when switching elements may be defective Gunawan (2014). Consequently, accurate reliability computation is a high priority for the development of reliable computational systems. Traditional

DOI: 10.1201/9781003787525-11

reliability analyses are conducted on binary-state systems, in which system components are described as either fully operating or entirely failed. The binary-state system model has limitations when applied to contemporary systems, whose components operate at multiple performance levels. The development of the multi-state system (MSS) theory was a more accurate technique for modelling degradable systems (Lisnianski et al., 2010). Sharma et al. (2010) studied the reliability index and communication path length in irregular fault-tolerant MINs to highlight the effects of structural irregularity on NR. The MSS system assumes that system components, as well as the system as a whole, can be found in any one of several states, each of which corresponds to a separate level of system performance. Rajkumar and Goyal (2016) conducted a thorough review of reliability and fault tolerance methods for MINs, discussing prominent models, performance indices, and design techniques for improving reliability. Qin and Li (2019) proposed an approach for reliability and sensitivity analysis of MSSs that accounts for common-cause failure, improving its accuracy with respect to system susceptibility and component values. Models for reliability assessment of multi-state k-out-of-n: F systems with multiple protective devices supporting m subsystems were proposed by Wang et al. (2022) to demonstrate improved reliability with redundancy and protective measures. The MSS can be applied to complex systems, including weighted k-out-of-n systems (Li & Zuo, 2008), phased-mission systems (Chen et al., 2025), and performance-sharing systems (Zhao et al., 2024).

The computational solution of MSS reliability was radically changed by the UGF method, which was developed in a mathematical formulation by Ushakov (1986). The UGF method provides an elegant algebraic solution by encoding the probability distributions of system components in polynomial form, enabling an effective derivation of the performance distribution of the entire system using combinatorial operators. Other authors have used the UGF method to calculate measures of importance and sensitivity for MSSs, for example, Levitin and Lisnianski (1999). Its applicability has been demonstrated for various complex system configurations, ranging from generalized models of k-out-of-n MSS with general configurations studied in Zuo and Tian (2006) to MSS with dependent components, as studied in Levitin (2004). Yeh proposed a set of efficient methods based on UGFs for evaluating the reliability of general multi-state node networks and for systematically specifying all minimal paths, which simplified the complexity of reliability analysis for networks (Yeh, 2008, 2009).

However, in general, the classical UGF method applies to non-repairable MSS. To cover the dynamics of repairable MSS with cyclic processes of failures and repairs in MSS components, a powerful method, the L_z-transform method, has been proposed by Lisnianski (2004). This method retains the combinatorial advantages of UGF but is tailored to solve Markov processes describing the dynamics of repairable MSS components. This enables the effective derivation of a unified Markov model of the entire MSS by combining models of individual components, greatly simplifying reliability evaluation. Recently, the L_z-transform has been successfully used to determine the reliability indices for multi-state weighted *k*-out-of-*n* systems (Bisht & Singh, 2023) and maintenance-dependent systems (Yi et al., 2022), thus proving its effectiveness compared with the traditional approach. Extensive research has been conducted on the dependability of MINs, with a specific focus on sensor event networks. Fard and Gunawan (2014) studied ways to improve terminal reliability (TR) for SEN systems based on the concept of fault-tolerant design. The primary investigations focused on combinatorial probability and Reliability Block Diagrams (RBDs; Blake & Trivedi, 1989; Bistouni & Jahanshah, 2014a). Ongoing research has progressed to encompass architectural enhancements for fault tolerance techniques, such as the Extra Group Network (Bistouni & Jahanshahi, 2014b). The emergence of the UGF technique marked a groundbreaking shift in research, with a systematic method for studying MINs. The concept of UGF was proposed by Gunawan (2014), followed by applications of this technique on individual networks, such as the reliability analysis of 4 × 4 and 8 × 8 SENs (Bisht & Singh, 2021, 2022). These analysis techniques have proven efficient for obtaining a closed-form expression for the reliability of static network topologies.

A literature survey of existing work shows a particular gap in the methodological aspects. Although the UGF approach has been employed for the analysis of SENs and the L_z-transform for the repairable MSSs, no study has applied the L_z-transform for the analysis of SENs. This survey is relevant for two reasons:

1. The partial failures of network switches and/or operations with degraded performance capabilities are quite in line with the MSS model. More significantly, a repairable system is a realistic scenario in practice. The L_z-transform is the most applicable method for modelling a repairable MSS network with dynamic and state-dependent system reliability measures.
2. The application of the L_z-transform to a 4 × 4 SEN would prove the concept of combining a state-of-the-art reliability analysis tool (the L_z-transform) with a classic network design. This would enable one to compute, in addition to reliability, an important measure for a repairable SEN that goes beyond the usual UGF calculations: availability.

The following part of the chapter goes as follows: Section 11.2 gives the L_z-transform and its properties; Section 11.3 presents the model description and proposes an algorithm for terminal, broadcast, and network performance assessment of 4 × 4 SEN, 4 × 4 SEN + 1, 4 × 4 SEN + 2; Section 11.4 provides a numerical illustration of determining the performance measures; Section 11.5 concludes the findings.

11.2 PRELIMINARIES

In this section the preliminaries such as basic definitions, results, and properties needed to determine the dynamic reliability indices using L_z-transform are provided in detail.

11.2.1 Markov process

Consider a Markov process $G(t)$ characterized by discrete state and continuous time (DSCT), defined by $G(t) \in \{g_{i1}, g_{i2}, \dots, g_{iN_i}\}; t \geq 0$, where g_{ij} indicates the performance level of each switch i in any state j $(j = 1, \dots, N_i)$, where N_i represent the total number of possible states. This Markov process is described by the set of possible states, $g = \{g_{i1}, g_{i2}, \dots, g_{iN_i}\}$, transition intensity matrix $X = [x_{ij}(t)]$, where $i, j = (1, \dots, N_i)$, and the initial state probability distribution $p_0 = [p_{10}, \dots, p_{N0}]$ where $p_{i0} = \text{Prob}\{G(0) = g_{i0}\}$. Hence, the DSCT Markov process can be defined as the triplet: $G(t) = \{g, X, p_0\}$.

11.2.2 L_z definition

L_z-transform of a DSCT Markov process $G(t) = \{g, X, p_0\}$ is a function of $f(z, t, p_0)$, which is defined as

$$L_z(E(t)) = f(z, t, p_0) = \sum_{j=1}^{N_i} p_{ij}(t) z^{g_{ij}} \tag{11.1}$$

where $p_{ij}(t)$ denotes a probability distribution that varies with time for a network comprising multiple switches, with the state of the switch i being in state j at time $t \geq 0$ and g_{ij} is the performance level of switch i in state j. The initial state probabilities are given by p_0 and this expression can be extended by incorporating the complex variable z.

11.2.3 Properties of L_z-transform

Property 1: When a constant a is applied to the DSCT Markov process, it will also be applied to the corresponding performance level for each state i.

$$L_z(aE(t)) = \sum_{j=1}^{N_i} p_{ij}(t) z^{a g_{ij}} \tag{11.2}$$

Property 2: L_z-transform of two DSCT Markov processes $R(t)$ and $S(t)$ from a function $f(R(t), S(t))$ can be calculated by using Ushakov's UGF operator Ω_f to $R(t)$ and $S(t)$ for $t \geq 0$ as:

$$L_z\{f[R(t),S(t)]\} = \Omega_f\left(L_z\{R(t)\}, L_z\{S(t)\}\right) \tag{11.3}$$

11.2.4 4 × 4 Shuffle Exchange Networks

The 4 × 4 SEN, 4 × 4 SEN + 1, and 4 × 4 SEN + 2 networks are shown in Figures 11.1–11.3.

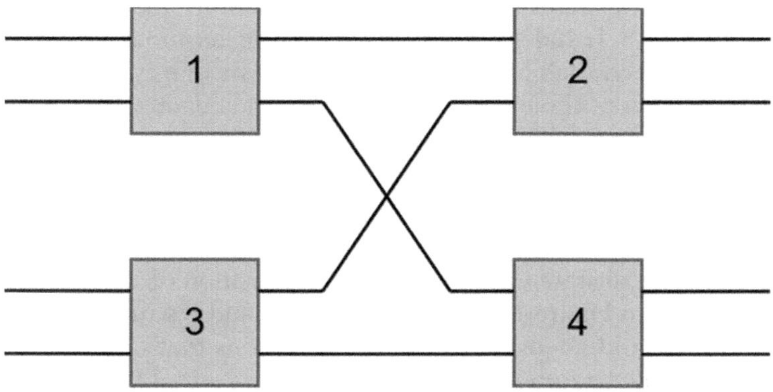

Figure 11.1 4 × 4 SEN.

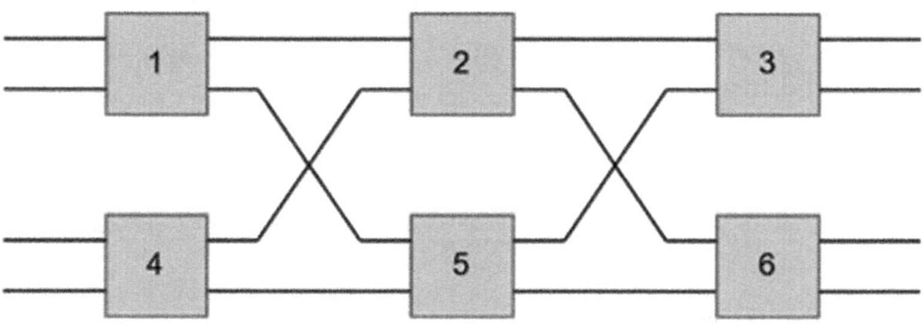

Figure 11.2 4 × 4 SEN + 1.

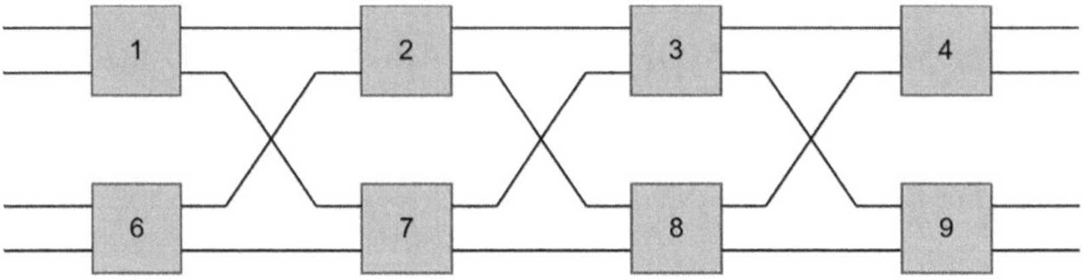

Figure 11.3 4 × 4 SEN + 2.

11.3 ALGORITHM TO EVALUATE TERMINAL, BROADCAST, NR MEASURES OF 4 × 4 SEN, 4 × 4 SEN + 1, 4 × 4 SEN + 2 NETWORKS USING L_z-TRANSFORM

In this section, the algorithm evaluates the reliability indices for the 4 × 4 SEN, 4 × 4 SEN + 1, and 4 × 4 SEN + 2, including terminal, broadcast, NR, and steady-state availability. The network consists of n switches, with each switch consisting of three performance levels and denoted by $g_{i1}, g_{i2}, \ldots g_{im_i}$. A Markov process $G(t)$ is constructed with states of the switches as discrete and continuous time which is defined as $G(t) \in \{g_{i1}, g_{i2}, g_{i3}\}$, with time $t \geq 0$ and g_{ij} denoting the performance level of ith switch at j^{th} state. The state transition diagram of varying degradation levels with a failure transition of $\lambda_{k,1}$ from state k to state 1 and repair of the transition of $\mu_{1,k}$ from state 1 to state k is given in Figure 11.4. Also, the failure and repair rates are categorized as minor and major based on the transitions that occur. The $j - i = 1$ transition is called a minor transition, and if the $j - i > 1$ transition is called a major transition. In the following algorithm, minor failures, major failures, and repairs are considered in the evaluation.

Step 1: Find the set of states of different performing levels $g = \{g_{i1}, g_{i2}, g_{i3}\}$, where g_{ij} denotes the performance level of ith switch at j^{th} state.

Step 2: Determine the initial probability p_0.

Step 3: Construct the differential equations for the transient-state probabilities.

$$\frac{dp_{i1}(t)}{dt} = -\mu_{13}^{(i)} p_{i1}(t) + \lambda_{21}^{(i)} p_{i2}(t) + \lambda_{31}^{(i)} p_{i3}(t)$$

$$\frac{dp_{i2}(t)}{dt} = -\lambda_{21}^{(i)} p_{i2}(t) + \lambda_{32}^{(i)} p_{i3}(t)$$

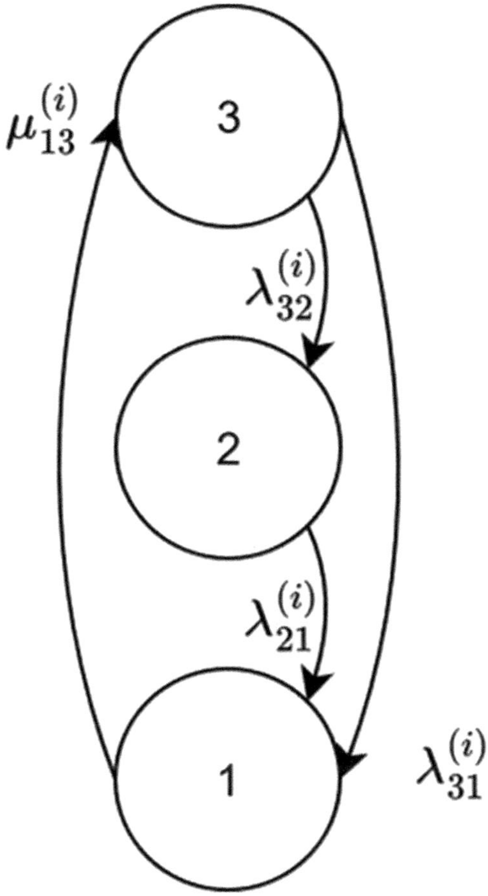

Figure 11.4 State transition diagram of a multi-state switch.

$$\frac{dp_{i3}(t)}{dt} = \mu_{13}^{(i)} p_{i1}(t) - \left(\lambda_{31}^{(i)} + \lambda_{32}^{(i)}\right) p_{i3}(t)$$

Step 4: Solve the differential equations with the initial probabilities p_0 using the Laplace transform and find the state probabilities $p_{ij}(t)$.
Step 5: Evaluate the L_z-transform for the multi-state switch.
Step 6: Repeat steps 1 to 5 for all the switches of the network to find the L_z-transform for each switch in the network.

Step 7: Let $L_z(e_1(t)), L_z(e_2(t)),\ldots, L_z(e_n(t))$ be the L_z-transform of the first, second, and n^{th} switch of the network from the definition of the L_z-transform of the ith switch given by

$$L_z(e_i(t)) = \sum_{j=1}^{3} p_{ij}(t) z^{g_{ij}} \tag{11.4}$$

Here $p_{ij}(t)$ is the probability of the ith switch in state j.

Step 8: Determine the L_z-transform of the network consisting of n multi-state switches connected as 4 × 4 SEN network using the Ushakov operator Ω_f. The L_z-transform of the network is evaluated by the property (11.2). The L_z-transform of the network is given by

$$L_z\left(E_{(i+1)-r}(t)\right) = \Omega_f\left(E_{i-r}(t), e_i(t)\right), \quad i = 1,2,\ldots, n$$

$$= \sum_{j_{i-r}=1}^{3} \sum_{j_i=1}^{3} \left(p_{ij_{i-r}}(t).p_{ij}(t) z^{\max\left(g_{ij_{i-r}}, g_{ij}\right)}\right)$$

$$= \sum_{j_{i-r}=1}^{3} \sum_{j_i=1}^{3} \left(p_{ij_{i-r}}(t).p_{ij}(t) z^{\min\left(g_{ij_{i-r}}, g_{ij}\right)}\right)$$

where Ω_f is the structure function of the respective network configuration. For a parallel configuration, the structure function is represented by the maximum operator, whereas for a series configuration, it is represented by the minimum operator. The L_z-transformation for the entire network can be obtained using the steps above, and the network's performance indices are determined in the section below.

11.4 PERFORMANCE ANALYSIS OF 4 × 4 SEN, 4 × 4 SEN + 1, 4 × 4 SEN + 2

The performance analysis of 4 × 4 SEN, 4 × 4 SEN + 1, and 4 × 4 SEN + 2 is predicated on three configurations: terminal, broadcast, and NR, availability, and steady-state availability. We propose to examine this using L_z-transform. Table 11.1 presents the transition rates between states and the performance rates for each state at every switch. The performance capacity of the switches in 4 × 4 SENs is quantified in kilobytes (KBs).

Table 11.1 State transition rates of the 4 × 4 SEN, 4 × 4 SEN + 1, 4 × 4 SEN + 2

	States	Transition rates s_1	s_2	s_3	Performance level (g_{ij})
Switch 1	s_1	0.0	0.0	4.1	0
	s_2	0.67	0.0	0.0	510
	s_3	1.2	0.89	0.0	1,100
Switch 2	s_1	0.0	0.0	4.7	0
	s_2	0.54	0.0	0.0	720
	s_3	1.5	0.79	0.0	1,050
Switch 3	s_1	0.0	0.0	3.9	0.0
	s_2	0.44	0.0	0.0	410
	s_3	1.1	0.92	0.0	950
Switch 4	s_1	0.0	0.0	5.3	0.0
	s_2	0.74	0.0	0.0	600
	s_3	1.8	1.2	0.0	1,020
Switch 5	s_1	0.0	0.0	3.2	0.0
	s_2	0.45	0.0	0.0	680
	s_3	0.98	0.75	0.0	1,120
Switch 6	s_1	0.0	0.0	6.2	0.0
	s_2	0.88	0.0	0.0	460
	s_3	2.3	1.5	0.0	990
Switch 7	s_1	0.0	0.0	2.0	0.0
	s_2	0.34	0.0	0.0	650
	s_3	0.7	0.5	0.0	1,200
Switch 8	s_1	0.0	0.0	1.2	0.0
	s_2	0.2	0.0	0.0	310
	s_3	0.5	0.3	0.0	920

11.4.1 Terminal performance

The TR of a network is defined as the probability that at least one fault-free path exists between the source and destination. Should the switch be repairable, we can ascertain the terminal availability (TA) at a given time t. As t approaches infinity, we can calculate the terminal steady-state availability (TSA).

$$\text{TR} = 1 - f_{ij}$$

f_{ij} is the sum of the probability terms, which is the coefficient of z^0

$$\text{TA} = 1 - f_{ij}$$

In TA calculations, each switch's repair rate is considered:

$$\text{TSA} = \lim_{t\to\infty} \text{TA}(t)$$

11.4.1.1 Terminal performance of 4 × 4 SEN

The RBD for evaluating the terminal performance of a 4 × 4 SEN is shown in Figure 11.5. The terminal performance of the 4 × 4 SEN can be computed with the L_z-transform as follows:

$$L_z(e_1(t)) = p_{11}(t)z^0 + p_{12}(t)z^{510} + p_{13}(t)z^{1,100}$$

$$L_z(e_2(t)) = p_{21}(t)z^0 + p_{22}(t)z^{720} + p_{23}(t)z^{1,050}$$

$$L_z(E(t)) = \Omega_f\left(L_z(e_1(z)),\, L_z(e_2(z))\right)$$

$$L_z(E(t)) = \left(L_z(e_1(z)) \underset{\min}{\otimes} L_z(e_2(z))\right)$$

$$\text{TR} = 1 - f_{ij}$$

f_{ij} is the sum of the probability terms, which is the coefficient of z^0.

The TA is the same as the TR, with repair rates of the switch considered, and TSA is given by

$$\text{TSA} = 0.6848993845$$

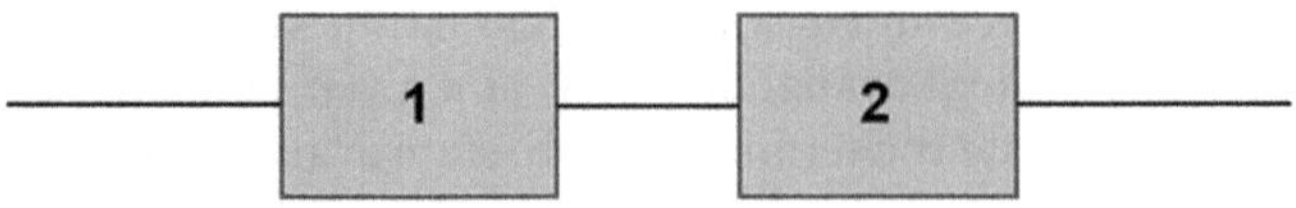

Figure 11.5 RBD of 4 × 4 SEN.

11.4.1.2 Terminal performance of 4 × 4 SEN + 1

The RBD for evaluating the terminal performance of 4 × 4 SEN + 1 is shown in Figure 11.6. The terminal performance of 4 × 4 SEN + 1 can be computed with the L_z-transform as follows:

$$L_z\left(e_i\left(t\right)\right) = p_{i1}\left(t\right)z^{g_{i1}} + p_{i2}\left(t\right)z^{g_{i2}} + p_{i3}\left(t\right)z^{g_{i3}}$$

$$L_z\left(E(t)\right) = \Omega_f\left(L_z\left(e_1\left(z\right)\right),\ L_z\left(e_2\left(z\right)\right), L_z\left(e_3\left(z\right)\right),\ L_z\left(e_4\left(z\right)\right)\right)$$

$$L_z\left(E(t)\right) = L_z\left(e_1\left(z\right)\right) \underset{\min}{\otimes} \left(L_z\left(e_2\left(z\right)\right) \underset{\max}{\otimes} L_3\left(e_2\left(z\right)\right)\right) \underset{\min}{\otimes} L_z\left(e_4\left(z\right)\right)$$

$$\text{TR} = 1 - f_{ij}$$

f_{ij} is the sum of the probability terms, which is the coefficient of z^0.

The TA is the same as the TR, with repair rates of the switch considered, and the TSA is given by

$$\text{TSA} = 0.6587186591$$

11.4.1.3 Terminal performance of 4 × 4 SEN + 2

The RBD for evaluating the terminal performance of 4 × 4 SEN + 2 is shown in Figure 11.7. The terminal performance of 4 × 4 SEN + 2 can be computed with the L_z-transform as follows:

$$L_z\left(e_i\left(t\right)\right) = p_{i1}\left(t\right)z^{g_{i1}} + p_{i2}\left(t\right)z^{g_{i2}} + p_{i3}\left(t\right)z^{g_{i3}}$$

$$L_z\left(E(t)\right) = \Omega_f\begin{pmatrix} L_z\left(e_1\left(z\right)\right),\ L_z\left(e_2\left(z\right)\right), L_z\left(e_3\left(z\right)\right),\ L_z\left(e_4\left(z\right)\right), \\ L_z\left(e_5\left(z\right)\right),\ L_z\left(e_6\left(z\right)\right) \end{pmatrix}$$

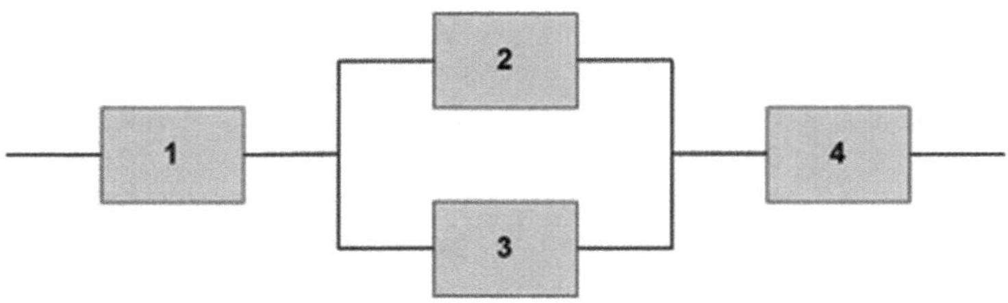

Figure 11.6 RBD of 4 × 4 SEN + 1.

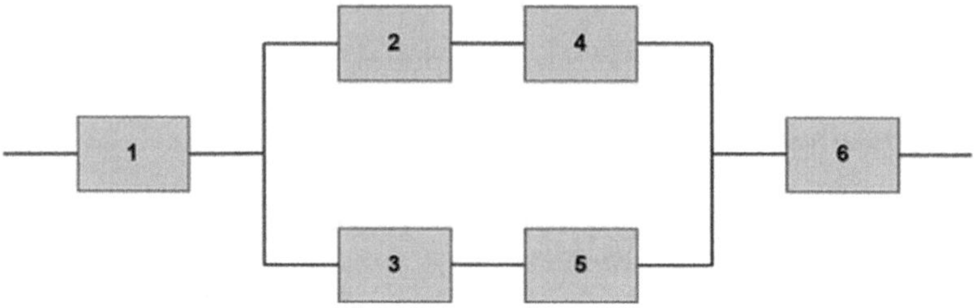

Figure 11.7 RBD of 4 × 4 SEN + 2.

$$L_z\left(E(t)\right)=L_z\left(e_1(z)\right)\underset{\min}{\otimes}\left(\left(L_z\left(e_2(z)\right)\underset{\min}{\otimes}L_z\left(e_3(z)\right)\right)\underset{\max}{\otimes}\left(L_z\left(e_4(z)\right)\underset{\min}{\otimes}L_z\left(e_5(z)\right)\right)\right)$$
$$\underset{\min}{\otimes}L_z\left(e_6(z)\right)$$

$$\text{TR}=1-f_{ij}$$

f_{ij} is the sum of the probability terms, which is the coefficient of z^0.

The TA is the same as the TR, with repair rates of the switch considered, and the TSA is given by

$$\text{TSA}=0.6084771622$$

From Table 11.2, throughout the entire observed time period ($t > 0$), the standard 4 × 4 SEN consistently exhibits the highest availability, followed by 4 × 4 SEN + 1 and 4 × 4 SEN + 2. It can be seen that 4 × 4 SEN + 1 and 4 × 4 SEN + 2 configurations likely introduce additional stages or switching

Table 11.2 Terminal availability of the 4 × 4 SEN, 4 × 4 SEN + 1, 4 × 4 SEN + 2

Unit time (t)	*TA of 4 × 4 SEN*	*TA of 4 × 4 SEN + 1*	*TA of 4 × 4 SEN + 2*
0	1	1	1
1	0.653201	0.630488	0.574219
2	0.675489	0.652074	0.6
3	0.682185	0.657158	0.606268
4	0.684108	0.658337	0.60786
5	0.684667	0.658622	0.608297
6	0.68483	0.658693	0.608423
7	0.684879	0.658712	0.60846
8	0.684893	0.658717	0.608472
9	0.684898	0.658718	0.608475
10	0.684899	0.658719	0.608477

elements to the base 4 × 4 SEN. Since availability decreases as complexity increases, these additions likely introduce more series-dependent points of failure without providing sufficient redundancy to offset the risk. Therefore, the baseline 4 × 4 SEN is the most reliable structure among the three (Figures 11.8 and 11.9).

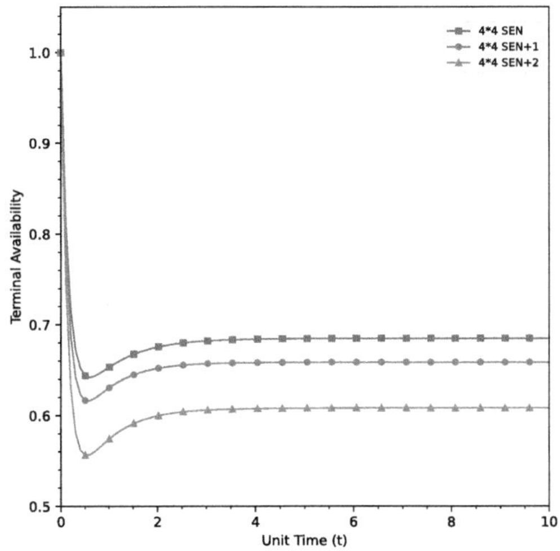

Figure 11.8 Terminal availability.

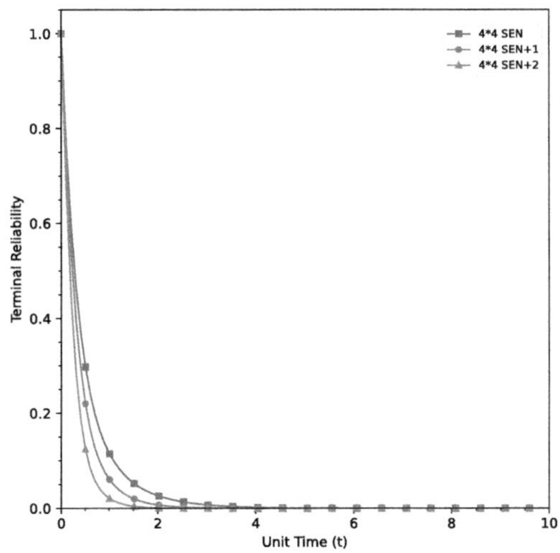

Figure 11.9 Terminal reliability.

11.4.2 Broadcast performance

The Broadcast Reliability (BR) of a network is the probability that a message gets transmitted from one source to all destination nodes. Should the switch be repairable, we can ascertain the Broadcast Availability (BA) at a particular moment t. As t approaches infinity, we can ascertain the Broadcast steady-state availability (BSA).

$$\text{BR} = 1 - f_{ij}$$

f_{ij} is the sum of the probability terms, which is the coefficient of z^0.

$$\text{BA} = 1 - f_{ij}$$

In BA calculations, each switch's repair rate is taken into account.

$$\text{BSA} = \lim_{t \to \infty} \text{BA}(t)$$

11.4.2.1 Broadcast performance of 4 × 4 SEN

The RBD for evaluating the broadcast performance of a 4 × 4 SEN is shown in Figure 11.10. The broadcast performance of the 4 × 4 SEN can be computed with the L_z-transform as follows:

$$L_z(e_1(t)) = p_{11}(t)z^0 + p_{12}(t)z^{510} + p_{13}(t)z^{1,100}$$

$$L_z(e_2(t)) = p_{21}(t)z^0 + p_{22}(t)z^{720} + p_{23}(t)z^{1,050}$$

$$L_z(E(t)) = \Omega_f\left(L_z(e_1(z)),\ L_z(e_2(z)), L_z(e_3(z))\right)$$

$$L_z(E(t)) = \left(L_z(e_1(z)) \underset{\min}{\otimes} L_z(e_2(z)) \underset{\min}{\otimes} L_z(e_3(z))\right)$$

$$\text{BR} = 1 - f_{ij}$$

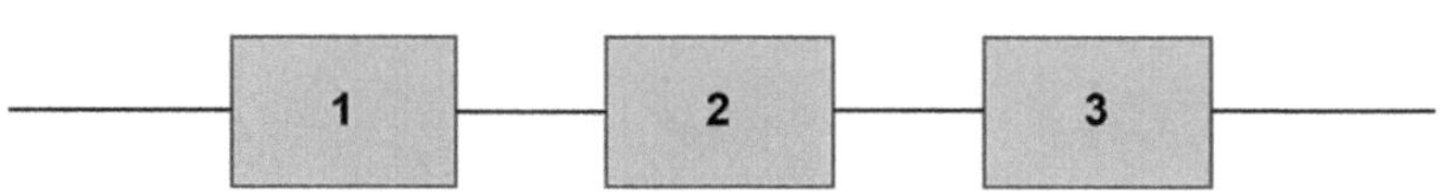

Figure 11.10 RBD of 4 × 4 SEN.

f_{ij} is the sum of the probability terms, which is the coefficient of z^0.

The BA is the same as the BR, with repair rates of the switch considered, and the BSA is given by

$$\text{BSA} = 0.5866015915$$

11.4.2.2 Broadcast performance of 4 × 4 SEN + 1

The RBD for evaluating the broadcast performance of a 4 × 4 SEN + 1 is shown in Figure 11.11. The broadcast performance of the 4 × 4 SEN + 1 can be computed with the L_z-transform as follows:

$$L_z\left(e_i(t)\right) = p_{i1}(t)z^{g_{i1}} + p_{i2}(t)z^{g_{i2}} + p_{i3}(t)z^{g_{i3}}$$

$$L_z\left(E(t)\right) = \Omega_f\left(L_z\left(e_1(z)\right),\, L_z\left(e_2(z)\right), L_z\left(e_3(z)\right),\, L_z\left(e_4(z)\right), L_z\left(e_5(z)\right)\right)$$

$$L_z\left(E(t)\right) = L_z\left(e_1(z)\right) \underset{\min}{\otimes} \left(L_z\left(e_2(z)\right) \underset{\max}{\otimes} L_3\left(e_2(z)\right)\right) \underset{\min}{\otimes} L_z\left(e_4(z)\right) \underset{\min}{\otimes} L_z\left(e_5(z)\right)$$

$$\text{BR} = 1 - f_{ij}$$

f_{ij} is the sum of the probability terms, which is the coefficient of z^0.

The BA is the same as the BR, with repair rates of the switch considered, and the BSA is given by

$$\text{BSA} = 0.5476842375$$

11.4.2.3 Broadcast performance of 4 × 4 SEN + 2

The RBD for evaluating the broadcast performance of a 4 × 4 SEN + 2 is shown in Figure 11.12. The broadcast performance of the 4 × 4 SEN + 2 can be computed with the L_z-transform as follows:

$$L_z\left(e_i(t)\right) = p_{i1}(t)z^{g_{i1}} + p_{i2}(t)z^{g_{i2}} + p_{i3}(t)z^{g_{i3}}$$

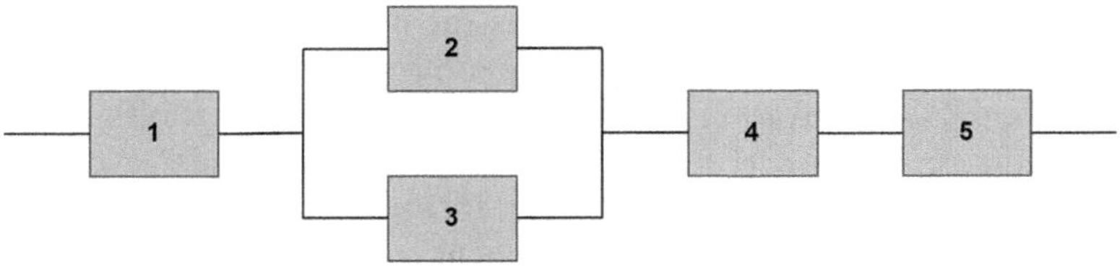

Figure 11.11 RBD of 4 × 4 SEN + 1.

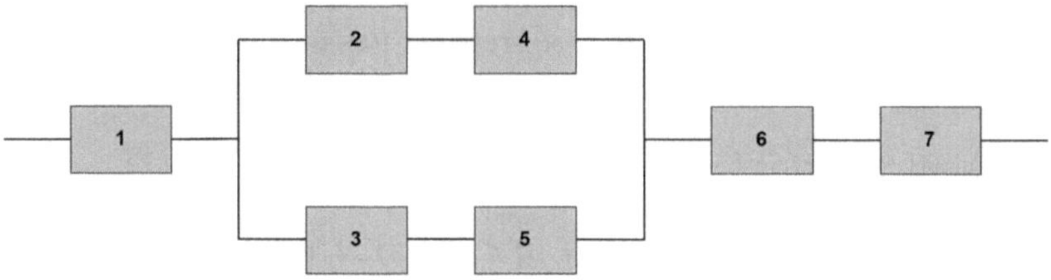

Figure 11.12 RBD of 4 × 4 SEN + 2.

$$L_z(e_i(t)) = p_{i1}(t)z^{g_{i1}} + p_{i2}(t)z^{g_{i2}} + p_{i3}(t)z^{g_{i3}}$$

$$L_z\left(E(t)\right) = \Omega_f\begin{pmatrix} L_z\left(e_1(z)\right), L_z\left(e_2(z)\right), L_z\left(e_3(z)\right), L_z\left(e_4(z)\right), \\ L_z\left(e_5(z)\right), L_z\left(e_6(z)\right), L_z\left(e_7(z)\right) \end{pmatrix}$$

$$L_z\left(E(t)\right) = L_z\left(e_1(z)\right) \underset{\min}{\otimes} \left(\left(L_z\left(e_2(z)\right) \underset{\min}{\otimes} L_z\left(e_3(z)\right) \right) \underset{\max}{\otimes} \left(L_z\left(e_4(z)\right) \underset{\min}{\otimes} L_z\left(e_5(z)\right) \right) \right)$$

$$\underset{\min}{\otimes} L_z\left(e_6(z)\right) \underset{\min}{\otimes} L_z\left(e_7(z)\right)$$

$$\text{BR} = 1 - f_{ij}$$

f_{ij} is the sum of the probability terms, which is the coefficient of z^0.

The BA is the same as the BR, with repair rates of the switch considered, and the BSA is given by

$$\text{BSA} = 0.4895793258$$

From Table 11.3, throughout the entire observed time period ($t > 0$), the standard 4 × 4 SEN consistently exhibits the highest availability, followed by 4 × 4 SEN + 1 and 4 × 4 SEN + 2. It can be seen that 4 × 4 SEN + 1 and 4 × 4 SEN + 2 configurations likely introduce additional stages or switching elements to the base 4 × 4 SEN. Since availability decreases as complexity increases, these additions likely introduce more series-dependent points of failure without providing sufficient redundancy to offset the risk. Therefore, the baseline 4 × 4 SEN is the most reliable structure among the three (Figures 11.13 and 11.14).

Table 11.3 Broadcast availability of the 4 × 4 SEN, 4 × 4 SEN + 1, 4 × 4 SEN + 2

Unit time (t)	*BA of 4 × 4 SEN*	*BA of 4 × 4 SEN + 1*	*BA of 4 × 4 SEN + 2*
0	1	1	1
1	0.542474	0.509378	0.45102
2	0.572986	0.536195	0.473214
3	0.582585	0.544319	0.482982
4	0.585413	0.546658	0.486779
5	0.586248	0.547361	0.488342
6	0.586496	0.54758	0.489018
7	0.58657	0.54765	0.489321
8	0.586592	0.547673	0.489459
9	0.586599	0.54768	0.489523
10	0.586601	0.547683	0.489553

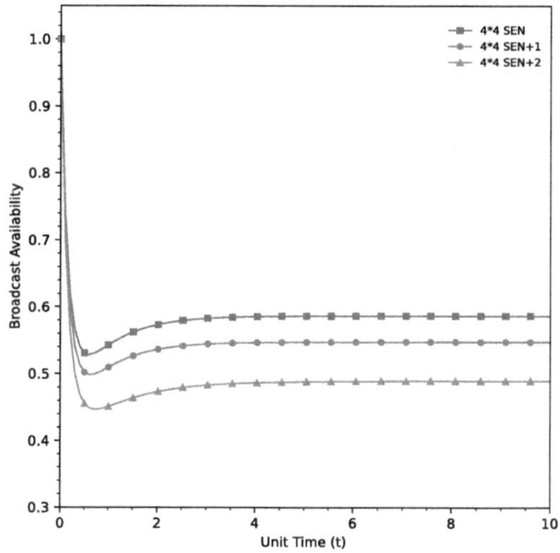

Figure 11.13 Broadcast availability.

11.4.3 Network performance

NR is the probability that a signal is successfully transmitted from all source nodes to all sink nodes. If the switch is repairable, we can determine the network availability (NA) at time t. As t approaches infinity, we can ascertain the network steady-state availability (NSA).

$$\mathrm{NR} = 1 - f_{ij}$$

f_{ij} is the sum of the probability terms, which is the coefficient of z^0.

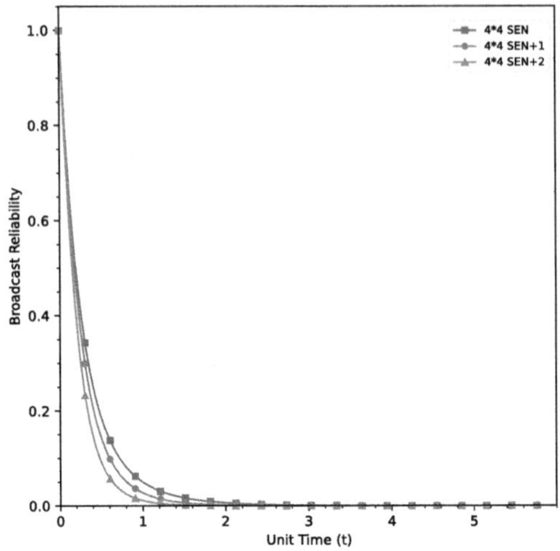

Figure 11.14 Broadcast reliability.

$$\mathrm{NA} = 1 - f_{ij}$$

In NA calculations, each switch's repair rate is taken into account.

$$\mathrm{NSA} = \lim_{t \to \infty} \mathrm{NA}(t)$$

11.4.3.1 Network performance of 4 × 4 SEN

The RBD for evaluating the network performance of a 4 × 4 SEN is shown in Figure 11.15. The network performance of the 4 × 4 SEN can be computed with the L_z-transform as follows:

$$L_z(e_1(t)) = p_{11}(t)z^0 + p_{12}(t)z^{510} + p_{13}(t)z^{1,100}$$

$$L_z(e_2(t)) = p_{21}(t)z^0 + p_{22}(t)z^{720} + p_{23}(t)z^{1,050}$$

$$L_z(E(t)) = \Omega_f\left(L_z(e_1(z)),\ L_z(e_2(z)), L_z(e_3(z)),\ L_z(e_3(z)), L_z(e_4(z))\right)$$

$$L_z(E(t)) = \left(L_z(e_1(z)) \underset{\min}{\otimes} L_z(e_2(z)) \underset{\min}{\otimes} L_z(e_3(z)) \underset{\min}{\otimes} L_z(e_4(z))\right)$$

$$\mathrm{NR} = 1 - f_{ij}$$

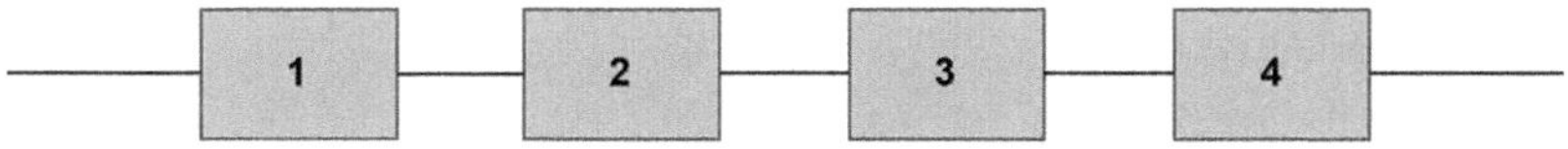

Figure 11.15 RBD of 4 × 4 SEN.

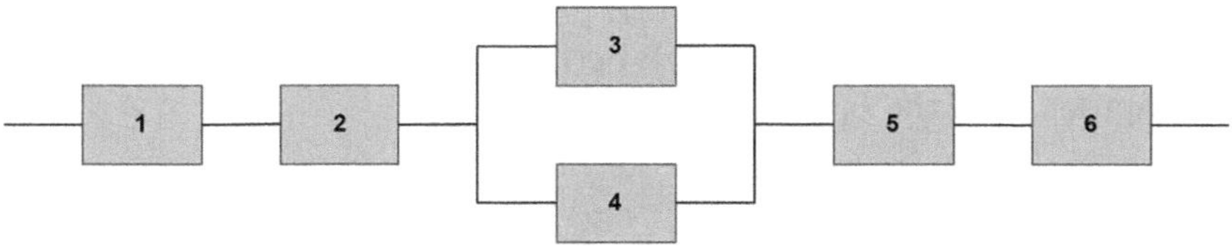

Figure 11.16 RBD of 4 × 4 SEN + 1.

f_{ij} is the sum of the probability terms, which is the coefficient of z^0.

The NA is the same as the NR, with repair rates of the switch considered, and the NSA is given by

$$\text{NSA} = 0.4824378150$$

11.4.3.2 Network performance of 4 × 4 SEN + 1

The RBD for evaluating the network performance of a 4 × 4 SEN + 1 is shown in Figure 11.16. The network performance of the 4 × 4 SEN + 1 can be computed with the L_z-transform as follows:

$$L_z\left(e_i\left(t\right)\right) = p_{i1}\left(t\right)z^{g_{i1}} + p_{i2}\left(t\right)z^{g_{i2}} + p_{i3}\left(t\right)z^{g_{i3}}$$

$$L_z\left(E(t)\right) = \Omega_f\begin{pmatrix} L_z\left(e_1\left(z\right)\right),\ L_z\left(e_2\left(z\right)\right), L_z\left(e_3\left(z\right)\right), \\ L_z\left(e_4\left(z\right)\right), L_z\left(e_5\left(z\right)\right), L_z\left(e_6\left(z\right)\right) \end{pmatrix}$$

$$L_z\left(E(t)\right) = L_z\left(e_1\left(z\right)\right) \underset{\min}{\otimes} \left(L_z\left(e_2\left(z\right)\right) \underset{\max}{\otimes} L_3\left(e_2\left(z\right)\right)\right)$$

$$\underset{\min}{\otimes} L_z\left(e_4\left(z\right)\right) \underset{\min}{\otimes} L_z\left(e_5\left(z\right)\right) \underset{\min}{\otimes} L_z\left(e_6\left(z\right)\right)$$

$$\text{NR} = 1 - f_{ij}$$

f_{ij} is the sum of the probability terms, which is the coefficient of z^0.

The NA is the same as the NR, with repair rates of the switch considered, and the NSA is given by

$$\text{NSA} = 0.5476842375$$

11.4.3.3 Network performance of 4 × 4 SEN + 2

The RBD for evaluating the network performance of a 4 × 4 SEN + 2 is shown in Figure 11.17. The network performance of the 4 × 4 SEN + 2 can be computed with the L_z-transform as follows:

$$L_z\left(e_i\left(t\right)\right) = p_{i1}\left(t\right)z^{g_{i1}} + p_{i2}\left(t\right)z^{g_{i2}} + p_{i3}\left(t\right)z^{g_{i3}}$$

$$L_z\left(E(t)\right) = \Omega_f\begin{pmatrix} L_z\left(e_1\left(z\right)\right),\ L_z\left(e_2\left(z\right)\right), L_z\left(e_3\left(z\right)\right),\ L_z\left(e_4\left(z\right)\right), \\ L_z\left(e_5\left(z\right)\right), L_z\left(e_6\left(z\right)\right), L_z\left(e_7\left(z\right)\right), L_z\left(e_8\left(z\right)\right) \end{pmatrix}$$

$$L_z\left(E(t)\right) = L_z\left(e_1\left(z\right)\right) \underset{\min}{\otimes} \left(\begin{array}{c} \left(L_z\left(e_2\left(z\right)\right) \underset{\min}{\otimes} L_z\left(e_3\left(z\right)\right)\right) \\ \underset{\max}{\otimes} \left(L_z\left(e_4\left(z\right)\right) \underset{\min}{\otimes} L_z\left(e_5\left(z\right)\right)\right) \end{array}\right)$$

$$\underset{\min}{\otimes} L_z\left(e_6\left(z\right)\right) \underset{\min}{\otimes} L_z\left(e_7\left(z\right)\right) \underset{\min}{\otimes} L_z\left(e_8\left(z\right)\right)$$

$$\text{NR} = 1 - f_{ij}$$

f_{ij} is the sum of the probability terms, which is the coefficient of z^0.

The NA is the same as the NR, with repair rates of the switch considered, and the NSA is given by

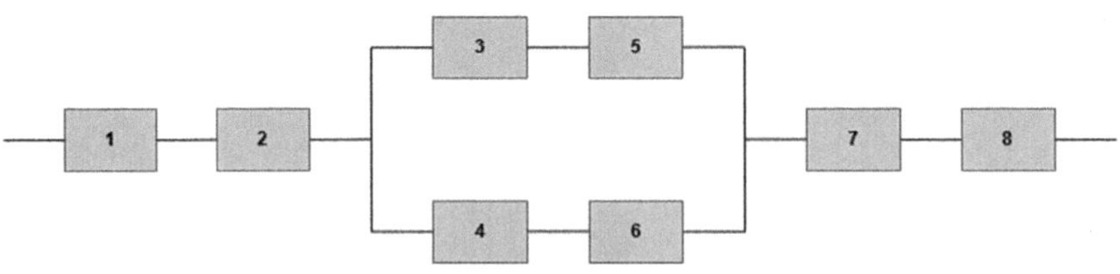

Figure 11.17 RBD of 4 × 4 SEN + 2.

NSA = 0.4895793258

From Table 11.4, throughout the entire observed time period ($t > 0$), the standard 4 × 4 SEN consistently exhibits the highest availability, followed by 4 × 4 SEN + 1 and 4 × 4 SEN + 2. It can be seen that 4 × 4 SEN + 1 and 4 × 4 SEN + 2 configurations likely introduce additional stages or switching elements to the base 4 × 4 SEN. Since availability decreases as complexity increases, these additions likely introduce more series-dependent points of failure without providing sufficient redundancy to offset the risk. Therefore, the baseline 4 × 4 SEN is the most reliable structure among the three (Figures 11.18 and 11.19).

Table 11.4 Network availability of the4 × 4 SEN, 4 × 4 SEN + 1, 4 × 4 SEN + 2

Unit time (t)	*NA of 4 × 4 SEN*	*NA of 4 × 4 SEN + 1*	*NA of 4 × 4 SEN + 2*
0	1	1	1
1	0.437524813	0.409660333	0.372246891
2	0.469565558	0.439543197	0.379822533
3	0.478825005	0.448560955	0.391982446
4	0.481404	0.451222775	0.399535086
5	0.482136947	0.452037542	0.403929177
6	0.482349138	0.452293109	0.406525008
7	0.482411439	0.452374545	0.408097166
8	0.482429919	0.452400777	0.409069223
9	0.48243544	0.452409294	0.409679333
10	0.482437098	0.452412077	0.410066325

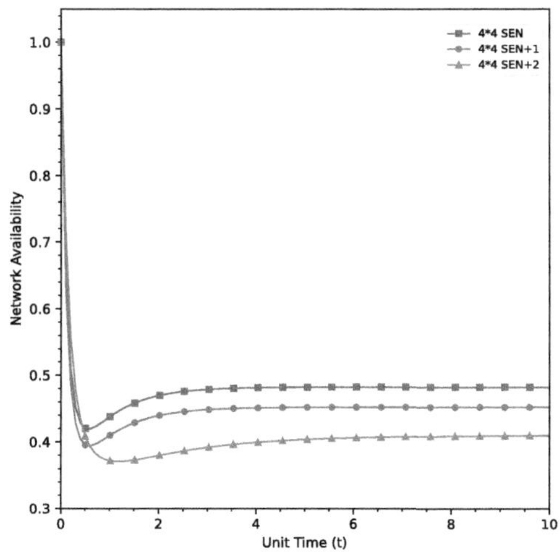

Figure 11.18 Network availability.

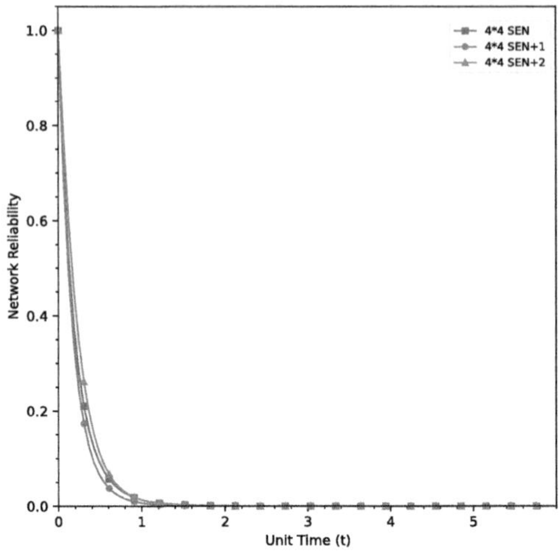

Figure 11.19 Network reliability.

11.5 CONCLUSION

This research successfully demonstrates the application of the L_z-transform method for the reliability and availability analysis of repairable MSSs within the context of SENs. Unlike permissible static models, this approach effectively integrates the dynamic behaviour of component failures and repairs into a unified Markov model, allowing for the precise computation of terminal, broadcast, and network performance indices. The comparative analysis of the 4×4 SEN, 4×4 SEN + 1, and 4×4 SEN + 2 topologies yielded significant findings regarding structural complexity and system dependability: Across all observed time periods ($t > 0$), the standard 4×4 SEN consistently demonstrated superior availability compared to both 4×4 SEN + 1 and 4×4 SEN + 2. The study identifies that the 4×4 SEN + 1 and 4×4 SEN + 2 configurations, by introducing additional stages or switching elements, inadvertently increase the system's susceptibility to failure. The analysis suggests that these additions create more series-dependent points of failure without providing sufficient redundancy to offset the associated risks. The results validate the L_z-transform as a powerful tool for evaluating complex, repairable networks where components operate at varying performance levels (measured here in KBs).

REFERENCES

Bisht, V., & Singh, S. B. (2021). Reliability estimation of 4 × 4 SENs using UGF method. *Journal of Reliability and Statistical Studies*, *14*, 173–198. https://doi.org/10.13052/jrss0974-8024.1418

Bisht, V., & Singh, S. B. (2022). Reliability analysis of 8 × 8 SEN using UGF method. In M. Ram & H. Pham (Eds.), *Reliability and maintainability assessment of industrial systems* (pp. 329–342). Springer. https://doi.org/10.1007/978-3-030-93623-5_15

Bisht, V., & Singh, S. B. (2023). Lz-transform approach to evaluate reliability indices of multistate repairable weighted *k*-out-of-*n* systems. *Quality and Reliability Engineering International,* 39(3), 1043–1057. https://doi.org/10.1002/qre.3279

Bistouni, F., & Jahanshahi, M. (2014a). Improved extra group network: A new fault-tolerant multistage interconnection network. *The Journal of Supercomputing,* 69(1), 161–199. https://doi.org/10.1007/s11227-014-1132-y

Bistouni, F., & Jahanshahi, M. (2014b). Analyzing the reliability of shuffle-exchange networks using reliability block diagrams. *Reliability Engineering & System Safety,* 132, 97–106. https://doi.org/10.1016/j.ress.2014.07.012

Blake, J. T., & Trivedi, K. S. (1989). Multistage interconnection network reliability. *IEEE Transactions on Computers,* 38(11), 1600–1604. https://doi.org/10.1109/12.42134

Chen, G., Hu, Y., Wang, C., Wu, Z., Rong, W., & Guan, Q. (2025). Efficient reliability analysis of generalized *k*-out-of-*n* phased-mission systems. *Reliability Engineering & System Safety,* 253, 110581. https://doi.org/10.1016/j.ress.2024.110581

Fard, N. S., & Gunawan, I. (2005). Terminal reliability improvement of shuffle-exchange network systems. *International Journal of Reliability, Quality and Safety Engineering,* 12(1), 51–60. https://doi.org/10.1142/S0218539305001677

Gunawan, I. (2014). *Fundamentals of Reliability Engineering: Applications in Multistage Interconnection Networks.* John Wiley & Sons.

Levitin, G. (2004). A universal generating function approach for the analysis of multi-state systems with dependent elements. *Reliability Engineering & System Safety,* 84(3), 285–292. https://doi.org/10.1016/j.ress.2003.12.002

Levitin, G., & Lisnianski, A. (1999). Importance and sensitivity analysis of multi-state systems using the universal generating function method. *Reliability Engineering & System Safety,* 65(3), 271–282. https://doi.org/10.1016/S0951-8320(99)00005-8

Li, W., & Zuo, M. J. (2008). Optimal design of multi-state weighted *k*-out-of-*n* systems based on component design. *Reliability Engineering & System Safety,* 93(11), 1673–1681. https://doi.org/10.1016/j.ress.2008.01.009

Lisnianski, A. (2004). Universal generating function technique and random process methods for multi-state system reliability analysis. In *Proceedings of the 2nd International Workshop in Applied Probability (IWAP2004). Piraeus, Greece* (pp. 237–242).

Lisnianski, A., Frenkel, I., & Ding, Y. (2010). *Multi-state System Reliability Analysis and Optimization for Engineers and Industrial Managers.* London: Springer London.

Qin, J., & Li, Z. (2019). Reliability and sensitivity analysis method for a multistate system with common cause failure. *Complexity, 2019*, 6535726. https://doi.org/10.1155/2019/6535726

Rajkumar, S., & Goyal, N. K. (2016). Review of multistage interconnection networks reliability and fault-tolerance. *IETE Technical Review,* 33(3), 223–230. https://doi.org/10.1080/02564602.2015.1102098

Sharma, S., Kahlon, K. S., & Bansal, P. K. (2010). Reliability and path length analysis of irregular fault tolerant multistage interconnection network. *ACM SIGARCH Computer Architecture News*, 37(5), 16–23. https://doi.org/10.1145/1755235.1755239

Ushakov, I. A. (1986). A universal generating function. *Soviet Journal of Computer and Systems Sciences,* 24(5), 118–129.

Wang, X., Ning, R., Zhao, X., & Zhou, J. (2022). Reliability evaluations for a multi-state *k*-out-of-*n*: F system with m subsystems supported by multiple protective devices. *Computers & Industrial Engineering,* 171, 108409. https://doi.org/10.1016/j.cie.2022.108409

Wu, C. L., & Feng, T. Y. (1980). On a class of multistage interconnection networks. *IEEE Transactions on Computers,* C-29(8), 694–702. https://doi.org/10.1109/TC.1980.1675651

Yeh, W. C. (2008). A simple universal generating function method for estimating the reliability of general multi-state node networks. *IIE Transactions,* 41(1), 3–11. https://doi.org/10.1080/07408170802322622

Yeh, W. C. (2009). A simple universal generating function method to search for all minimal paths in networks. *IEEE Transactions on Systems, Man, and Cybernetics – Part A,* 39(6), 1247–1254. https://doi.org/10.1109/TSMCA.2009.2026209

Yi, X. J., Xu, C. H., Liu, S. L., Xing, M. X., & Mu, H. N. (2022). Failure analysis of repairable systems with *k*-out-of-*m* configuration considering common cause failure and maintenance correlation based on goal-oriented method. *Engineering Failure Analysis,* 140, 106615. https://doi.org/10.1016/j.engfailanal.2022.106615

Zhao, X., Han, H., Jiao, C., & Qiu, Q. (2024). Reliability modeling of *k*-out-of-*n*: F balanced systems with common bus performance sharing. *Reliability Engineering & System Safety,* 248, 110144. https://doi.org/10.1016/j.ress.2024.110144

Zuo, M. J., & Tian, Z. (2006). Performance evaluation of generalized multi-state *k*-out-of-*n* systems. *IEEE Transactions on Reliability,* 55(2), 319–327. https://doi.org/10.1109/TR.2006.874916

Chapter 12

Mathematical modeling and reliability attributes of feeding system in sugarcane industry

Geetanjali Rajput, Soni Bisht, Manjinder Kaur, and Surendra Vikram Singh Padiyar

12.1 INTRODUCTION: BACKGROUND AND RESEARCH MOTIVATION

Sugar is one of the most essential food commodities consumed worldwide and plays a vital role in the agro-based economy of India. According to the NITI Aayog (2020) report, the sugar industry in India comprises more than 700 licensed factories, with a collective annual turnover of approximately Rs. 80,000 crore. It is important to note that these numbers represent aggregate figures for the entire industry rather than output from individual factories. Of these licensed units, only around 500 are currently operational, reflecting constraints in production and management. This sector directly employs over 5 lakh people through sugar mills and indirectly supports nearly 10 lakh individuals, while the livelihood of around 5 crore farmers and associated workers depends on sugarcane cultivation and related industrial activities. These mills collectively have a crushing capacity of about 340 lakh metric tons, highlighting the scale and importance of sugar production in India.

This sector is closely linked with several allied industries. For example, the jute industry benefits from sugar packaging needs, molasses is used in alcohol production and animal feed, bagasse is utilized in paper and biofuel production. This industry also contributes to the automobile and energy sectors. As a result, any disruptions in sugar production can have cascading effects on these dependent industries and the broader national economy.

Despite its significance, sugar sector faces multiple challenges, such as shortage of sugarcane, inadequate power supply, outdated technology, low-capacity utilization, and insufficient financial support for farmers (Dahiya et al., 2019a). These issues raise operational costs, reduce productivity, and results in persistent losses. Therefore, assessing the feeding system of sugar mills and evaluating its reliability are essential for improving operational performance and boosting economic efficiency.

Several studies have examined the reliability of the manufacturing systems of the sugar industry with various methods such as supplementary variable technique (SVT), Markov process, fuzzy approach, genetic

DOI: 10.1201/9781003787525-12

algorithm, Runge–Kutta (R–K) method, and Laplace transform (LT) to evaluate availability, reliability, profit, and sensitivity of various subsystems (Kumar et al., 1988; Sharma & Vishwakarma, 2014a, 2014b; Kadyan & Kumar, 2017, 2018, 2019; Saini & Kumar, 2019; Dahiya et al., 2019a, 2019b). Similar approaches have also been applied to thermal power plants, manufacturing systems, butter oil production units, and cloud computing systems (Kumar, 2019b; Kumari et al., 2019; Manglik et al., 2020; Mehta et al., 2018; Shakuntla et al., 2011). Kumar (2019a) developed the mathematical modeling of the network, which comprises three main units: Automated Teller Machine (ATM), host computer, and bank computer, using Markov process. They considered that the bank computer and power supply have standby redundancy. The authors used SVT and LT to compute numerous reliability attributes. Manglik and Ram (2013) analyzed several reliability attributes of a two-unit cold standby system with the Markov process, under the assumption that the repair follows general distribution while all the failure rates are constant. Sharifi et al. (2021) studied a load-sharing system in which identical components possess time-dependent failure rates. They used the Markov process to compute system reliability and applied simulation technique to compare the obtained results. Finally, they concluded that the computing efficiency of the suggested model is superior to the simulation method for tackling large-scale problems.

Mostly, researchers have used a single distribution in failure or repair analysis. However, the copula, which connects or couples a multivariate distribution function to its one-dimensional marginal distribution functions, can be used if two different distributions are to be applied simultaneously in repair or failure. Through the use of copulas, numerous researchers have examined systems with various failure modes. Nailwal and Singh (2011) examined numerous reliability characteristics of complex system with three types of repair facilities using Gumbel–Hougaard copula (G–H copula). Nautiyal et al. (2020) worked on the reliability evaluation of k-out-of-n network. The authors incorporated minimal cuts and G–H copula approach to compute the reliability characteristics of the network. Ram and Singh (2009) applied G–H copula in computing the reliability attributes of complex system with two types of repair facilities and highlighted the important results. Abubakar and Singh (2019) evaluated reliability, availability, mean time to failure (MTTF) and profit analysis of African textile manufacturer using copula approach. Chantola and Singh (2019) calculated the reliability, sensitivity, and MTTF of the transformer subject to a health index with the application of copula. Chopra and Ram (2019) studied two dissimilar parallel systems under constant failure rates and evaluated various reliability indices with the help of G–H copula. Bisht et al. (2020) analyzed the reliability and various transition state probabilities of a bleaching unit of a paper mill with the application of G–H copula approach. Lado and Singh (2019) investigated the reliability indices of a complex repairable system with two subsystems in parallel configuration by employing the copula methodology. Goyal et al. (2021) focused on the modeling and performance evaluation of windmill water pumping system employing copula approach. Maihulla et al. (2022) discussed the mathematical modeling and analyzed the availability of the solar photovoltaic (PV) system with the application of the G–H copula. These studies collectively demonstrate how copula-based approach can enhance the performance and efficiency of complex industrial systems.

Numerous researchers have made significant contributions toward improving the efficiency and performance of various subsystems in the sugar industry. However, the reliability metrics used to assess durability, potency, and efficiency of the feeding system are not well understood. More research is needed on the effectiveness of sugar industry's feeding system, the reliability of assessing system strength, and strategies to improve feeding system performance.

Motivated by the economic significance of sugar and the operational importance of the feeding system, this chapter focuses on feeding systems with slightly modified configuration consisting of seven subsystems, namely: unloaders, cane carrier, cutter unit, crushing unit, bagasse carrier unit, boilers, and turbines. The following subsections present the research gap, novelty, and detailed description of the considered system.

12.1.1 Research gap

The feeding system of sugar mills has been investigated using different methods such as the fuzzy approach, SVT, and R–K method. However, the copula-based approach has not yet been applied to model dependencies within the repair processes of subsystems. In this study, two types of repairs are considered: a general repair distribution and an exponential repair distribution. The copula-based approach is used to couple these repair types, providing a framework to capture interdependencies among subsystems during repairs, which has not been previously explored in sugar mill feeding system.

12.1.2 Novelty

This study presents assessment of the sugar mill feeding system, an essential subsystem in the industry, by modeling its real-world structure, which comprises seven interdependent subsystems arranged in a hybrid configuration. It introduces a copula-based repair modeling approach, combining both general and exponential repair times, to more accurately reflect practical maintenance scenarios. Furthermore, this is the first study to apply the combined copula approach to capture interdependencies among subsystem repairs in the sugar mill feeding system. By integrating the Markov process,

SVT, LT, and copula methodologies, the chapter derives key performance measures such as reliability, MTTF, sensitivity analysis, and expected profit, thereby offering a holistic framework that significantly enhances existing modeling approaches.

12.1.3 Feeding system of sugar industry

In the feeding system, sugarcane is brought from the field (through trucks via roads, etc.) and unloaded by the unloaders, and then sugarcane carrier supplies these sugarcanes to the cutter unit. These canes are then cut into fine pieces and supplied to crusher to extract juice, and the leftover (or squeezed cane, molasses) is used as fuel in boilers. Further, turbines use the steam from the boiler to produce electricity (Kadyan & Kumar, 2017). The feeding system considered in this chapter is based on a real-life industrial system from the sugar industry to facilitate its study. The considered feeding system is slightly updated with minor assumptions to enable mathematical modeling, and the G–H copula methodology has been applied for its analysis. The system comprises seven subsystems, which are connected in series. These subsystems are represented in the reliability block diagram shown in Figure 12.1, while the transition diagram in Figure 12.2 provides a mathematical representation of the system, illustrating the assumed state-to-state transitions among fully functional, partially functional, and completely failed states. A detailed description of these states is provided in Table 12.1. The subsystems of the considered feeding system are discussed in more detail below:

- **Subsystem A (Unloaders)**: It consists of three identical unloaders in a parallel configuration. Failure of any one unloader in this subsystem results in degradation of system performance. Here, failure of all the unloaders causes system failure.
- **Subsystem B (Cane carrier)**: Subsystem B is a chain conveyor, it carriers unloaded cane and transfers it to cutter unit. Operating in series with the other subsystems, its failure causes the entire system to stop.

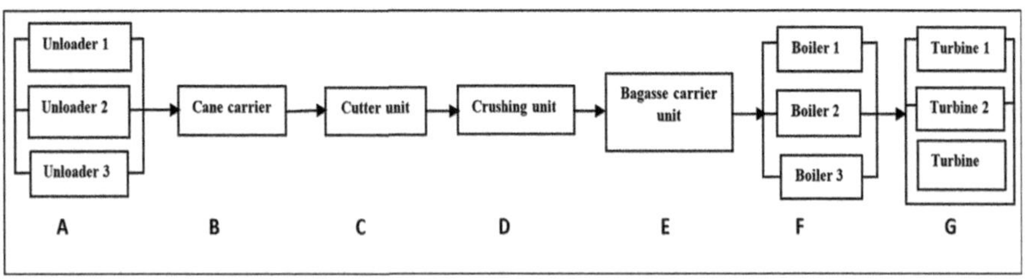

Figure 12.1 Block diagram of the feeding system.

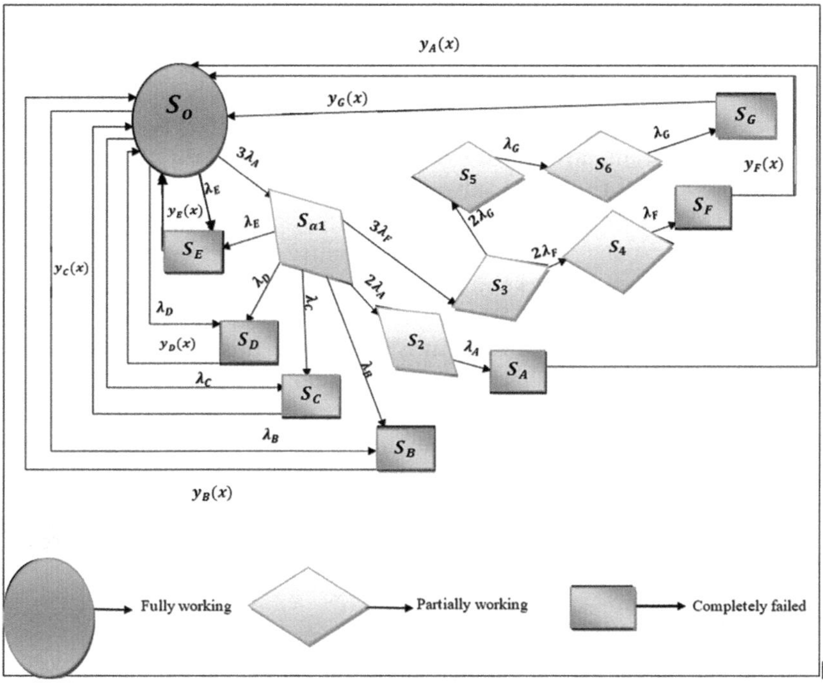

Figure 12.2 Transition diagram of the feeding system.

Table 12.1 State specification of transition diagram

States	*Number of good units in subsystem*			*States of subsystem*				*System state*
	A	*F*	*G (1 unit in cold standby)*	*B*	*C*	*D*	*E*	
S_o	3	3	2	O	O	O	O	O
S_{al}	2	3	2	O	O	O	O	P_w
S_2	1	3	2	O	O	O	O	P_w
S_A	0	3	2	O	O	O	O	N_w
S_3	2	2	2	O	O	O	O	P_w
S_4	2	1	2	O	O	O	O	P_w
S_F	2	0	2	O	O	O	O	N_w
S_5	2	2	1	O	O	O	O	P_w
S_6	2	2	1 (standby functioning)	O	O	O	O	P_w
S_G	2	2	0	O	O	O	O	N_w
S_B	3 or 2	3	2	N_w	O	O	O	N_w
S_C	3 or 2	3	2	O	N_w	O	O	N_w
S_D	3 or 2	3	2	O	O	N_w	O	N_w
S_E	3 or 2	3	2	O	O	O	N_w	N_w

O, fully operational; P_w, partially working; N_w, failed state.

- **Subsystem C (Cutter)**: This subsystem cuts the supplied sugarcane into small pieces. A failure in this unit leads to the shutdown of the entire sugar plant.
- **Subsystem D (Crusher)**: From subsystem C, the chopped sugarcane is crushed in subsystem D for maximum juice extraction. A failure in this subsystem can shut down the entire sugar plant, as it operates in series with all the other subsystems.
- **Subsystem E (Bagasse carrier)**: This subsystem transfers crushed cane from the crusher to subsystem F. Failure of this unit causes system failure because it works in series with all other subsystems.
- **Subsystem F (Boilers)**: It consists of three identical boilers in parallel configuration. This subsystem is connected in series with subsystems F and G. Failure of this subsystem results in system failure.
- **Subsystem G (Turbines)**: Two identical turbines are arranged in parallel as part of subsystem G, and one turbine is kept in cold standby. Failure of all the turbines results in a complete breakdown of the system.

12.2 METHODOLOGY

12.2.1 Mathematical modeling

Mathematical modeling converts real-life problems into mathematical equations (or forms). One can easily calculate real-life problems with mathematical modeling. Figure 12.3 shows the flowchart of mathematical modeling of a real-life problem.

12.2.2 Mathematical tools used to measure reliability indices

12.2.2.1 Markov process

A Markov process is a type of process where the future state depends only on the current state and not on the sequence of events that came before. This approach is commonly used to analyze decision-making problems in various fields (Ebeling, 2004). The process starts with creating a state transition diagram based on the arrangement of the system's components. Then, differential equations are set up to describe failures and repairs in the system. The governing equations are solved by applying the LT to obtain the probabilities of each system state. These probabilities are then used to evaluate the reliability indices of the system.

12.2.2.2 Supplementary variable technique

The SVT is commonly used to analyze systems in which failure or repair rates change over time, resulting in non-Markovian behavior. This technique

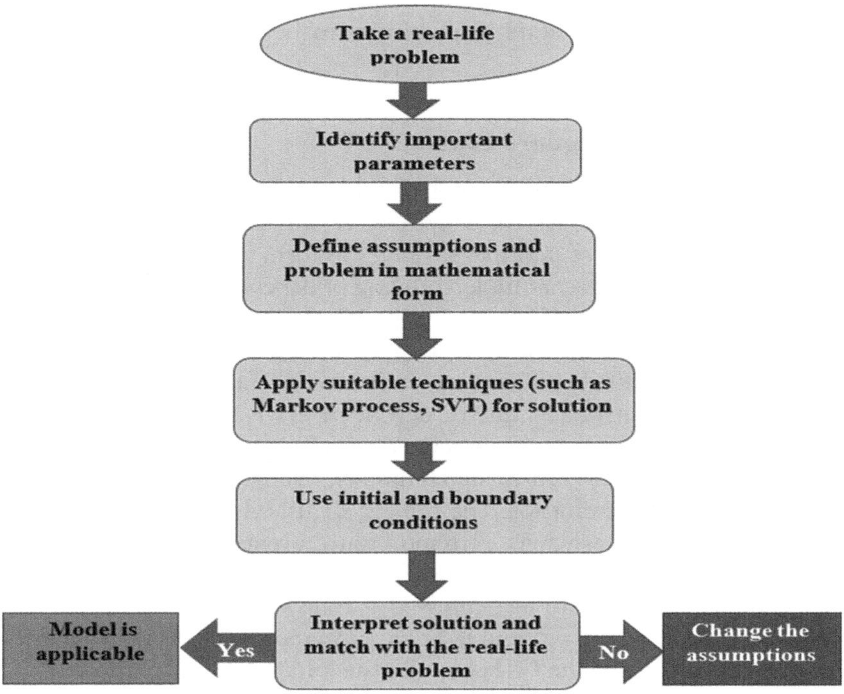

Figure 12.3 Flowchart of mathematical modeling in a real-life problem.

became widely accepted in stochastic modeling through the work of Cox (1955), who applied elapsed time as a supplementary variable. The SVT enables a non-Markovian system to be treated as Markovian by incorporating one or more supplementary variables, such as the elapsed repair time x. In this framework, when the system is in a failed state at time t, the probability of returning to a functional state is computed using the elapsed repair time x.

12.2.2.3 Laplace transform

The LT is a widely used mathematical tool for solving both ordinary and partial differential equations under specific initial or boundary conditions. By converting differential equations into algebraic equations, it simplifies these problems, which are generally easier to solve. Once the algebraic equation is solved, the inverse LT (ILT) is applied to recover the solution in time domain. Formally, for a function $M(t)$ defined for all $t \geq 0$, its LT $M(s)$ is given as (Balagurusamy, 1984):

$$M(s) = \int_0^\infty e^{-st} M(t)\, dt$$

This approach is particularly useful in Markov analysis, as it allows the system's time-dependent state probabilities to be assessed in a clear and systematic manner.

12.2.2.4 Gumbel–Hougaard copula

The G–H copula is used to model the dependence between component lifetimes in a repairable system. Copulas are statistical tools that link the marginal distributions of multiple variables to form a joint multivariate distribution, providing a deeper understanding of dependencies than what simple correlation measures. The term "copula", meaning a "link" or "bond" together, was derived from Latin (Nelsen, 2006) and was first formally introduced by Sklar (1959). Among various types, the G–H copula is particularly effective for representing asymmetric dependence, where extreme high values of one component are strongly associated with high values of another, while low values remain weakly correlated (Gumbel, 1960; Hougaard, 1986). This property is highly relevant for repairable systems, where early failures may occur independently, whereas extreme failure events often exhibit strong correlation. The joint cumulative distribution function (CDF) of two variables under this copula can be expressed as follows:

For instance, if $u_a(x) = e^x$ and $u_b(x) = y_i(x)$ represent a general distribution, then according to the G–H copula, the joint CDF is given by:

$$y(x) = \exp\left(-\left((-x)^{\beta} + \left(-\log y_i(x)\right)^{\beta}\right)^{\frac{1}{\beta}}\right)$$

where $\beta \in [1, \infty)$ is the copula parameter controlling the strength of dependence ($\beta = 1$ represents independence) (Gumbel, 1960; Hougaard, 1986).

This formulation offers a reliable framework for analyzing component dependencies and system behavior under repairable conditions.

12.2.3 Assumptions

a. At the start, the feeding system under study is fully functional, with each component performing flawlessly.
b. The system under study exhibits three possible states: fully functional (S_0), partially operational, and entirely failed.
c. System repair rates follow two types of distribution, i.e., exponential and general.
d. Repair facility will be provided only if the system stopped working (completely failed).
e. Repaired subsystems are restored to their original state, making the system fully operational again.

The notations used in this chapter are listed below:

λ_i; where $i = A, B, C, D, E, F, G$	Failure rates of units of subsystems
y_i; where $i = A$ to G	Repair rates of subsystems
$y(x)$	Copula repair rate
x	Elapsed repair time
s	LT variable
$P_o(t)$	Probability of the system in its fully functional state
$P_j(t)$	Probability that the system is in jth state at time t where $t > 0$ and j = a1, 2, 3, 4, 5, 6, A, B, C, D, E, F, G
$\overline{P}_j(t)$	LT of $P_j(t)$
$P_i(x,t)$; where $i = A$ to G	PDF associated with the system in failure state S_i at time t while under repair, with elapsed repair time x
$P_{up}(t)$	Probability of working state at time t
$P_{down}(t)$	Probability of failed state at time t
$E_P(t)$	Expected profit for the interval $(0,t]$
C_R and C_S	Revenue and service cost per unit time

12.3 MATHEMATICAL MODEL FORMULATION

Applying the Markov process, the model can be expressed mathematically using the differential-difference equations as given below:

$$\left[\frac{d}{dt} + 3\lambda_A + \lambda_B + \lambda_C + \lambda_D + \lambda_E\right] P_o(t)$$

$$= \int_0^\infty P_A(x,t)\, y_A(x)dx + \int_0^\infty P_B(x,t)\, y_B(x)dx$$

$$+ \int_0^\infty P_C(x,t)\, y_C(x)dx + \int_0^\infty P_D(x,t)\, y_D(x)dx$$

$$+ \int_0^\infty P_E(x,t)\, y_E(x)dx + \int_0^\infty P_F(x,t)\, y_F(x)dx$$

$$+ \int_0^\infty P_G(x,t)\, y_G(x)dx \tag{12.1}$$

$$\left[\frac{d}{dt}+2\lambda_A+\lambda_B+\lambda_C+\lambda_D+\lambda_E+3\lambda_F\right]P_{a1}(t)=3\lambda_A\ P_o(t) \tag{12.2}$$

$$\left[\frac{d}{dt}+\lambda_A\right]P_2(t)=3\lambda_A P_{a1}(t) \tag{12.3}$$

$$\left[\frac{d}{dt}+2\lambda_F+2\lambda_G\right]P_3(t)=3\lambda_F P_{a1}(t) \tag{12.4}$$

$$\left[\frac{d}{dt}+\lambda_F\right]P_4(t)=2\lambda_F P_3(t) \tag{12.5}$$

$$\left[\frac{d}{dt}+\lambda_G\right]P_5(t)=2\lambda_G P_3(t) \tag{12.6}$$

$$\left[\frac{d}{dt}+\lambda_G\right]P_6(t)=\lambda_G P_5(t) \tag{12.7}$$

$$\left[\frac{\partial}{\partial t}+\frac{\partial}{\partial x}+y_A(x)\right]P_A(x,t)=0 \tag{12.8}$$

$$\left[\frac{\partial}{\partial t}+\frac{\partial}{\partial x}+y_B(x)\right]P_B(x,t)=0 \tag{12.9}$$

$$\left[\frac{\partial}{\partial t}+\frac{\partial}{\partial x}+y_C(x)\right]P_C(x,t)=0 \tag{12.10}$$

$$\left[\frac{\partial}{\partial t}+\frac{\partial}{\partial x}+y_D(x)\right]P_D(x,t)=0 \tag{12.11}$$

$$\left[\frac{\partial}{\partial t}+\frac{\partial}{\partial x}+y_E(x)\right]P_E(x,t)=0 \tag{12.12}$$

$$\left[\frac{\partial}{\partial t}+\frac{\partial}{\partial x}+y_F(x)\right]P_F(x,t)=0 \tag{12.13}$$

$$\left[\frac{\partial}{\partial t}+\frac{\partial}{\partial x}+y_G(x)\right]P_G(x,t)=0 \tag{12.14}$$

Boundary conditions:

$$P_A(0,t) = \lambda_A \; P_2(t) \tag{12.15}$$

$$P_B(0,t) = \lambda_B \; P_o(t) + \lambda_B \; P_{a1}(t) \tag{12.16}$$

$$P_C(0,t) = \lambda_C \; P_o(t) + \lambda_C \; P_{a1}(t) \tag{12.17}$$

$$P_D(0,t) = \lambda_D \; P_o(t) + \lambda_D \; P_{a1}(t) \tag{12.18}$$

$$P_E(0,t) = \lambda_E \; P_o(t) + \lambda_E \; P_{a1}(t) \tag{12.19}$$

$$P_F(0,t) = \lambda_F \; P_4(t) \tag{12.20}$$

$$P_G(0,t) = \lambda_G \; P_6(t) \tag{12.21}$$

Initial condition:
$P_O(0) = 1$; otherwise zero

12.3.1 Model solution

On applying SVT and taking LT, we get

$$\begin{aligned}[s + \lambda_A + \lambda_B + \lambda_c + \lambda_D + \lambda_E]\overline{P}_o(s) = 1 &+ \int_0^\infty \overline{P}_A(x,s)\, y_A(x)dx + \int_0^\infty \overline{P}_B(x,s)\, y_B(x)dx \\ &+ \int_0^\infty \overline{P}_C(x,s)\, y_C(x)dx + \int_0^\infty \overline{P}_D(x,s)\, y_D(x)dx \\ &+ \int_0^\infty \overline{P}_E(x,s)\, y_E(x)dx + \int_0^\infty \overline{P}_F(x,s)\, y_F(x)dx \\ &+ \int_0^\infty \overline{P}_G(x,s)\, y_G(x)dx \end{aligned} \tag{12.22}$$

$$[s + 2\lambda_A + \lambda_B + \lambda_c + \lambda_D + \lambda_E + 3\lambda_F]\overline{P}_{a1}(s) = 3\lambda_A \overline{P}_o(s) \tag{12.23}$$

$$[s+\lambda_A]\bar{P}_2(s)=2\lambda_A\bar{P}_{a1}(s) \tag{12.24}$$

$$[s+2\lambda_F+2\lambda_G]\bar{P}_3(s)=3\lambda_F\bar{P}_{a1}(s) \tag{12.25}$$

$$[s+\lambda_F]\bar{P}_4(s)=2\lambda_F\bar{P}_3(s) \tag{12.26}$$

$$[s+\lambda_G]\bar{P}_5(s)=2\lambda_G\bar{P}_3(s) \tag{12.27}$$

$$[s+\lambda_G]\bar{P}_6(s)=\lambda_G\bar{P}_5(s) \tag{12.28}$$

$$\left[s+\frac{d}{dx}+y_A(x)\right]\bar{P}_A(x,s)=0 \tag{12.29}$$

$$\left[s+\frac{d}{dx}+y_B(x)\right]\bar{P}_B(x,s)=0 \tag{12.30}$$

$$\left[s+\frac{d}{dx}+y_C(x)\right]\bar{P}_C(x,s)=0 \tag{12.31}$$

$$\left[s+\frac{d}{dx}+y_D(x)\right]\bar{P}_D(x,s)=0 \tag{12.32}$$

$$\left[s+\frac{d}{dx}+y_E(x)\right]\bar{P}_E(x,s)=0 \tag{12.33}$$

$$\left[s+\frac{d}{dx}+y_F(x)\right]\bar{P}_F(x,s)=0 \tag{12.34}$$

$$\left[s+\frac{d}{dx}+y_G(x)\right]\bar{P}_G(x,s)=0 \tag{12.35}$$

Boundary conditions:

$$\bar{P}_A(0,s)=\lambda_A\bar{P}_2(s) \tag{12.36}$$

$$\bar{P}_B(0,s)=\lambda_B\bar{P}_o(s)+\lambda_B\bar{P}_{a1}(s) \tag{12.37}$$

$$\bar{P}_C(0,s) = \lambda_C \bar{P}_o(s) + \lambda_C \bar{P}_{a1}(s) \tag{12.38}$$

$$\bar{P}_D(0,s) = \lambda_D \bar{P}_o(s) + \lambda_D \bar{P}_{a1}(s) \tag{12.39}$$

$$\bar{P}_E(0,s) = \lambda_E \bar{P}_o(s) + \lambda_E \bar{P}_{a1}(s) \tag{12.40}$$

$$\bar{P}_F(0,s) = \lambda_F \bar{P}_4(s) \tag{12.41}$$

$$\bar{P}_G(0,s) = \lambda_G \bar{P}_6 \tag{12.42}$$

Using boundary conditions, and solving equations from (12.22) to (12.28), the state transition probabilities are given as:

$$\bar{P}_o(s) = \frac{1}{X(s)} \tag{12.43}$$

$$\bar{P}_{a1}(s) = \frac{3\lambda_A \bar{P}_o(s)}{(s+c)} \tag{12.44}$$

$$\bar{P}_2(s) = \frac{6\lambda_A^2 \bar{P}_o(s)}{(s+c)(s+\lambda_A)} \tag{12.45}$$

$$\bar{P}_3(s) = \frac{9\lambda_A \lambda_F \bar{P}_o(s)}{(s+c)(s+2\lambda_F+2\lambda_G)} \tag{12.46}$$

$$\bar{P}_4(s) = \frac{18\lambda_A \lambda_F^2 \bar{P}_o(s)}{(s+c)(s+\lambda_F)(s+2\lambda_F+2\lambda_G)} \tag{12.47}$$

$$\bar{P}_5(s) = \frac{18\lambda_A \lambda_F \lambda_G \bar{P}_o(s)}{(s+c)(s+\lambda_G)(s+2\lambda_F+2\lambda_G)} \tag{12.48}$$

$$\bar{P}_6(s) = \frac{18\lambda_A \lambda_F \lambda_G^2 \bar{P}_0(s)}{(s+c)(s+\lambda_G)^2(s+2\lambda_F+2\lambda_G)} \tag{12.49}$$

From Figure 12.2, the probability of the system under study being in a functional state is given as:

$$\bar{P}_{\text{up}}(s) = \bar{P}_o(s) + \bar{P}_{a1}(s) + \bar{P}_2(s) + \bar{P}_3(s) + \bar{P}_4(s) + \bar{P}_5(s) + \bar{P}_6(s) \tag{12.50}$$

Using equations (12.43)–(12.49), we get

$$\bar{P}_{up}(s)=\begin{bmatrix}1+\dfrac{3\lambda_A}{(s+c)}+\dfrac{6\lambda_A^2}{(s+c)(s+\lambda_A)}+\dfrac{9\lambda_A\lambda_F}{(s+c)(s+2\lambda_F+2\lambda_G)}\\ +\dfrac{18\lambda_A\lambda_F^2}{(s+c)(s+\lambda_F)(s+2\lambda_F+2\lambda_G)}+\dfrac{18\lambda_A\lambda_F\lambda_G}{(s+c)(s+\lambda_G)(s+2\lambda_F+2\lambda_G)}\\ +\dfrac{18\lambda_A\lambda_F\lambda_G^2}{(s+c)(s+\lambda_G)^2(s+2\lambda_F+2\lambda_G)}\end{bmatrix}\bar{P}_o(s) \tag{12.51}$$

Using boundary conditions from (12.36) to (12.42) and solving equations from (12.29) to (12.35), the probabilities of the failed states are as:

$$\bar{P}_A(s)=\frac{6\lambda_A^3\bar{P}_0(s)}{s(s+c)(s+\lambda_A)}\left[1-A(s)\right] \tag{12.52}$$

$$\bar{P}_B(s)=\frac{\lambda_B}{s}\left[\frac{3\lambda_A}{(s+c)}+1\right]\bar{P}_0(s)\left[1-B(s)\right] \tag{12.53}$$

$$\bar{P}_C(s)=\frac{\lambda_C}{s}\left[\frac{3\lambda_A}{(s+c)}+1\right]\bar{P}_0(s)\left[1-C(s)\right] \tag{12.54}$$

$$\bar{P}_D(s)=\frac{\lambda_D}{s}\left[\frac{3\lambda_A}{(s+c)}+1\right]\bar{P}_0(s)\left[1-D(s)\right] \tag{12.55}$$

$$\bar{P}_E(s)=\frac{\lambda_E}{s}\left[\frac{3\lambda_A}{(s+c)}+1\right]\bar{P}_0(s)\left[1-E(s)\right] \tag{12.56}$$

$$\bar{P}_F(s)=\frac{18\lambda_A\lambda_F^3\bar{P}_0(s)}{s(s+c)(s+\lambda_F)(s+2\lambda_F+2\lambda_G)}\left[1-F(s)\right] \tag{12.57}$$

$$\bar{P}_G(s)=\frac{18\lambda_A\lambda_F\lambda_G^3\bar{P}_0(s)}{s(s+c)(s+\lambda_G)^2(s+2\lambda_F+2\lambda_G)}\left[1-G(s)\right] \tag{12.58}$$

From transition diagram, i.e., Figure 12.2, the probability of the considered system being in a failed state is presented below:

$$\overline{P}_{\text{down}}(s)=\overline{P}_A(s)+\overline{P}_B(s)+\overline{P}_C(s)+\overline{P}_D(s)+\overline{P}_E(s)+\overline{P}_F(s)+\overline{P}_G(s) \tag{12.59}$$

From equations (12.52) to (12.58), equation (12.59) becomes

$$\overline{P}_{\text{down}}(s)=\begin{bmatrix} \frac{6\lambda_A^3}{s(s+c)(s+\lambda_A)}\left[1-A(s)\right] \\ +\frac{\lambda_B}{s}\left[\frac{3\lambda_A}{(s+c)}+1\right]\left[1-B(s)\right]+\frac{\lambda_C}{s}\left[\frac{3\lambda_A}{(s+c)}+1\right]\left[1-C(s)\right] \\ +\frac{\lambda_D}{s}\left[\frac{3\lambda_A}{(s+c)}+1\right]\left[1-D(s)\right]+\frac{\lambda_E}{s}\left[\frac{3\lambda_A}{(s+c)}+1\right]\left[1-E(s)\right] \\ +\frac{18\lambda_A\lambda_F^3}{s(s+c)(s+\lambda_F)(s+2\lambda_F+2\lambda_G)}\left[1-F(s)\right] \\ +\frac{18\lambda_A\lambda_F\lambda_G^3}{s(s+c)(s+\lambda_G)^2(s+2\lambda_F+2\lambda_G)}\left[1-G(s)\right] \end{bmatrix}\overline{P}_o(s) \tag{12.60}$$

where

$$X(s)=\begin{bmatrix} s+c_a-\frac{6\lambda_A^3}{(s+c)(s+\lambda_A)}A(s)-\lambda_B\left[\frac{3\lambda_A}{(s+c)}+1\right]B(s) \\ -\lambda_C\left[\frac{3\lambda_A}{(s+c)}+1\right]C(s)-\lambda_D\left[\frac{3\lambda_A}{(s+c)}+1\right]D(s) \\ -\lambda_E\left[\frac{3\lambda_A}{(s+c)}+1\right]E(s)-\frac{18\lambda_A\lambda_F^3}{(s+c)(s+\lambda_F)(s+2\lambda_F+2\lambda_G)}F(s) \\ -\frac{18\lambda_A\lambda_F\lambda_G^3}{(s+c)(s+\lambda_G)^2(s+2\lambda_F+2\lambda_G)}G(s) \end{bmatrix}$$

$$c_a=3\lambda_A+\lambda_B+\lambda_c+\lambda_D+\lambda_E$$

$$c=2\lambda_A+\lambda_B+\lambda_c+\lambda_D+\lambda_E+3\lambda_F$$

$$A(s)=\frac{\exp\left(-\left((-x)^{\beta}+(-\log y_A(x))^{\beta}\right)^{\frac{1}{\beta}}\right)}{s+\exp\left(-\left((-x)^{\beta}+(-\log y_A(x))^{\beta}\right)^{\frac{1}{\beta}}\right)}$$

$$B(s)=\frac{\exp\left(-\left((-x)^{\beta}+(-\log y_B(x))^{\beta}\right)^{\frac{1}{\beta}}\right)}{s+\exp\left(-\left((-x)^{\beta}+(-\log y_B(x))^{\beta}\right)^{\frac{1}{\beta}}\right)}$$

$$C(s)=\frac{\exp\left(-\left((-x)^{\beta}+(-\log y_C(x))^{\beta}\right)^{\frac{1}{\beta}}\right)}{s+\exp\left(-\left((-x)^{\beta}+(-\log y_C(x))^{\beta}\right)^{\frac{1}{\beta}}\right)}$$

$$D(s)=\frac{\exp\left(-\left((-x)^{\beta}+(-\log y_D(x))^{\beta}\right)^{\frac{1}{\beta}}\right)}{s+\exp\left(-\left((-x)^{\beta}+(-\log y_D(x))^{\beta}\right)^{\frac{1}{\beta}}\right)}$$

$$E(s)=\frac{\exp\left(-\left((-x)^{\beta}+(-\log y_E(x))^{\beta}\right)^{\frac{1}{\beta}}\right)}{s+\exp\left(-\left((-x)^{\beta}+(-\log y_E(x))^{\beta}\right)^{\frac{1}{\beta}}\right)}$$

$$F(s)=\frac{\exp\left(-\left((-x)^{\beta}+(-\log y_F(x))^{\beta}\right)^{\frac{1}{\beta}}\right)}{s+\exp\left(-\left((-x)^{\beta}+(-\log y_F(x))^{\beta}\right)^{\frac{1}{\beta}}\right)}$$

$$G(s)=\frac{\exp\left(-\left((-x)^{\beta}+(-\log y_G(x))^{\beta}\right)^{\frac{1}{\beta}}\right)}{s+\exp\left(-\left((-x)^{\beta}+(-\log y_G(x))^{\beta}\right)^{\frac{1}{\beta}}\right)}$$

On adding equations (12.51) and (12.60), we then get

$$\bar{P}_{\text{up}}(s) + \bar{P}_{\text{down}}(s) = \frac{1}{s} \tag{12.61}$$

12.4 RESULTS

This section presents the results derived from the analysis of the considered feeding system using copula approach.

12.4.1 Reliability analysis

Reliability of a system or component is defined as the probability that it performs its intended function without failure for a specified period of time under stated operating conditions (Balagurusamy, 1984). Mathematically, reliability can be expressed as

$$R(t) = P_{\text{up}}(t)$$

In other words, reliability measures how likely it is that a system will work correctly for a given duration under the conditions for which it was designed.

Setting repair rates $y = 0$, $x = 1$, $\beta = 1$ and failure rates $\lambda_A = 0.1$, $\lambda_B = 0.05$, $\lambda_C = 0.25$, $\lambda_D = 0.15$, $\lambda_E = 0.2$, $\lambda_F = 0.01$, and $\lambda_G = 0.01$ in equation (12.51), and then applying ILT, the equation of reliability of the model is obtained as:

$$R(t) = -2.142e^{-0.95t} + 3.04074e^{-0.88t} + 0.0905e^{-0.1t} \tag{12.62}$$

$$-0.00131e^{-0.04t} + 0.01207e^{-0.01t} + 0.00007te^{-0.01t}$$

Using equation (12.62), the reliability of the considered system has been calculated and is displayed in Table 12.2 and Figure 12.4.

12.4.2 MTTF analysis

It represents the average period during which the system remains operational before its first failure. Mathematically, it is obtained as

$$\text{MTTF} = \lim_{s \to 0} \bar{P}_{\text{up}}(s) \tag{12.63}$$

where s is the Laplace variable and $\bar{P}_{\text{up}}(s)$ represents the working states of the considered system. Consider the failure rates as $\lambda_A = 0.1$, $\lambda_B = 0.05$, $\lambda_C = 0.25$, $\lambda_D = 0.15$, $\lambda_E = 0.2$, $\lambda_F = 0.01$, and $\lambda_G = 0.01$. For the analysis,

Table 12.2 Time versus reliability

t (in years)	*R (t)*
0	1
1	0.5255
2	0.28762
3	0.1709
4	0.11351
5	0.08444
6	0.06872
7	0.059298
8	0.05298
9	0.0482
10	0.0443

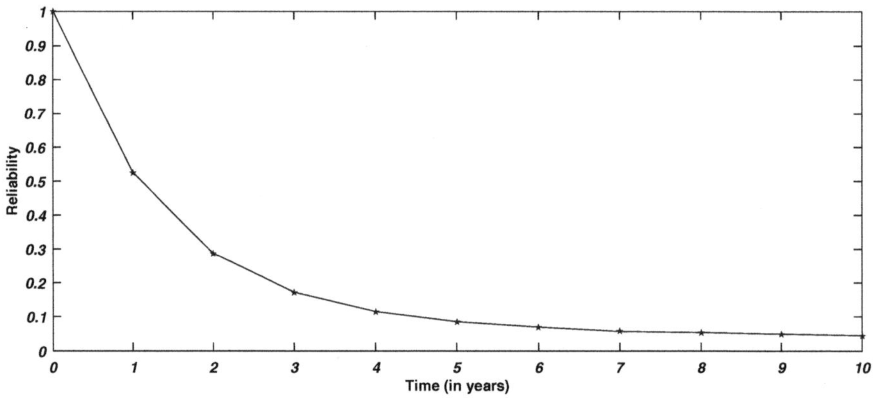

Figure 12.4 Time versus reliability.

each parameter is independently varied from 0.1 to 1, while keeping all other parameters constant in equation (12.63). The corresponding numerical results are presented in Table 12.3, and their graphical representations is shown in Figure 12.5, which illustrates how MTTF changes with respect to different parameters.

12.4.3 Sensitivity analysis

Sensitivity analysis is employed to identify the components of the model that have the greatest impact on system reliability, which helps engineers enhance overall system performance. Mathematical equation for the sensitivity analysis is given below:

Table 12.3 Variation in MTTF w.r.t. the failure rates

λ_i	$MTTF_{\lambda_A}$	$MTTF_{\lambda_B}$	$MTTF_{\lambda_C}$	$MTTF_{\lambda_D}$	$MTTF_{\lambda_E}$	$MTTF_{\lambda_F}$	$MTTF_{\lambda_G}$
0.1	4.0132	3.6613	5.4880	4.4244	4.9095	4.4976	2.4717
0.2	4.4667	3.0936	4.4244	3.6613	4.0132	3.8838	2.3086
0.3	4.3876	2.6586	3.6613	3.0936	3.3576	3.4276	2.2507
0.4	4.1563	2.3171	3.0936	2.6586	2.8623	3.0928	2.2211
0.5	3.8912	2.0435	2.6586	2.3171	2.4780	2.8391	2.2031
0.6	3.6322	1.8205	2.3171	2.0435	2.1730	2.6409	2.1910
0.7	3.3925	1.6360	2.0435	1.8205	1.9265	2.4819	2.1823
0.8	3.1752	1.4814	1.8205	1.6360	1.7241	2.3518	2.1757
0.9	2.9797	1.3504	1.6360	1.4814	1.5554	2.2433	2.1706
1	2.8041	1.2381	1.4814	1.3504	1.4133	2.1514	2.1665

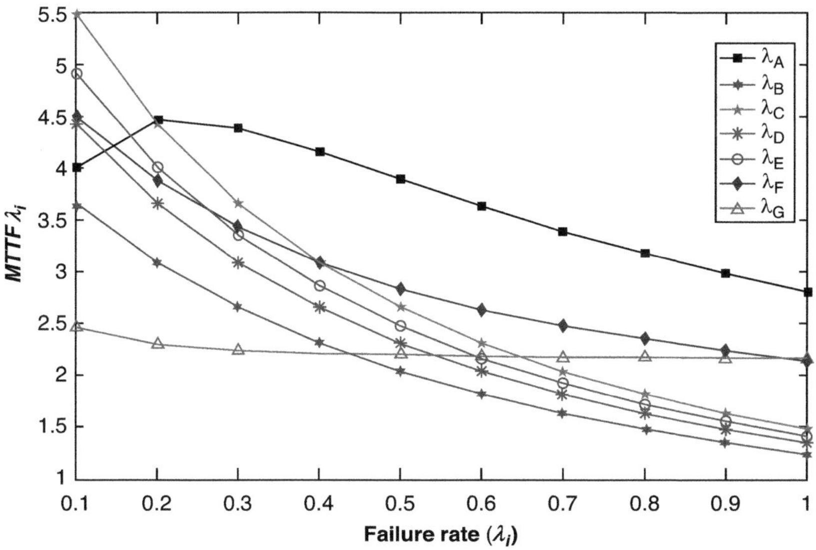

Figure 12.5 Failure rate versus MTTF.

$$S = \frac{\partial R(t)}{\partial \lambda_i} \tag{12.64}$$

where λ_i is parameter and i varies from A to G. Putting $\lambda_A = 0.1$, $\lambda_B = 0.05$, $\lambda_C = 0.25$, $\lambda_D = 0.15$, $\lambda_E = 0.2$, $\lambda_F = 0.01$, $\lambda_G = 0.01$ in equation (12.64), we get Table 12.4 and their corresponding graph (Figure 12.6).

Table 12.4 Time versus sensitivity

t (years)	S_{λ_A}	S_{λ_B}	S_{λ_C}	S_{λ_D}	S_{λ_E}	S_{λ_F}	S_{λ_G}
0	0	0	0	0	0	0	0
1	0.0535	−0.5184	−0.5184	−0.5184	−0.5184	0.0507	-1.1123×10^{-08}
2	0.2136	−0.5391	−0.5391	−0.5391	−0.5391	0.2192	-2.7035×10^{-07}
3	0.3687	−0.4321	−0.4321	−0.4321	−0.4321	0.4163	-1.5935×10^{-06}
4	0.4589	−0.3242	−0.3242	−0.3242	−0.3242	0.5800	-5.3149×10^{-06}
5	0.4836	−0.2447	−0.2447	−0.2447	−0.2447	0.6964	-1.3065×10^{-05}
6	0.4630	−0.1920	−0.1920	−0.1920	−0.1920	0.7732	-2.6597×10^{-05}
7	0.4180	−0.1580	−0.1580	−0.1580	−0.1580	0.8224	-4.7685×10^{-05}
8	0.3631	−0.1355	−0.1355	−0.1355	−0.1355	0.8541	-7.8073×10^{-05}
9	0.3071	−0.1198	−0.1198	−0.1198	−0.1198	0.8752	-1.1944×10^{-04}
10	0.2546	−0.1082	−0.1082	−0.1082	−0.1082	0.8897	-1.7341×10^{-04}

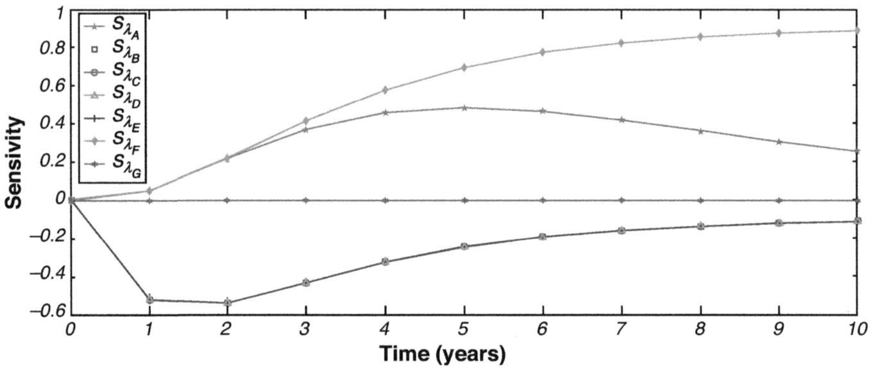

Figure 12.6 Time versus sensitivity.

12.4.4 Expected profit analysis

The mathematical formulation of the profit analysis is given as follows:

$$E_P = C_R \int_0^t P_{\text{up}}(t)dt - C_S t \tag{12.65}$$

where C_R and C_S are revenue and service cost, respectively.

Substituting $y = 1$, $x = 1$, $\beta = 1$, $\lambda_A = 0.1$, $\lambda_B = 0.05$, $\lambda_C = 0.25$, $\lambda_D = 0.15$, $\lambda_E = 0.2$, $\lambda_F = 0.01$, $\lambda_G = 0.01$ in equation (12.51), and applying ILT, the resulting $P_{\text{up}}(t)$ is placed into equation (12.65), which, upon further evaluation, yields:

$$E_P = (C_R 0.8005t + 0.0273e^{-1.1946t} - 0.0679e^{-0.0396t}$$
$$+4.1737 \times 10^{-05} e^{-0.01t} - 0.2768e^{-1.5940t} + 0.6905e^{-0.1380t}$$
$$+4.9276e^{-0.0119t} \cos(0.0044t) + 0.1063e^{-0.0119t} \sin(0.0044t)$$
$$+4.9276 \cos(0.0044t) e^{-0.0119t} + 0.1063 \sin(0.0044t) e^{-0.0119t} - 10.2284 - C_S t) \tag{12.66}$$

Keeping $C_R = 1$ and varying C_S from 0.1 to 0.5 in equation (12.66) yields the expected profit of the considered feeding system as shown in Table 12.5 and demonstrated in Figure 12.7.

Table 12.5 Time versus expected profit

	E_P at C_S				
t (years)	*0.1*	*0.2*	*0.3*	*0.4*	*0.5*
0	0	0	0	0	0
1	0.6998	0.5998	0.4998	0.3998	0.2998
2	1.2497	1.0497	0.8497	0.6497	0.4497
3	1.7789	1.4789	1.1789	0.8789	0.5789
4	2.3119	1.9119	1.5119	1.1119	0.7119
5	2.8524	2.3524	1.8524	1.3524	0.8524
6	3.4004	2.8004	2.2004	1.6004	1.0004
7	3.9551	3.2551	2.5551	1.8551	1.1551
8	4.5159	3.7159	2.9159	2.1159	1.3159
9	5.0821	4.1821	3.2821	2.3821	1.4821
10	5.6532	4.6532	3.6532	2.6532	1.6532

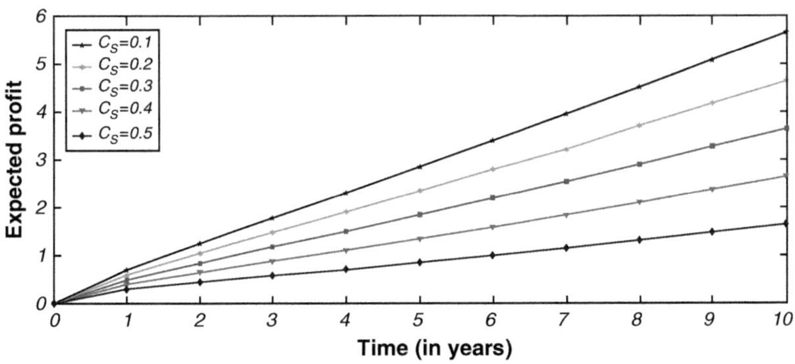

Figure 12.7 Time versus expected profit.

12.5 CONCLUSION

In this chapter, several reliability attributes of a sugar mill feeding system, including system reliability, sensitivity, MTTF, and expected profit, are examined using the G–H copula approach. Based on analytical and numerical evaluations, the following observations are obtained:

- From Figure 12.4 and Table 12.2, one can easily find out that system reliability declines rapidly as time goes on and eventually reaches a value of 0.04 after a considerable amount of time.
- Figure 12.5 and Table 12.3 show the variation of MTTF in context to different failure rates. From Figure 12.5 we concluded that MTTF decreases continuously with respect to the failure rate λ_B, λ_C, λ_D, λ_E, λ_F, λ_G, respectively, in the same manner for the considered values of failure rates but MTTF with respect to λ_A first increases from 4.0132 to 4.4667 then goes on decreasing up to 2.8041. It is also concluded that MTTF of the system with respect to failure rate of turbines remains approximately the same and it does not affect the system performance completely as the failure rate of turbines increases.
- Sensitivity of the feeding system with respect to different parameters is given in Table 12.4 and demonstrated in Figure 12.6. From Table 12.4, it is concluded that the sensitivity of the feeding system to the failure rates λ_A, λ_F, and λ_G continuously increases over time from 0.0535 to 0.2546, 0.0507 to 0.8897, and -1.1123×10^{-08} to -1.7341×10^{-04}, respectively. Surprisingly, the system is equally sensitive with respect to failure rate λ_B, λ_C, λ_D, λ_E, which ranges from −0.5184 to 0.1198 and is highly sensitive with respect to failure rate of subsystems A and F while less sensitive to the failure rate of subsystem G. So, if we reduce the failure rates of subsystems A and F, then system become less sensitive. Hence, it is concluded from the graph that redundancy in subsystem G significantly improves the system's reliability.
- From the critical examination of Figure 12.7, we conclude that the expected profit of the considered feeding system increases with time but decreases continuously as the service cost increases. From Table 12.5, it is also noted that the highest and lowest values of expected profit are 5.6532 and 0.2998, respectively, for assumed values of cost. Therefore, controlling service costs is essential for determining the maximum profit.

12.5.1 Managerial applications

i The results obtained in this study may help managers and engineers in the sugar industry to develop practical Standard Operating Procedures (SOPs) related to equipment handling, workforce allocation, and maintenance scheduling. Such procedures can be particularly useful during peak operational periods, especially the harvesting season, when system load is high.

ii By identifying the most critical subsystems and their failure characteristics, management is able to focus on preventive maintenance efforts on key components, such as unloaders and boilers, which significantly influence system reliability and downtime.

iii Including expected profit in the proposed model allows management to evaluate the trade-offs between operational efficiency and repair cost, thereby supporting more realistic and cost-effective decisions. This also aids in budgeting for maintenance and assessing the financial impact of equipment failures.

12.5.2 Potential sources of error or uncertainty in the model

i Component-level failure and repair rates may be estimated from historical or assumed data, not actual site-specific observations.

ii Factors such as temperature, humidity, dust, and cane quality affect the wear and efficiency of mechanical components (e.g., cutters, crushers), but these are not explicitly modeled.

iii Power supply fluctuations or human operator behavior may also impact system reliability but are not part of the model.

iv The division into fully working, partially failed, and completely failed states may be subjective. Minor changes in these definitions could alter reliability or availability calculations significantly.

REFERENCES

Aayog, N. I. T. I. (2020). *Final Report of the Task Force on Sugarcane and Sugar Industry*. NITI Aayog, Delhi.

Abubakar, M. I., & Singh, V. V. (2019). Performance assessment of industrial system (African Textile Manufacturers, LTD) through copula linguistic approach. *Operations Research and Decisions*, 29(4), 5–22.

Balagurusamy, E. (1984). *Reliability Engineering*. New Delhi, India. Tata McGraw-Hill Education, New Delhi.

Bisht, S., Kumar, A., Goyal, N., & Ram, M. (2020). Reliability characteristics of industrial system using copula in repair. In *2020 8th International Conference on Reliability, Infocom Technologies and Optimization (Trends and Future Directions) (ICRITO)* (pp. 1265–1267). IEEE, Noida.

Chantola, N., & Singh, S. B. (2019). The effect of transformer parameters on reliability assessment with the help of health index. *International Journal of Quality & Reliability Management*, 36(6), 895–916.

Chopra, G., & Ram, M. (2019). Reliability measures of two dissimilar units parallel system using Gumbel–Hougaard family copula. *International Journal of Mathematical, Engineering and Management Sciences*, 4(1), 116.

Cox, D. R. (1955, July). The analysis of non-Markovian stochastic processes by the inclusion of supplementary variables. In B. J. Green (ed.), *Mathematical Proceedings of the Cambridge Philosophical Society* (Vol. 51, No. 3, pp. 433–441). Cambridge University Press, Cambridge.

Dahiya, O., Kumar, A., & Saini, M. (2019a). An analysis of feeding system of sugar plant subject to coverage factor. *International Journal of Mechanical Production Engineering Research Development (IJMPERD)*, 9(1), 495–508.

Dahiya, O., Kumar, A., & Saini, M. (2019b). Mathematical modeling and performance evaluation of A-pan crystallization system in a sugar industry. *SN Applied Sciences*, 1(4), 339.

Ebeling, C. E. (2004). *An Introduction to Reliability and Maintainability Engineering*. McGraw-Hill, New York.

Goyal, N., Ram, M., Kumar, A., Bisht, S., & Klochkov, Y. (2021). Reliability measures and profit exploration of windmill water-pumping systems incorporating warranty and two types of repair. *Mathematics*, 9(8), 822.

Gumbel, E. J. (1960). Bivariate exponential distributions. *Journal of the American Statistical Association,* 55(292), 698–707. https://doi.org/10.1080/01621459.1960.10483368

Hougaard, P. (1986). A class of multivariate failure time distributions. *Biometrika,* 73(3), 671–678. https://doi.org/10.1093/biomet/73.3.671

Kadyan, M. S., & Kumar, R. (2017). Availability and profit analysis of a feeding system in sugar industry. *International Journal of System Assurance Engineering and Management*, 8(1), 301–316.

Kumar, A. (2019a). ATM modeling through probabilistic and reliability approach subject to different failures. *Journal of Engineering Science & Technology Review*, 12(3), 145–154.

Kumar, A. (2019b). Sensitivity and reliability evaluation for a thermal power plant subject to complex failures and human error. *Multidiscipline Modeling in Materials and Structures,* 15(5), 895–912.

Kumar, D., Singh, J., & Singh, I. P. (1988). Availability of the feeding system in the sugar industry. *Microelectronics Reliability*, 28(6), 867–871.

Kumar, R., & Kadyan, M. S. (2018). Performance analysis and maintenance planning of evaporation system in the sugar industry by using supplementary variable technique. *Journal of Industrial Integration and Management*, 3(01), 1850004.

Kumar, R., & Kadyan, M. S. (2019). Improving industrial systems reliability—An application in sugar industry. *Journal of Industrial Integration and Management*, 4(04), 1950011.

Kumari, P., Kadyan, M. S., & Kumar, J. (2019). Profit analysis of butter-oil (ghee) producing system of milk plant using supplementary variable technique. *International Journal of System Assurance Engineering and Management*, 10(6), 1627–1638.

Lado, A. K., & Singh, V. V. (2019). Cost assessment of complex repairable system consisting two subsystems in series configuration using Gumbel–Hougaard family copula. *International Journal of Quality & Reliability Management*, 36(10), 1683–1698.

Maihulla, A. S., Yusuf, I., & Salihu Isa, M. (2022). Reliability modeling and performance evaluation of solar photovoltaic system using Gumbel–Hougaard family copula. *International Journal of Quality & Reliability Management*, 39(8), 2041–2057.

Manglik, M., & Ram, M. (2013). Reliability analysis of a two unit cold standby system using Markov process. *Journal of Reliability and Statistical Studies*, 6, 65–80.

Manglik, M., Rawat, N., & Ram, M. (2020). Reliability and availability analysis of a cloud computing transition system under multiple failures. *International Journal of Quality & Reliability Management*, 37(6/7), 823–835.

Mehta, M., Singh, J., & Sharma, S. (2018). Availability analysis of an industrial system using supplementary variable technique. *Jordan Journal of Mechanical & Industrial Engineering*, 12(4), 245–251.

Nailwal, B., & Singh, S. B. (2011). Performance evaluation and reliability analysis of a complex system with three possibilities in repair with the application of copula. *International Journal of Reliability and Applications*, 12(1), 15–39.

Nautiyal, N., Singh, S. B., & Bisht, S. (2020). Analysis of reliability and its characteristics of a *k*-out-of-*n* network incorporating copula. *International Journal of Quality & Reliability Management*, 37(4), 517–537.

Nelsen, R. B. (2006). *An Introduction to Copulas*. Springer, New York.

Ram, M., & Singh, S. B. (2009). Analysis of reliability characteristics of a complex engineering system under copula. *Journal of Reliability and Statistical Studies*, 2(1), 91–102.

Saini, M., & Kumar, A. (2019). Performance analysis of evaporation system in sugar industry using RAMD analysis. *Journal of the Brazilian Society of Mechanical Sciences and Engineering*, 41(4), 175.

Shakuntla, S., Lal, A. K., Bhatia, S. S., & Singh, J. (2011). Reliability analysis of polytube industry using supplementary variable technique. *Applied Mathematics and Computation*, 218(8), 3981–3992.

Sharifi, M., Pourkarim Guilani, P., Zaretalab, A., & Abhari, A. (2021). Reliability evaluation of a system with active redundancy strategy and load-sharing time-dependent failure rate components using Markov process. *Communications in Statistics-Theory and Methods*, 52(13), 4514–4533.

Sharma, S. P., & Vishwakarma, Y. (2014a). Application of Markov process in performance analysis of feeding system of sugar industry. *Journal of Industrial Mathematics*, 2014(1), 593176.

Sharma, S. P., &Vishwakarma, Y. (2014b). Availability optimization of refining system of Sugar Industry by Markov process and Genetic Algorithm. In *2014 International Conference on Reliability Optimization and Information Technology (ICROIT)* (pp. 29–33). IEEE, Faridabad.

Sklar, A. (1959). Fonctions de répartition à *n* dimensions et leurs marges [Functions of *n*-dimensional distribution functions and their margins]. *Publications de l'Institut de Statistique de l'Universite de Paris*, 8(3), 229–231.

Chapter 13

AI-driven UAV solutions for sustainable and reliable infrastructure monitoring in Industry 5.0

Irem Bayraktar and Murat Bakirci

13.1 INTRODUCTION

Since the early 2010s, industrial systems have undergone significant technological and conceptual changes [1]. Automation, digitalization, and data-driven production have become core components of Industry 4.0. Cyber-physical systems, artificial intelligence (AI)-based decision-making processes, and smart sensors have become fundamental elements of industrial infrastructure [2,3]. As a result of this transformation, productivity and speed have increased significantly. Nevertheless, fundamental engineering and societal issues such as sustainability, system reliability, and human-centricity have only been partially addressed [4]. Industrial systems are no longer just smart. It is now clear that industrial systems also need to be resilient, human-friendly, and environmentally responsible [5]. This requirement has led to the concept of Industry 5.0, offering a new industrial vision that places people, the environment, and long-term system health ahead of technology. Unlike previous paradigms, Industry 5.0 views automation as a tool that collaborates with humans rather than an ultimate goal [6]. This strategy requires rethinking the interaction between humans and machines in the design and operation of industrial infrastructures, prioritizing system resilience, and incorporating environmental impacts into engineering decisions [7,8]. This new paradigm does not emerge merely as a vision. It arises as a practical necessity, especially when dealing with complex and interdependent structures such as energy grids, manufacturing facilities, transportation infrastructures, and critical industrial communication systems [9]. These infrastructures are systems where human safety is at risk due to maintenance and monitoring procedures, environmental costs become more pronounced, and even a minor malfunction can trigger chain reactions [10].

Fixed sensor networks, recurring field inspections, and frequent reactive interventions are fundamental elements of traditional infrastructure monitoring and maintenance techniques [11,12]. Especially for infrastructures spread over large areas or are difficult to access, physically sending human teams to the field leads to high costs, time loss, and safety hazards [13]. Fixed

 DOI: 10.1201/9781003787525-13

sensor solutions often require maintenance. They can only collect data from pre-defined locations. They struggle to adapt to changing environmental conditions [14,15]. This situation can hinder early fault detection, make it difficult to monitor the overall and real-time condition of infrastructure systems, and lead to operational burdens that conflict with sustainability goals. In the field of industrial infrastructure monitoring, unmanned aerial vehicles (UAVs) hold an important position [16–18]. Drones surpass the limitations of traditional approaches due to their high maneuverability, ability to rapidly scan large areas, and capacity to be equipped with a wide range of sensors. High-resolution data obtained from optical, thermal, or multispectral sensors allows the physical condition of infrastructure components to be evaluated remotely and without contact [19]. The use of AI-based analysis techniques to process this data enables the early detection of both potential future problems and existing damage [20,21].

The aerial data collection capability of UAV-based surveillance systems is only one aspect of their true power. Another is their capacity to transform this data into useful and actionable information [22]. In complex industrial contexts, advanced object detection and anomaly detection algorithms can reliably capture information that the human eye might miss [23]. UAVs highlight their strength and their speed under challenging conditions. AI-supported analytical techniques, especially considering the inherent challenges of aerial imaging such as changing lighting conditions, varying scales, and complex backgrounds, are transforming UAV systems from passive observation tools into active decision-making components [24,25]. This shift enables maintenance and operational procedures to move from reactive to proactive and even predictive. From an Industry 5.0 perspective, the combination of UAVs and AI is not merely a technical advancement. It is also a tactical tool that supports human-centered and sustainable objectives [26,27]. Deploying UAVs to workplaces reduces the number of human teams sent to the field and the risks they encounter. It also reduces carbon emissions resulting from operations and transportation, thereby improving workplace safety [28]. Early fault detection extends the lifespan of infrastructure. It improves resource utilization by preventing extensive damage and unplanned shutdowns. In this context, UAV-based monitoring systems have two advantages: one is reducing their negative impact on the environment, and the other is enhancing system reliability [29,30].

This study has been conducted at a timely juncture because we are at a critical point in transitioning to UAV systems. Expectations for durability and sustainability are increasing alongside the complexity of industrial systems. In the long term, faster or cheaper alternatives are not sufficient. There is a need for reliable, adaptable, and human-friendly systems [31]. In the context of Industry 5.0, there is a lack of a comprehensive approach toward these solutions. UAV-based, AI-supported monitoring systems have reached a level of maturity that can directly address these objectives [32].

This study aims to fill this gap by providing a methodical perspective on how UAV technology can be strategically integrated into reliable and sustainable industrial infrastructures. In addition to technical performance, the study examines UAV-based infrastructure monitoring systems from the perspectives of sustainability, reliability, and human-centered operational principles [33]. It explains how AI-powered object detection and analysis procedures support proactive maintenance plans, reduce operational and environmental costs, and improve industrial network continuity. Additionally, it aims to demonstrate the new engineering perspective required by Industry 5.0 through a practical technological application. Besides showcasing the capabilities of UAV and AI technologies, this study also attempts to explain why these technologies are necessary today and specifically for these systems in this context [34]. Reliability, sustainability, and human connection are as important as automation and speed for the future of industrial infrastructures. At the intersection of these complex objectives, UAV-based smart monitoring solutions provide a powerful tool for the real-world implementation of the Industry 5.0 vision [15].

13.2 INDUSTRY 5.0 AND INFRASTRUCTURE MONITORING PARADIGM

Industry 5.0 is no longer just a progression of industrial systems with the addition of new technologies. It has become a radical paradigm shift in how infrastructures are perceived, used, and maintained. This approach does not treat infrastructures merely as functional components. Instead, it views them as living systems constantly interacting with humans, the environment, and technology. The monitoring and management of critical industrial infrastructures, due to their increasing complexities and interconnections, has transformed from a simple automation or data collection problem into a multifaceted technical and management challenge. In Industry 5.0 infrastructure systems, there has been a shift from performance-focused designs to designs centered on resilience and adaptability. This is one of the most significant impacts of Industry 5.0 on infrastructure systems. The expectations for modern infrastructures have greatly increased. While this increase continues, previous industrial paradigms have mostly focused on maximum efficiency, minimum cost, and rapid production speed. The performance of an industrial system under nominal conditions is no longer the sole factor used to determine its success. Other factors include how effectively it handles unforeseen situations, how quickly it can detect errors, and how much it reduces its environmental impact. As a result of this situation, infrastructure monitoring procedures are becoming a fundamental part of system design.

In the context of Industry 5.0, reliability, continuity, and sustainability are not general objective criteria. They are specific engineering criteria. Reliable infrastructure systems operate flawlessly, and deliver continuous and predictable performance over time. Continuity minimizes service interruptions. The reason for this is to prevent systems from going offline due to regular maintenance, repairs, or environmental impacts. Sustainability, on the other hand, encompasses a broader range. Sustainability covers usage from the materials used and operational procedures to energy consumption and maintenance strategies. In the design of modern infrastructure monitoring solutions, these three concepts are not seen as separate. They are regarded as interconnected objectives. Monitoring systems are now much more than passive structures that simply report the current status. According to Industry 5.0, monitoring is an active process aimed at continuously understanding system behavior, detecting changes early, and providing informative feedback to human operators. Monitoring bridges the gap between data collection and decision-making. In this way, it aims to enhance the knowledge, adaptability, and predictability of strategic infrastructure management. Such a monitoring strategy provides comprehensive awareness, which is particularly difficult to achieve with traditional approaches, especially in large-scale and geographically distributed systems.

Industry 5.0 places human-AI-robot collaboration at the center of system design. This is another distinguishing feature of Industry 5.0. Here, AI and autonomous systems do not replace humans. They are positioned as tools that enhance and support human decision-making processes. This approach, when it comes to infrastructure monitoring, requires the use of automated data collection and analysis procedures alongside the knowledge and intuition of human experts. Large amounts of data make it extremely difficult for human effort and time to identify patterns and anomalies. AI algorithms can be used for this task. Human operators should interpret the results using contextual information before making judgments about the data. The system moves away from being entirely dependent on humans or completely autonomous. It transforms into a hybrid model that combines the advantages of both. In facilities where physical access is dangerous or difficult, relying solely on human labor is a significant risk. In such a situation, the importance of robotic and autonomous platforms is particularly evident. The ability of autonomous systems to perform hazardous monitoring tasks is crucial for operational efficiency and from an ethical and human perspective in the Industry 5.0 approach, which prioritizes human safety.

The need for monitoring in Industry 5.0 is particularly apparent in critical systems such as energy generation and transmission networks, transportation infrastructure, large-scale manufacturing facilities, and industrial communication networks. These infrastructures exhibit complex behaviors due to their interrelationships, are often spread across vast geographical areas, and consist of various sub-components. A performance degradation or physical damage to a small component can trigger a series of events affecting the overall functionality of the system. Therefore, instead of focusing solely on specific components, monitoring solutions should reveal the entire state and trends of the system. Fault detection is not the only reason for monitoring such infrastructures. Monitoring data also forms the basis for long-term maintenance planning, resource efficiency, and environmental impact reduction. The Industry 5.0 methodology encourages the use of this data as technical indicators, and as inputs supporting strategic decision-making processes. Consequently, monitoring systems transform from a passive component of infrastructure management to active components that drive system development. This conceptual framework clearly demonstrates that infrastructure monitoring solutions are essential for the Industry 5.0 goal: only human-AI-robot collaborative models addressing operational and environmental aspects together can achieve sustainability, continuity, and reliability. This background further indicates that UAV-based and AI-powered monitoring techniques discussed in later chapters are an inherent extension and technological advancement corresponding to the new infrastructure concept required by Industry 5.0.

13.3 UAV-BASED MONITORING SYSTEMS: ARCHITECTURE AND CAPABILITIES

In the context of Industry 5.0, UAV-based monitoring systems are much more than a single aircraft or a specific detection algorithm. These systems should be seen as a system-of-systems structure, where elements from various disciplines come together with loose but significant connections, functioning individually but generating new capabilities when considered as a whole. This strategy ensures that UAVs become active system components that support the continuity and reliability of industrial infrastructures, rather than merely serving as data collection platforms. From an Industry 5.0 perspective, the multi-layered architecture of UAV-based infrastructure monitoring systems is shown in Figure 13.1. The lowest layer of this design, the physical infrastructure, includes assets that are constantly changing and directly affected by environmental conditions, such as electrical transmission lines, industrial facilities, transportation networks, and communication systems. These infrastructures are dynamic elements that continuously activate the rest of the system rather than being passive components in the monitoring process. The sensing and decision-making layers receive new information whenever there is a change in the physical world.

Atop this infrastructure layer, the system's mobile and adaptive sensing elements are represented by UAV platforms. The UAVs are equipped with various sensor sets and can simultaneously acquire data in multiple modalities, such as RGB, thermal, or multi-band imaging, while being able to adapt to the environment during flight. These platforms rely on edge hardware to generate raw data, and to perform situational awareness analysis and first-level preprocessing. This feature enhances the overall system's resilience by reducing reliance on centralized systems, especially in

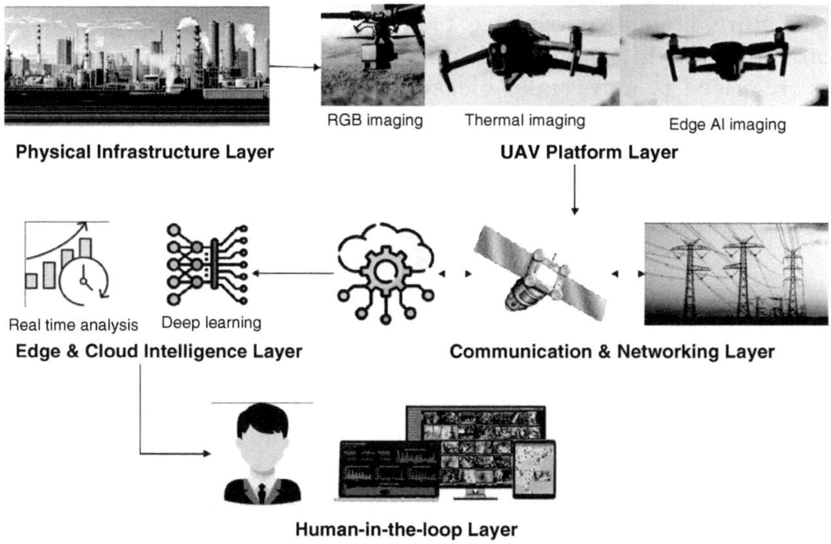

Figure 13.1 System-of-systems architecture of UAV-based infrastructure monitoring in the context of Industry 5.0. Multi-layer, loosely coupled architecture where UAV platforms, communication networks, edge–cloud intelligence, and human operators act as distinct yet collaborating subsystems is presented to support distributed decision-making, adaptability, and explicit integration of human input, thereby clearly demonstrating the system-of-systems nature of monitoring with UAVs.

latency-sensitive infrastructure monitoring scenarios. The communication and network architecture create a sensing layer composed of UAVs that connect with edge and cloud-based intelligence components. This communication layer does not use a static and continuous structure but rather adopts an adaptive and context-aware network logic, as depicted in Figure 13.1. Data transmission can occur more slowly and efficiently in terms of energy consumption during usual monitoring tasks, while in critical situations low-latency data transmission is prioritized. This approach allows for the monitoring of infrastructures distributed over large geographical areas and facilitates scalability.

The edge and cloud intelligence layers take part in the decision-making process of the UAV-based monitoring system. In scenarios with real-time constraints, AI models at the edge deliver fast inferences, while comprehensive analyses, model updates, and long-term learning take place in the cloud environment. This distributed intelligence architecture decouples the decision-making process from a single point of failure and allows for dynamic adaptation to diverse operational conditions. Consequently, UAV-based monitoring evolves from a static data processing pipeline into a learning and adaptive system. At the top layer of the UAV-based monitoring architecture, human-centricity as one of the pillars of Industry 5.0 is very clear. In Figure 13.1, the human-loop layer indicates that human operators are placed as active stakeholders informed by AI-powered outputs and able to intervene if necessary rather than an autonomous structure where human operators are completely excluded from the system. This enhances operational reliability and allows for designing systems with accountability and ethics considerations.

13.4 AI-DRIVEN OBJECT DETECTION AND ANALYTICS FOR INFRASTRUCTURE INSPECTION

Transforming high-resolution data into accurate and reliable information under uncertainty conditions represents an equally essential dimension in the sustainable and reliable monitoring of industrial infrastructure. The primary role of AI in UAV-based monitoring systems is to provide insightful details on infrastructure conditions from complex aerial images that are frequently noisy. An object and anomaly detection task should not be considered as a visual classification problem but rather as an analysis process with multidimensional technical criteria such as environmental variability, system reliability, and decision tolerance. The complete AI-based detection and analysis pipeline for UAV-based infrastructure assessment is illustrated in Figure 13.2. It is this pipeline that comprehensively takes care of the process from multimodal data taken from the aircraft to its transformation into outputs for decision-making. The process starts with

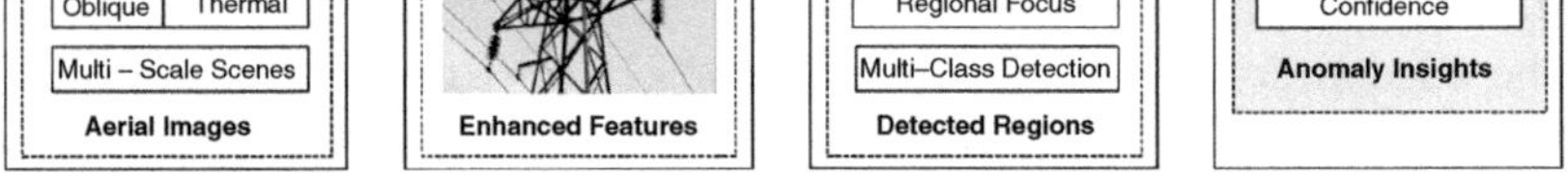

Figure 13.2 AI-based pipeline for top-down object detection and any anomaly reasoning regarding infrastructure inspection. The image shows the complete process from data collection through various aerial modalities to outputs ready for decision-making highlighting deep learning detection and reasoning about anomalies in a pipeline that highly emphasizes the production of results actionable and confidence-aware as well as robust against unpredictable environments.

the data acquisition phase carried out by sensors mounted on the UAV; different types of data including RGB and thermal images are captured at different scales and oblique viewing angles. This variety constitutes one of the key elements that complicate the investigation process while permitting more thorough physical condition documentation about infrastructure assets. Consequently, preprocessing and normalization steps become important rather than directly feeding raw aerial photographs into deep learning models. Through correcting scale differences, denoising, and lighting adjustment, the model isolates infrastructure components from any environmental influence. This stage depicted in Figure 13.2 acts as a preliminary preparation layer improving generalizability together with stability within AI-based analyses.

The activities in the object detection phase based on deep learning include the positioning, classification, and identification of areas of interest of infrastructure components. This phase is not limited to the mere confirmation of object existence. An object's actual state and its expected state as described may differ; interpreting such a difference becomes crucial within the context of infrastructure inspection. Hence, besides confirming an object's presence, the anomaly inference layer depicted in Figure 13.2 also evaluates the condition and risk levels for any newly discovered structures. When these outputs are generated with confidence levels and priority information, it further enhances AI's decision support role while increasing its interaction with human operators. In this context, one of the greatest challenges to aerial imaging-based infrastructure inspection relates to limitations inherent in traditional computer vision techniques. The importance of deep learning to solving problems in aerial vision and the insufficiency of traditional methods is illustrated by Figure 13.3. The reliability falls sharply for fixed feature-based approaches under dramatically changed lighting conditions, motion blur due to industrial activity backgrounds varying from one moment to another, and drastically different scales at height and viewing angles where the same object appears. This unpredictability is not addressed by deep learning-based methods as a direct problem but rather as a learnable feature domain. As shown in the figure, deep networks can recognize infrastructure components context-independently by jointly learning visual differences that happen under diverse environmental conditions. In this way, AI can work reliably under unpredictable conditions with much higher accuracy. Such generalizability becomes very important when fault tolerance is less, like in industrial infrastructure monitoring situations.

The performance of AI is significantly influenced by the latency and reliability of the predictions in addition to model performance. The real-world performance of UAV-based monitoring systems is currently determined by the trade-off between edge and cloud-based inference techniques. The pros and cons of edge and cloud-based AI inference are presented in Figure 13.4, which also explains why a hybrid approach is more suitable for Industry 5.0.

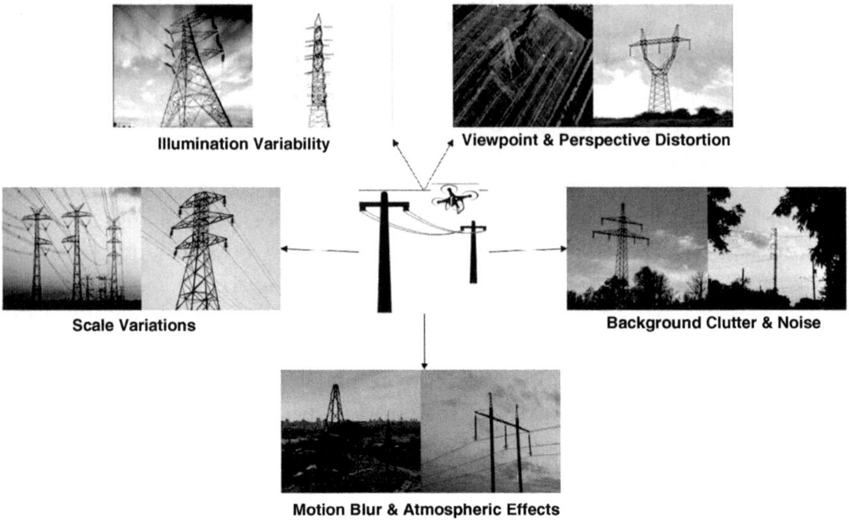

Figure 13.3 Main challenges in aerial vision and deep learning involvement in infrastructure inspection. This figure presents the major sources of visual uncertainty when using UAVs for inspection: background clutter; illumination variation; scale variation; and perspective distortion. It compares traditional vision methods' shortcomings with deep learning models' strengths to adapt to challenging aerial imaging conditions.

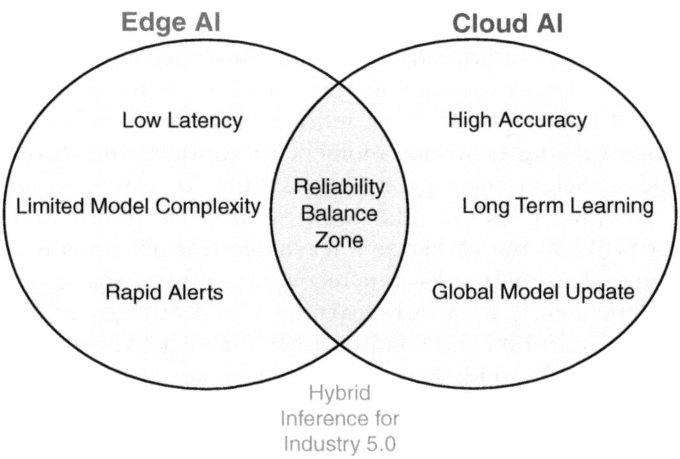

Figure 13.4 Edge–cloud AI inference trade-off for reliable UAV-based infrastructure monitoring. The figure illustrates a hybrid reliability balance zone latency-accuracy-learning capability comparison between edge inference and cloud inference paradigms under different scenarios. This hybrid approach ensures fault tolerance, business continuity, and scalable intelligence in Industry 5.0-oriented inspection systems.

AI models that run at the edge are essential for low-latency applications where immediate critical alerts are needed, but these models typically have a low level of complexity due to technology constraints. Cloud-based models, while facing latency and connectivity issues, offer high accuracy and long-term learning with deeper and wider network structures, but these two strategies do not rule each other out; rather, their synergy leads to a more reliable and flexible monitoring system as illustrated in the reliability equilibrium region in Figure 13.4. This hybrid inference strategy increases the system's fault tolerance and supports the continuity of infrastructure monitoring activities. Cloud-based deeper studies, model updates, and long-term trend inferences increase the reliability of the system over time, while edge-based fast and rough inferences provide emergency awareness. Thus, AI becomes an element that makes instantaneous decisions, and an element that continuously learns and improves infrastructure health over time.

13.5 SUSTAINABILITY AND RELIABILITY IMPACTS OF UAV-BASED INSPECTION

In the perspective of Industry 5.0, one of the most interesting aspects that can be found in UAV-based infrastructure inspection systems is their ability to achieve both technical performance enhancements as well as sustainability and reliability objectives. The UAV-assisted methods provide an opportunity for dealing with infrastructure management within a broader context, whereas conventional inspection methods are limited to short-term maintenance needs and do not take into consideration high operating costs, human safety hazards, and environmental impacts. This section reviews the effects of UAV-based inspection on sustainability and system reliability based on quantitative comparisons as well as scenario-based inferences. The first major effect from a sustainability point of view is the drastic decrease in carbon footprint and operating costs. It is common practice for traditional inspections of infrastructures to necessitate ground vehicle mobilization together with heavy machinery and several personnel to the site; apart from raising fuel consumption, these processes can render the asset fully or partially unoperational during inspection time. On the contrary, drone-based inspections can cover similar grounds much quicker using less energy without hindering any ongoing activities related to assets' operation. A comparison between traditional means of inspection vis-à-vis UAV-based inspection in terms of sustainability as well as operational cost has been encapsulated in Table 13.1. From Table 13.1, it can be observed that drone-based inspections drastically cut down carbon emissions plus total operational costs more so when undertaken at remote locations plus facilities having challenging entry points. Emissions per inspection are lowered by using electric or low-fuel drones; rapid data collection also reduces personnel requirement plus inspection time significantly.

Table 13.1 Comparative sustainability and operational impact of traditional and UAV-based infrastructure inspection

Metric	*Traditional inspection*	*UAV-based inspection*	*Observed impact*
Carbon emissions per inspection	High (vehicle- and crew-intensive)	Low (electric or low-fuel UAVs)	Significant reduction
Average inspection duration	Long (hours to days)	Short (minutes to hours)	Faster completion
Workforce requirement	Multiple field personnel	Minimal field presence	Reduced labor demand
Operational cost	High	Moderate to low	Cost-efficiency

Table 13.2 Impact of inspection approach on reliability, safety, and infrastructure lifespan

Aspect	*Traditional inspection*	*UAV-based inspection*
Fault detection timing	Reactive	Proactive
Inspection frequency	Periodic, infrequent	High frequency, flexible
Human safety risk	High in hazardous sites	Significantly reduced
Infrastructure lifespan	Limited by late interventions	Extended through early detection

UAV-based inspections have the potential to shift maintenance strategies from a reactive approach to a proactive model, thereby having a major effect on sustainability. In conventional inspections, infrastructure elements are generally examined at fixed intervals and failures are usually detected after considerable physical damage has occurred. This approach may allow small defects to develop into large failures over time, leading to increased costs for repairs. UAV and AI-enabled inspection systems, on the other hand, identify early-stage deteriorations and anomalies in infrastructure components so that immediate and targeted maintenance interventions can be undertaken. Such an increase in the failure rate is reduced and increases the overall reliability of the infrastructure. With continuous high-frequency monitoring, time-dependent performance trends of the infrastructure become clearer and maintenance schedules can be adjusted accordingly. The effects of traditional inspection methods versus UAV-based inspection techniques on reliability and long-term asset life are summarized in Table 13.2. From an evaluation of human safety and workload, UAV-based inspections fit very well with Industry 5.0's human-centered perspective. Manual inspections in dangerous environments, such as those involving high-voltage power lines, tall structures, or hazardous industrial areas, have a high risk of accidents for workers. UAVs assume these tasks so that human operators are not directly exposed to physical threats and can make the inspection process safer; furthermore,

AI analytics reduce operator workload enabling human intervention only when truly necessary and critical.

The early identification of minor damages, as one of the major benefits brought by UAV-based inspection systems, prevents systematic deterioration from taking place. Regular and thorough inspections allow infrastructure elements to be safely utilized for durations that more closely approach their designed service life. This enhances the financial and ecological returns on large infrastructure investments, and reduces maintenance and replacement costs. UAV-based infrastructure inspection does not see sustainability and reliability goals as separate benefits but rather as two sides of the same coin; it views them as mutually reinforcing elements that create a monitoring paradigm which, when integrated with proactive maintenance strategies, reduces carbon footprint and operating costs, improves human safety, prolongs infrastructure lifespan, and fortifies system continuity. In this context, UAV-based inspection is a practical engineering solution in line with the environmentally friendly, resilient, and human-centric principles of Industry 5.0.

13.6 INDUSTRIAL USE CASES AND DEPLOYMENT SCENARIOS

One aspect where there is direct implementation of the Industry 5.0 vision is UAV-based infrastructure inspection because this methodology offers an adaptable common monitoring core for various industrial domains rather than being limited to one particular sector. It can be seen in Figure 13.5 that other industries such as manufacturing, energy, transportation, and communications can contextually reconfigure the same UAV plus AI-based inspection infrastructure. The fact that these UAV-based monitoring systems can be applied across multiple domains clearly proves their scalability and system-wide nature. Energy transmission lines represent one industry where UAV-based inspection has reached advanced stages and offers substantial added value; manually inspecting transmission lines over vast geographic distances poses significant limitations both in terms of time and human safety. As indicated in Figure 13.5a, structural defects can be detected from a distance by UAVs equipped with thermal sensors and RGB cameras as well as abnormal heating at connection points and initial stages of insulator damage. This approach allows for maintenance before failures become critical, thereby reducing power outages. From the point of view of engineering, the biggest benefit is that we can keep track of very big, long structures all the time without any human help. Inspections using UAVs help industrial places because they lower the chances of accidents and unplanned stops in work. In places like chemical factories, oil refineries, or huge production sites, pipes, tanks, and flue systems are always at danger from rust, leaking fluids, and weak structures. In the case shown in

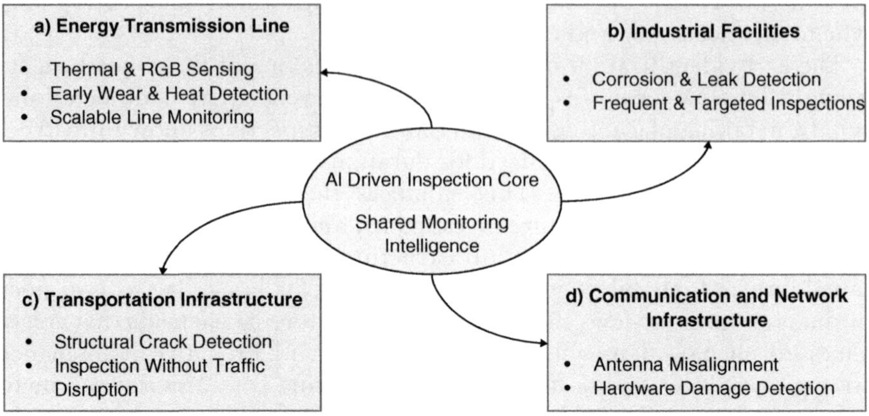

Figure 13.5 Inspection of industrial infrastructures using UAVs in different deployment domains. The figure shows the use of a common UAV and an AI inspection core in diverse industrial sectors, including communication networks, energy transmission, industrial facilities, and transportation infrastructure. Each domain employs a common monitoring intelligence to discover inspection objectives specific to that sector.

Figure 13.5b, the ability of UAVs to do frequent and focused checks allows for early finding of small problems; this cuts down maintenance costs and offers a key advantage in keeping operations going smoothly in places where stopping work costs much money. Transportation infrastructure is another area where UAV-based monitoring systems provide maximum benefits in terms of service continuity. Traditional inspections of railway systems, viaducts, and bridges often result in operational or traffic disruptions. UAVs, as shown in Figure 13.5c, can detect surface deteriorations and structural failures in such infrastructure without disrupting traffic flow. This method minimizes the societal impact on public infrastructure by providing a monitoring paradigm that simultaneously considers the availability and safety of transportation systems. Another area where physical degradation has an indirect but critical impact on service quality is communication and network infrastructures. Minor misalignment errors or hardware damage in base stations, antenna masts, and connection equipment can lead to widespread service disruptions. The situation shown in Figure 13.5d demonstrates how UAV-based monitoring can quickly and reliably inspect such components. Increased visibility into the physical layer of network infrastructures and the ability to identify problems before they affect users are the most significant technical contributions.

The implementation sequence depicted in Figure 13.6 further illustrates the practical realization of these multi-sector use cases. It represents UAV-based inspection not merely as a technological concept, but rather as an end-to-end mission-planned and operationally engineered process. This process begins with the mission planning stage, wherein flight parameters and inspection

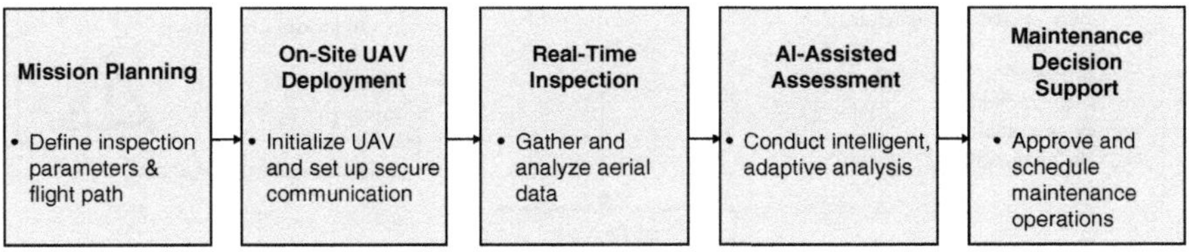

Figure 13.6 Workflow for the deployment of UAV-based inspection in industrial settings. It shows the entire operational workflow of UAV-based inspection, including real-time analysis and maintenance decision support, mission planning, and on-site deployment. The workflow highlights seamless integration into industrial maintenance procedures, minimal human exposure, and AI-assisted assessment.

goals are established. It proceeds through the deployment of the UAV asset to the field, followed by the establishment of secure communications. At this point, physical hazards are largely mitigated since human operators are confined to system initialization and monitoring roles. During the real-time inspection phase, UAVs can prioritize emergencies and provide instant awareness with end-based AI. This architecture, as illustrated in Figure 13.6, shows that field data is not merely recorded but also processed and interpreted in near real time. The subsequent evaluation phase, augmented by AI, ensures that the data collected undergoes further detailed analysis and produces outputs that support maintenance decisions. Based on these outputs, human operators plan and approve maintenance and intervention options in the final phase. Overall, as illustrated in Figures 13.5 and 13.6, UAV-based infrastructure monitoring provides an inherently scalable and flexible solution both vertically (along the on-site operational workflow) and horizontally (across different industrial sectors). This architecture shifts UAV-based monitoring from being a technology for a given use case into becoming a key enabling technology for flexible, resilient, and human-centric infrastructure management demanded by Industry 5.0.

13.7 CHALLENGES, LIMITATIONS, AND ETHICAL CONSIDERATIONS

While UAV-based infrastructure inspection has substantial technical and operational benefits within the context of the Industry 5.0 vision, its real-world implementation must be assessed against the backdrop of specific limitations, hazards, and ethical considerations. The objective here is not to assert that the proposed system is without fault; instead, it seeks to establish a credible and accountable engineering solution by unambiguously articulating the conditions under which it is robust and those under which it is not. This viewpoint aligns with both the technical maturity of the system and the level of managerial cognizance necessary for industrial

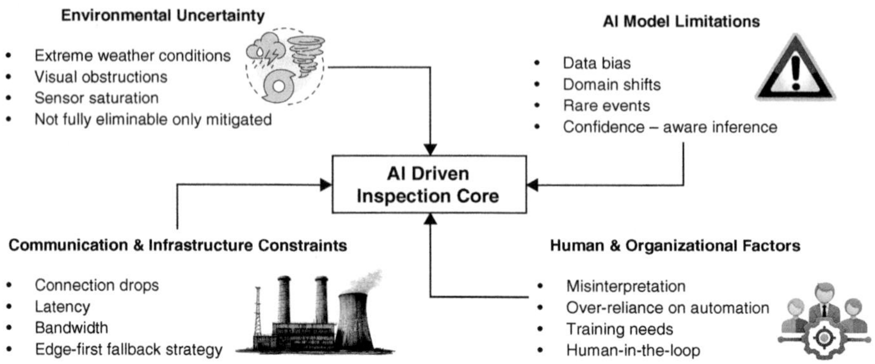

Figure 13.7 Risk-aware framework for UAV-based inspection toward Industry 5.0-oriented infrastructures. The figure presents a critical perspective on UAV-based inspection by environmental uncertainties, AI model limitations, communication constraints, and human factors around an operational intelligence core centered more on the promotion of risk mitigation and mandatory human-in-the-loop decision-making rather than hazard elimination.

deployment. Figure 13.7 illustrates the risk-aware conceptual framework of UAV-based inspection systems. The main decision-making architecture where UAV platforms and AI components work in synergy is represented by the operational intelligence core at the center of the image, while surrounding risk areas clearly identify uncertainties in which this core functions. This method places high importance on treating risks as not being outside exceptions but as an inherent part of design. One unavoidable disadvantage associated with using UAVs for inspection relates to environmental uncertainty. Data quality may be directly influenced by severe weather conditions, visibility-restricting obstacles, and sensor saturation. This risk area is particularly highlighted in Figure 13.7 with a warning that it is a risk that cannot be completely eliminated but can be minimized. This statement indicates that the system does not claim absolute resilience to environmental conditions but rather aims to manage risks through task planning, multimodal sensing, and operational flexibility.

The limitations of the AI model are also explicitly described. Data bias, field shift, and poor representation of rare events are known issues in deep learning-based systems. These limitations are acknowledged with human oversight triggered at low confidence levels, as seen in the confidence-sensitive inference approach shown in Figure 13.7. Therefore, AI is placed as a decision support tool that quantitatively represents risks rather than an ultimate decision-maker. Operational reliability is also affected by infrastructure and communication limitations particularly in large-scale industrial systems. Centralized cloud-based solutions have critical weaknesses because of bandwidth constraint delays and connection interruptions; therefore, the goal of the endpoint-prioritized fallback technique described in Figure 13.7 is to keep the system from becoming entirely inoperable under these conditions. Basic awareness and alert generation can be maintained operationally without dependence on connectivity due to local inference capability. The organizational and psychological aspects that go beyond just technical limitations play a vital role for safe UAV-based inspection system deployment since over-reliance on automation, misreading of outputs, or poor training can have field consequences even if the system works perfectly from a technical standpoint. The approach proposed here clearly states that critical decisions should not be automated without some form of human intervention and as seen in Figure 13.7 treats the principle of human intervention as one design element that must be included.

A systematic summary of these risks and limitations is presented in Table 13.3. The table lists the main hazards, causes, and impacts associated with UAV-based inspection. It also shows how the proposed framework attempts to mitigate these risks. Specifically, the tactics in the "mitigation" column demonstrate a realistic engineering approach aimed at reducing hazards to manageable levels rather than eliminating them entirely. This approach bases the system's reliability on transparency and predictability, rather than making exaggerated claims. The human-centered approach of Industry 5.0 places ethical and social considerations at the forefront of this technological discourse. UAV-based inspection systems do not intend to supplant human labor entirely but rather to allocate hazardous and repetitive tasks to autonomous systems, thereby allowing human experts to

Table 13.3 Major limitations, threats, and mitigation strategies in infrastructure inspection using UAVs

Limitation/risk	*Source*	*Potential impact on inspection*	*Mitigation strategy in proposed framework*
Illumination variability (day/night, shadows, glare)	Environmental conditions	Reduced detection accuracy, false negatives in visual inspection	Multimodal sensing (RGB + thermal), illumination normalization, confidence-aware inference
Weather effects (rain, fog, wind)	Environmental uncertainty	Flight instability, degraded image quality, incomplete data acquisition	Mission planning with weather constraints, adaptive flight control, fallback inspection scheduling
Scale variation due to altitude changes	Aerial imaging geometry	Missed small defects or inconsistent object representation	Multi-scale deep learning models, adaptive feature pyramids
Viewpoint and perspective distortion	UAV motion and oblique views	Geometric misinterpretation of infrastructure components	Data augmentation with diverse viewpoints, perspective-aware training

(*Continued*)

Table 13.3 (Continued) Major limitations, threats, and mitigation strategies in infrastructure inspection using UAVs

Limitation/risk	*Source*	*Potential impact on inspection*	*Mitigation strategy in proposed framework*
Background clutter in industrial scenes	Complex environments	Increased false positives, reduced precision	Region-of-interest focus, contextual feature learning
Rare failure modes	Data scarcity	Undetected critical anomalies	Confidence scoring, human review for low-confidence predictions
Dataset bias	Limited training diversity	Reduced generalization to unseen sites	Continuous dataset expansion, cross-domain validation
Model drift over time	Aging infrastructure and changing environments	Gradual performance degradation	Periodic retraining, cloud-based model updates
Communication loss	Network instability	Delayed alerts, incomplete data transfer	Edge-first inference, local decision buffering
Latency in cloud processing	Infrastructure constraints	Slow response to critical events	Hybrid edge–cloud inference architecture
Over-reliance on automation	Human factors	Unsafe decisions without human verification	Enforced human-in-the-loop approval mechanisms
Insufficient operator training	Organizational factors	Misinterpretation of AI outputs	Explainable AI interfaces, operator training protocols

This table captures the major technical, environmental, and organizational hazards associated with inspection via UAVs and describes the mitigation techniques adopted in the proposed framework. These prioritize transparency, risk awareness, and practical engineering trade-offs rather than total risk avoidance.

devote their efforts toward higher-level decision-making processes. This view seeks to lessen the negative consequences that automation could have on employment and enhance collaboration between humans and machines. Furthermore, compliance with privacy and accountability regulations in visual data collection processes is essential for the acceptability of such systems within industrial and societal settings.

13.8 CONCLUSION

This study presents UAV-based infrastructure monitoring as an intrinsic aspect of Industry 5.0, which envisions sustainable, reliable, and human-centric industrial systems rather than merely a technological advancement.

The architectural framework introduced here conceptualizes UAV platforms, AI-enabled sensing, endpoint-cloud intelligence deployment, and human-machine collaboration not as one-size-fits-all solutions but rather as a system-of-systems approach comprising distinct yet collectively significant subsystems. This viewpoint shifts the emphasis from enhancing sensing capabilities toward solving the broader problem of infrastructure monitoring—specifically concerning long-term system health, operational continuity, and environmental sustainability. The AI-based object detection and anomaly analysis articulated herein substantiate why UAV-based surveillance should be considered a pragmatic and dependable solution even amid real-world uncertainties. Aerial vision challenges such as dimension shift, field-of-view distortion, and environmental noise are addressed by deep learning-based technologies, and handled with trust awareness and human supervision. The balanced mix of endpoint inference with cloud support allows real-time reaction capability while being able to evolve over time. In conclusion, AI is evolving from an element that automates judgments to a reliable decision support system that collaborates with human experts.

Sustainability and reliability advances indicate that UAV-based monitoring is not just an operational improvement but also a strategic transformation in industrial infrastructure. Carbon footprint and operating cost reductions, coupled with proactive maintenance measures, extend infrastructure lifespan, reduce unplanned downtime, and enhance human safety. Use cases from varied sectors like energy, manufacturing, transportation, and communications prove this approach not as limited to a specific field but rather how common monitoring intelligence can be contextually reconfigured. One of the biggest contributions of this work is that it does not present UAV-based monitoring as an idealized, perfect solution but rather proposes a mature engineering framework recognizing its limitations by hazard identification and risk mitigation. Environmental uncertainties, AI model limitations, communication constraints, and human factors are not viewed as problems to be avoided in system design but rather consciously accepted inputs. This approach underlines the fact that reliability can be attained through transparency, predictability, and human-centered decision-making processes plus accuracy measurements.

REFERENCES

1. Yin, R., Rahman, M. N. A., Hishamuddin, H., Ikram, I., & , Sabtu, M. I. (2026). Towards Industry 5.0: Integrating Technical Optimization of Automated Guided Vehicles with Human-Centricity in Sustainable Production Systems. *Journal of Mechanical Science and Technology*, 40, 619–632. https://doi.org/10.1007/s12206-025-1252-8

2. Lazarević, M., & Obrecht, M. (2026). Artificial Intelligence and Sustainability in Industry 4.0 and 5.0: Trends, Networks of Leading Countries and Evolution of the Research Focus. *Sustainability*, 18(2), 877. https://doi.org/10.3390/su18020877
3. Terziyska, M., Ilieva, I., Terziyski, Z., & Komitov, N. (2026). A Review of the Transition from Industry 4.0 to Industry 5.0: Unlocking the Potential of TinyML in Industrial IoT Systems. Science, 8(1), 10. https://doi.org/10.3390/sci8010010
4. Hammad, M., Ali, H., Hussain, G., & Wang, X. (2026). Sustainability and Circular Economy in Industry 5.0. In: *Industry 5.0: The Human-Centric Future of Industry. Springer Series in Advanced Manufacturing.* Springer, Cham. https://doi.org/10.1007/978-3-032-12622-1_6
5. Majeed, H., & Iftikhar, T. (2026). Industrial Revolution 4.0, 5.0 Sustainable Transformation into Industrial Revolution 6.0. In: *Intelligent Manufacturing in Industry 6.0.* Springer, Cham. https://doi.org/10.1007/978-3-032-07278-8_3
6. Bucci, I., Fani, V., Rossi, M., Sassanelli, C., & Bandinelli, R. (2026). Product Lifecycle Management Towards Industry 5.0: A Descriptive Analysis. In: Mas, F., Del Valle, C., Eynard, B., Rivest, L., Bouras, A. (eds) *Product Lifecycle Management. PLM in the Age of Model-Based Engineering in Industry. PLM 2025.* IFIP Advances in Information and Communication Technology, vol 773. Springer, Cham. https://doi.org/10.1007/978-3-032-09704-0_30
7. Bartolomei, M., Barravecchia, F., Mastrogiacomo, L., Gatta, D. M., & Franceschini, F. (2026). A Multidimensional Framework for Modelling and Assessing Cognitive Assistance Systems for Industry 5.0. *The International Journal of Advanced Manufacturing Technology*, 142, 3053–3073. https://doi.org/10.1007/s00170-025-17203-1
8. Chauhan, A., Pandey, G., Attri, S., Sethi, M., Singh, M. V., Barrili, S., & Malimba, P. M. (2026). Revolutionizing Waste Management: Harnessing AI and Industry 5.0 Applications for Sustainable Solutions. In: Choudhury, M., Majumdar, S., Sillanpää, M. (eds) *The Circular Path. Waste as a Resource.* Springer, Cham. https://doi.org/10.1007/978-3-032-05302-2_10
9. Bhatia, M., & Kumar, V. (2026). The Role of Digital Twin in Urban Infrastructure and Building Management: A Global Scientometric Perspective. *Cluster Computing*, 29, 38. https://doi.org/10.1007/s10586-025-05813-z
10. Okeyia, C., & Marques de Almeida, N. (2026). The Impact of Blockchain Technology for Sustainable Asset Maintenance in Telecoms Domain in Emerging Markets. In: Crespo Márquez, A., Seecharan, T., Abdul-Nour, G., Amadi-Echendu, J., Lee, J. (eds) *Case Studies in Digital Transformation. Engineering Asset Management Review*, vol 4. Springer, Cham. https://doi.org/10.1007/978-3-032-05592-7_4
11. Zhou, H., Xu, C., Tang, X., Wang, S., & Zhang, Z. (2022). A Review of Vision-Laser-Based Civil Infrastructure Inspection and Monitoring. *Sensors*, 22(15), 5882. https://doi.org/10.3390/s22155882
12. Solla, M., Pérez-Gracia, V., & Fontul, S. (2021). A Review of GPR Application on Transport Infrastructures: Troubleshooting and Best Practices. *Remote Sensing*, 13(4), 672. https://doi.org/10.3390/rs13040672
13. López-González, P. J., Reyes-González, D., Moreno-Vázquez, O., Vivar-Ocampo, R., Zamora-Castro, S. A., Santos Cortés, L. d. C., Trujillo-García, B. S., & Sangabriel-Lomelí, J. (2026). Inspection and

Evaluation of Urban Pavement Deterioration Using Drones: *Review of Methods, Challenges, and Future Trends. Future Transportation*, 6(1), 10. https://doi.org/10.3390/futuretransp6010010
14. Meenakshi, K., Singh, K., & Bhargava, D. (2026). Digital Twin Models in the Blockchain Metaverse Framework: Convergence of Virtual Replicas and Decentralized Ecosystems. In: Kumar, A., Rathore, P. S., Moreira, F. (eds) *Blockchain Enabled Metaverse for Smart Wireless Sensor Networks. Information Systems Engineering and Management*, vol 72. Springer, Cham. https://doi.org/10.1007/978-3-032-07305-1_6
15. Elkhodr, M., & Al-Sabbagh, A. (2026). An Integrative Framework for Flood Monitoring and Water Quality Management Enabled by Emerging IoT, Drone, Blockchain, and AI Technologies. *Cluster Computing*, 29, 97. https://doi.org/10.1007/s10586-025-05812-0
16. Zhang, Z., Xue, Y., Zhang, P., Yang, X., Wang, X., Wang, C., Chen, H., Zheng, X., Yin, X., & Zhang, T. (2026). Thermally Drawn Flexible Fiber Sensors: Principles, Materials, Structures, and Applications. *Nano-Micro Letters*, 18, 4. https://doi.org/10.1007/s40820-025-01840-y
17. Bayraktar, I., & Bakirci, M. (2025). Performance Evaluation and Challenges in Automated Oil Tank Detection in Remote Sensing Images. In: Seyman, M. N. (eds) *3rd International Congress of Electrical and Computer Engineering. ICECENG 2024. EAI/Springer Innovations in Communication and Computing*. Springer, Cham. https://doi.org/10.1007/978-3-031-88999-8_15
18. Aishwarya, N., Bharani, H. D., Dhanushree, Y., Sameera Muskan, I., Dwivedi, A. K., & Singh, V. (2026). Literature Survey on Designing and Application of Unmanned Aerial Vehicles for Surveillance Systems. In: Uddin, M. S., Singh, D., Ganokratanaa, T., Kumawat, G. (eds) *Proceedings of International Conference on Innovations in Data Science. ICIDS 2024*. Lecture Notes in Networks and Systems, vol 1367. Springer, Singapore. https://doi.org/10.1007/978-981-96-6329-3_20
19. Ouyang, H., Liu, J., Liang, L., Wang, J., Li, S., Leng, J., & Ding, W. (2026). A Survey of Intelligent Optimization Algorithms for Unmanned Aerial Vehicle-Assisted Mobile Edge Computing. *Archives of Computational Methods in Engineering*, 33, 4419–4440. https://doi.org/10.1007/s11831-026-10496-9
20. Zia, M. U., Xiang, W., Huang, T., Ahmad, J., Chattha, J. N., Naqvi, I. H., & Butt, F. A. (2026). Unifying Ground and Air: A Comprehensive Review of Deep Learning-Enabled CAVs and UAVs. *Artificial Intelligence Review*, 59, 19. https://doi.org/10.1007/s10462-025-11425-1
21. Hu, J., Wang, W., Liu, Y., & Zhang, J. (2026). Large Model in Low-Altitude Economy: Applications and Challenges. *Big Data and Cognitive Computing*, 10(1), 33. https://doi.org/10.3390/bdcc10010033
22. Ali, M. A. H., Zulkifle, M. Z. A., Nik Ghazali, N. N., Apsari, R., Zulkifli, M. M. F. M., & Alkhedher, M. (2025). AI-Driven UAV with Image Processing Algorithm for Automatic Visual Inspection of Aircraft External Surface. *Scientific Reports*, 15, 19581. https://doi.org/10.1038/s41598-025-02902-2
23. Reckling, W., Mitasova, H., Wegmann, K., Kauffman, G., & Reid, R. (2021). Efficient Drone-Based Rare Plant Monitoring Using a Species Distribution Model and AI-Based Object Detection. *Drones*, 5(4), 110. https://doi.org/10.3390/drones5040110

24. Shin, H., Kim, J., Kim, K., & Lee, S. (2023). Empirical Case Study on Applying Artificial Intelligence and Unmanned Aerial Vehicles for the Efficient Visual Inspection of Residential Buildings. *Buildings*, 13(11), 2754. https://doi.org/10.3390/buildings13112754
25. Abdusalomov, A., Umirzakova, S., Tashev, K., Egamberdiev, N., Belalova, G., Meliboev, A., Atadjanov, I., Temirov, Z., & Cho, Y. I. (2025). AI-Driven UAV Surveillance for Agricultural Fire Safety. *Fire*, 8(4), 142. https://doi.org/10.3390/fire8040142
26. Panduman, Y. Y. F., Husna, R., Noprianto, Funabiki, N., Sakamaki, S., Sukaridhoto, S., Syaifudin, Y. W., & Rahmadani, A. A. (2025). An Application of SEMAR IoT Application Server Platform to Drone-Based Wall Inspection System Using AI Model. *Information*, 16(2), 91. https://doi.org/10.3390/info16020091
27. Ayoub, N., & Schneider-Kamp, P. (2021). Real-Time On-Board Deep Learning Fault Detection for Autonomous UAV Inspections. *Electronics*, 10(9), 1091. https://doi.org/10.3390/electronics10091091
28. Pal, O. K., Shovon, M. S. H., Mridha, M. F., & Shin, J. (2024). In-Depth Review of AI-Enabled Unmanned Aerial Vehicles: Trends, Vision, and Challenges. *Discover Artificial Intelligence*, 4, 97. https://doi.org/10.1007/s44163-024-00209-1
29. Lemos, R., Cabral, R., Ribeiro, D., Santos, R., Alves, V., & Dias, A. (2023). Automatic Detection of Corrosion in Large-Scale Industrial Buildings Based on Artificial Intelligence and Unmanned Aerial Vehicles. *Applied Sciences*, 13(3), 1386. https://doi.org/10.3390/app13031386
30. Hoang, M. L. (2023). Smart Drone Surveillance System Based on AI and on IoT Communication in Case of Intrusion and Fire Accident. *Drones*, 7(12), 694. https://doi.org/10.3390/drones7120694
31. Zhou, W., Zhou, L., Yuan, T., Chen, R., & Liu, D (2026). Robust Performance Optimization of UAV Dynamic Systems Using MPC-PID Hybrid Control. *Science Reports*, 16, 2585. https://doi.org/10.1038/s41598-025-32436-6
32. Hammad, S. Y., Kovács, G. P., & Milics, G. (2026). Ambrosia artemisiifolia in Hungary: A Review of Challenges, Impacts, and Precision Agriculture Approaches for Sustainable Site-Specific Weed Management Using UAV Technologies. *AgriEngineering*, 8(1), 30. https://doi.org/10.3390/agriengineering8010030
33. Zhang, Y., Luo, Z., Wu, W., & Tian, W. (2026). RIS-UAV Cooperative ISAC Technology for 6G: Architecture, Optimization, and Challenges. *Algorithms*, 19(1), 65. https://doi.org/10.3390/a19010065
34. Zhou, H., Duan, X., Li, X., Luo, J., Zhang, B., Guo, X., & Sun, L. (2026). HAARN: A Deep Neural Network-Based Intelligent Control Method for High-Altitude Adaptability of Heavy-Load UAV Power Systems. *Sensors*, 26(2), 389. https://doi.org/10.3390/s26020389

Chapter 14

Sustainable availability and MTTF estimation of machine repair systems under fuzzy conditions

Amit Kumar

14.1 PRELIMINARY AND LITERATURE REVIEW

Reliability analysis of Machine Repair System (MRS) has long been a key area of research, with extensive applications in manufacturing, telecommunications, and, more recently, cloud computing. Traditional MRS models typically assume exponential distributions for failure and repair processes, and often treat system parameters as deterministic. However, in complex, real-world systems particularly in cloud-based infrastructures, uncertainty in operational data arising from system complexity, environmental variations, or limited observations necessitates the adoption of fuzzy logic to more accurately capture the inherent system behavior. Effective management of preventive and corrective actions plays a crucial role in enhancing the performance and efficiency of various sectors, including industries, power plants, manufacturing, and engineering. This study investigates a redundant repairable system consisting of two active units and two warm standby units, integrating additional factors such as heterogeneous repairmen. The concept of standby redundancy has been extensively studied in reliability engineering, where various researchers (cf. [1–6]) have explored reliability, standby configurations, and heterogeneous systems to improve performance and reliability characteristics. A review of machining systems further emphasizes the need for deeper investigation into machine repair issues, particularly within uncertain or imprecise environments, in order to foster continued advancements in this field.

In recent years, the increasing complexity and scale of modern industrial and computing systems have placed growing emphasis on the reliability and maintainability of such systems. Machine repair models play a critical role in analyzing the performance and dependability of systems subject to failures and repair operations. A particularly important variant is the warm standby MRS, wherein standby units remain partially active and can quickly replace failed units, thereby enhancing system availability and reducing downtime. Another practical consideration in real-world repair systems is the heterogeneity of repairmen. In many industrial and cloud-based environments, repair personnel differ in skill levels, efficiency,

DOI: 10.1201/9781003787525-14

availability, and repair rates. Such heterogeneity introduces significant variability into system performance, making traditional crisp or deterministic modeling approaches insufficient for capturing the full range of uncertainties present. This study investigates a redundant repairable system consisting of two active units and two warm standby units, integrating additional factors such as heterogeneous repairmen. Standby redundancy is a foundational concept in reliability engineering, and its performance implications have been rigorously analyzed under various system configurations. Prior studies (cf. [1–6]) have addressed multiple aspects of standby systems, including cold, warm, and hot standby modes, as well as the incorporation of heterogeneous repair facilities. These investigations demonstrate that system reliability and availability are significantly influenced by the nature of redundancy and the repair policies employed. Additionally, reviews of repairable and machining systems underscore the necessity for analytical models capable of handling real-world uncertainties. In particular, the integration of fuzzy set theory into reliability analysis has gained traction as a means to manage imprecise failure and repair data, thus providing a more robust framework for evaluating complex systems under uncertainty.

Redundancy is a critical design strategy for ensuring uninterrupted operation and improving the reliability and availability of machining and repairable systems. It is generally classified into two broad categories: active and passive redundancy, each warranting detailed analysis due to their performance implications and cost considerations. Passive redundancy, which involves standby units not operating under normal conditions, is further categorized into cold, warm, and hot standby configurations based on the failure characteristics of the standby components. The foundational concept of standby redundancy was introduced by Srinivasan [7], laying the groundwork for modeling backup components in repairable systems. Subsequently, Barlow and Proschan [8] provided a comprehensive mathematical treatment of redundant repairable systems, offering significant insights into their reliability structures and operational dynamics. Building on this foundation, extensive research has addressed diverse system configurations, optimization techniques, and practical applications (e.g., Gizatullin and Yushkova [9]; Bhandari et al. [10]; Zhong et al. [11]). In a recent advancement, Zhang et al. [12] proposed a novel Reliability and Redundancy Allocation Problem (RRAP) model for complex systems. Their approach leverages graph theoretic concepts, specifically functionality multigraphs and spanning trees to compute system reliability. The proposed method demonstrated superior performance compared to existing models, as validated through benchmark problem sets and a real-world security system case study.

When an active unit fails, an available warm standby unit is promptly switched on to maintain uninterrupted operation of the machining system, with minimal delay. This switchover can be either perfect or imperfect. In cases of imperfect switching, the process is retried iteratively until a successful replacement occurs, or all standby units are exhausted. The concept of imperfect switching was first introduced by Breneman et al. [13]. In a related study, Wang et al. [14] analyzed a warm standby M/M/R machine repair problem incorporating switching failures, balking, and reneging. They developed a profit optimization model for spares and repairmen under availability constraints, utilizing direct search and steepest descent methods, alongside sensitivity analysis. More recently, Shekhar et al. [15] examined the increasing prevalence of finite queueing and reliability models in machining systems, emphasizing the effectiveness of Markov models in reducing costs and enhancing service. These models are particularly useful for finite population queueing systems in industrial machine repair contexts. In real-world computing and repair systems, service agents exhibit variability in expertise, speed, and reliability, leading to heterogeneous repair dynamics. Modeling these heterogeneous repairmen is essential for accurately evaluating distributed computing systems, cloud maintenance, and fault-tolerant infrastructures. Failure to account for this variability can result in suboptimal resource allocation and diminished system resilience. Consequently, integrating heterogeneous repair facilities improves the accuracy and realism of performance analysis for complex computer-based infrastructures. Yue [16] studied a machine repair problem (MRP) involving warm spares and two heterogeneous repairmen. The study derived steady-state probabilities and reliability measures, developed a cost model to optimize system performance, and conducted sensitivity analysis. Similarly, Ke et al. [17] investigated a MRP with queue-dependent heterogeneous repairmen, formulating a profit optimization model that adjusted service thresholds and rates using PGSL, direct search, and Quasi-Newton methods. Their approach was validated through numerical experiments.

In complex and uncertain operational environments, traditional system reliability measures such as availability and mean time to failure (MTTF) become inadequate due to the inherent imprecision and variability in real-world data. To address these limitations, fuzzy set theory has been integrated into reliability modeling, enabling the representation of uncertain parameters—such as failure and repair rates—as fuzzy numbers instead of fixed deterministic values. Fuzzy availability quantifies the likelihood or degree of possibility that a system will remain operational at a given time under uncertain conditions, providing a more flexible and realistic performance evaluation. Similarly, fuzzy MTTF extends the traditional concept by associating a membership function with a range of possible failure times, thus accommodating the vagueness and ambiguity commonly observed in practical maintenance and repair scenarios. Given the critical need for precise reliability assessments in uncertain environments, considerable research has focused on the development of fuzzy-based models for evaluating availability and MTTF. These models, which are discussed in the following literature review, offer significant advantages in capturing the uncertainty and imprecision that characterize real-world system behavior.

Kumar [18] investigated a two-unit repairable system operating under constant coverage and detection delays in a fuzzy environment, where failure and repair times follow fuzzified exponential distributions. The system characteristics were derived using the α-cut method, facilitating a more flexible reliability analysis under uncertainty. Mahendra et al. [19] developed a fuzzy-based fault-tolerant machining model involving two active units, a warm standby, and a single repair facility. Their model accounted for switching failures and imperfect repairs, with reliability measures such as availability and MTTF analyzed using the α-cut method and Zadeh's extension principle through nonlinear parametric programming, under various design constraints. Huang [20] proposed a fuzzy framework for a repairable system consisting of two primary units, one standby unit, and a repair facility, with the possibility of switching failure. Failure and repair times were fuzzified and analyzed using the α-cut method and nonlinear parametric programming, offering deeper insights through membership functions, which aided in both design and practical application. Similarly, Ke et al. [21] modeled a fuzzy repairable system with switching failures, applying α-cut and nonlinear programming (NLP) to construct membership functions based on fuzzified exponential failure and repair distributions.

Recent advancements in reliability analysis have introduced various innovative approaches to assess system performance and optimize complex systems. Malik et al. [22] presented a comprehensive reliability assessment of a non-series-parallel (NSP) system with constant triangular fuzzy failure rates, emphasizing the critical role of fuzzy set theory in analyzing complex system behavior. In a similar vein, Yadav et al. [23] employed a Markov approach to study the reliability, availability, and maintainability of a three-unit repairable system, contributing significantly to the understanding of dynamic system behavior. Kadyan et al. [24] utilized picture fuzzy set theory to evaluate the reliability and risk in complex industrial systems, highlighting the effective application of fuzzy logic in risk management. Additionally, Ram et al. [25] performed sensitivity and reliability analyses on a maintainable energy system with priority repair, providing valuable insights for optimizing energy systems. A complementary study by Ram et al. [26] focused on cost and sensitivity analyses in standby system reliability modeling, further deepening the understanding of system maintenance in uncertain conditions. Jain et al. [27] examined a MRP involving unreliable servers and controlled arrivals of failed machines, offering critical insights for enhancing industrial system operations. Singh and Singh [28] conducted dynamic reliability analyses of a weighted (k, r)-out-of-n cold standby system, incorporating multi-performance, multi-state components to better address the complexity of real-world systems. Finally, Levitin et al. [29] investigated standby systems where activation moments depend on component operation time limits, providing essential insights into system configuration and operational efficiency under varying conditions.

Extensive research has been conducted on redundant two-unit repairable systems under varying failure and service mechanisms within fuzzy environments, as highlighted in the preceding literature review. However, critical research gaps remain, specifically: (1) the absence of studies addressing redundant multi-unit machining systems incorporating heterogeneous repairmen and warm standby units, and (2) the lack of models considering real-time vagueness, ambiguity, imprecision, and uncertainty in failure behaviors. The proposed framework distinguishes itself from existing models by introducing: (1) the modeling of repair facilities operated by heterogeneous repairmen within an MRP, (2) an effective estimation approach under vague and imprecise conditions, (3) the integration of fuzzy theory with conventional reliability methodologies, and (4) the deployment of warm standby units to bolster system reliability. Notably, this chapter utilizes fuzzy numbers for the first time in modeling availability and MTTF, a methodology not previously explored in existing research. By leveraging trapezoidal and triangular fuzzy numbers, this study can accurately represent the vagueness and uncertainty inherent in real-world failure and repair behaviors, marking a significant advancement in the field. Given the critical need for high-precision and sensitive applications, the primary objective of this study is to manage the inherent imprecision, vagueness, and uncertainty associated with system performance. The major contribution lies in the comprehensive analysis of switching failures and the operational impact of warm standby and heterogeneous repairmen in redundant machining systems with both active and passive redundancy under a fuzzy environment. This study is pivotal for enhancing preventive, corrective, and predictive maintenance strategies in advanced machining systems.

This study is structured into sections that develop and analyze a fuzzy-based machine repair model. Section 14.3 introduces key notations, assumptions, and formulates the redundant machining system using transient-state Chapman–Kolmogorov equations. Sections 14.4 and 14.5 apply Laplace transforms, linear algebra, and hazard rate analysis to evaluate reliability metrics like MTTF and availability for systems with heterogeneous repairmen. Section 14.6 incorporates fuzzy logic to handle uncertainty in failure and repair parameters, deriving membership functions for fuzzy MTTF and availability. Section 14.7 uses γ-cut methods and Zadeh's extension principle within parametric NLP to compute fuzzy reliability characteristics. Section 14.8 validates the model through numerical examples, triangular fuzzy membership functions for MTTF and availability. Finally, Sections 14.9, 14.10, and 14.11 discuss findings, summarize contributions, and suggest future research directions, offering a comprehensive approach to modeling repairable system reliability under uncertainty.

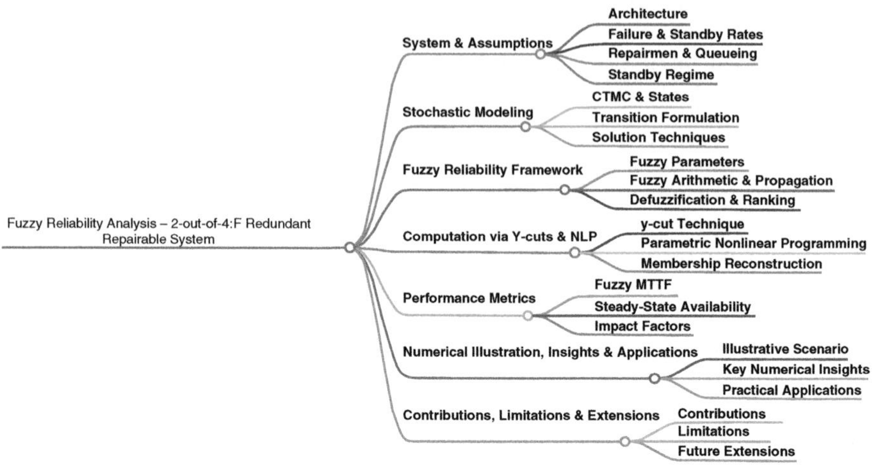

Figure 14.1 Roadmap of the chapter.

14.2 ORGANIZATION OF THE RESEARCH

This section outlines the structure and main contributions of the present study. The chapter is organized as follows:

- The study examines a 2-out-of-4: *F* redundant repairable system with two active and two warm standby units in a fuzzy environment.
- It incorporates fuzzy set theory to model uncertainties in failure and repair data, enhancing system performance evaluations.
- The repair process involves two heterogeneous repairmen, with priority given to the faster one, and failed units are queued when both are busy.
- The model evaluates critical performance metrics such as MTTF and steady-state availability using fuzzy parameters.
- Numerical solutions and parametric NLP are used to derive fuzzy reliability measures and assess system performance under uncertainty.
- The research highlights the importance of redundancy strategies and the impact of repair rate heterogeneity on system reliability.
- Future work may explore extensions like imperfect switching mechanisms and economic cost constraints in fuzzy reliability models (Figure 14.1).

14.3 MODEL DESCRIPTION

To address the increasing complexity and inherent uncertainty in cloud-based infrastructures, this study proposes a fuzzy reliability and availability framework for MRSs featuring heterogeneous repairmen and warm standby units. By embedding fuzzy logic into the modeling of failure and repair rates, the proposed approach captures the vagueness and imprecision often observed in real-world data. This enables a more robust and realistic assessment of system performance, particularly in environments characterized by incomplete information and stochastic variability. The model is especially suited for modern cloud-based systems, where reliability must be ensured despite fluctuating operational conditions and heterogeneous maintenance capabilities.

14.3.1 Failure process

- The time-to-failure of each operating machine is assumed to follow an exponential distribution with constant failure rate λ.
- Upon the failure of an operating machine, an available standby unit if present is immediately switched into operation with negligible delay. The standby unit assumes the full functionality of the failed machine and is subject to the same failure dynamics post-activation.
- The standby units are of warm type, meaning they remain partially energized and monitored prior to activation. These units are susceptible to failure even while in standby mode. The time-to-failure for a warm standby unit also follows an exponential distribution with failure rate α, where $0 < \alpha < \lambda$.
- Failures of operating and standby machines occur independently of each other. All machines are repairable, and failed units are immediately directed to the repair facility. If all repairmen are occupied, newly failed machines enter a queue and await service based on availability.

14.3.2 Repair process

- The repair times of the two heterogeneous repairmen are exponentially distributed with respective rate parameters μ and ν, where $\nu < \mu$. Repair times are assumed to be independent of the system state. The repairman with the higher repair rate (i.e., faster repair) is always prioritized when available.
- When no machines are under repair (i.e., the system is empty), any newly failed machine is assigned to the faster repairman by default.
- If a failed machine arrives and only one repairman is available, it is immediately assigned to the idle repairman, irrespective of the repair rate or how long the other repairman has been occupied.
- When a repairman completes a repair task and there is a queue of failed machines awaiting service, the next machine in line is immediately assigned to the available repairman. Thus, repairmen remain continuously engaged as long as the queue is non-empty.
- Once a failed machine is assigned to a repairman, it remains with that repairman until the repair process is completed.

- Repair jobs are non-preemptive and cannot be divided or shared between the two repairmen. Additionally, once a repair has commenced, a failed machine cannot be transferred from the slower to the faster repairman.

To formulate the forward Chapman–Kolmogorov differential-difference equations for the proposed redundant repairable system, we define the system state at any time t using the following state variables:

- $W(t)$: the number of active (operating) units in the system at time t,
- $S(t)$: the number of standby units available in the system at time t.

The stochastic process $\{W(t), S(t)\}$ forms a continuous-time Markov chain (CTMC), where the system's state at any time t is described by the pair (i, j), with i representing the number of active (operating) units and j the number of available standby units. The corresponding state space is given by:

$$\mathcal{F} = \{(i, j) \mid i = 0, 1, 2;\ j = 0, 1, 2\}$$

This finite two-dimensional state space captures all possible combinations of operating and standby units in the system, enabling the construction of the transition rate diagram and derivation of the system's transient and steady-state behavior through Chapman–Kolmogorov differential-difference equations. The system is assumed to fail completely when there are no active units, i.e., $S(t) = 0$. Let the time-dependent state probabilities be defined as:

- $P_{22}(t)$ = Prob {At any given time t, the storage system consists of two active units and two warm standby units}
  i.e., $P_{22}(t) = \Pr[W(t) = 2,\ S(t) = 2]$
- $P_{21}(t)$ = Prob {At any given time t, the storage system consists of two active units and one warm standby unit}
  i.e., $P_{21}(t) = \Pr[W(t) = 2,\ S(t) = 1]$
- $P_{20}(t)$ = Prob {At any given time t, the storage system consists of two active units and no warm standby units}
  i.e., $P_{20}(t) = \Pr[W(t) = 2,\ S(t) = 0]$
- $P_{10}(t)$ = Prob {At any given time t, the storage system consists of one active unit and no warm standby units}
  i.e., $P_{10}(t) = \Pr[W(t) = 1,\ S(t) = 0]$
- $Q_{21}(t)$ = Prob {At any given time t, the storage system consists of two active units, one warm standby unit, and one repaired unit currently under inspection}
  i.e., $Q_{21}(t) = \Pr[W(t) = 2,\ S(t) = 1,\ \text{one unit under inspection}]$

Thus, the forward Chapman–Kolmogorov differential-difference equations, formulated in terms of the parameters λ, α, μ, and v, govern the balance between the inflow and outflow rates of the system states in the reliability analysis of the repairable redundant storage system. These equations can be compactly represented in matrix form as follows:

$$\begin{bmatrix} \frac{dP_{22}(t)}{dt} \\ \frac{dP_{21}(t)}{dt} \\ \frac{dQ_{21}(t)}{dt} \\ \frac{dP_{20}(t)}{dt} \\ \frac{dP_{10}(t)}{dt} \end{bmatrix} = \begin{bmatrix} -(2\lambda+2\alpha) & \mu & v & 0 & 0 \\ 2\lambda+2\alpha & -(2\lambda+\alpha+\mu) & 0 & v & 0 \\ 0 & 0 & -(2\lambda+\alpha+v) & \mu & 0 \\ 0 & 2\lambda+\alpha & 2\lambda+\alpha & -(2\lambda+\mu+v) & v+\mu \\ 0 & 0 & 0 & 2\lambda & -(\mu+v) \end{bmatrix}$$

$$\cdot \begin{pmatrix} P_{22}(t) \\ P_{21}(t) \\ Q_{21}(t) \\ P_{20}(t) \\ P_{10}(t) \end{pmatrix}$$

At time $t = 0$, the system is initialized with no failed units. Specifically, both primary units are operational, and the two warm standby units are functioning correctly. Accordingly, the initial state probabilities are given by: $P_{2,2}(0)=1$; $P_{2,1}(0)=0$; $P_{2,0}(0)=0$; $P_{1,0}(0)=0$; $Q_{2,1}(0)=0$

14.4 THE SYSTEM RELIABILITY CHARACTERISTICS

14.4.1 Mean time to failure

We define the Laplace transforms of the state probabilities and transition functions as $\mathcal{L}\{P_{ij}(t)\} = \widehat{P_{ij}}(s) = P_{ij}(s)$ and $\mathcal{L}\{Q_{21}(t)\} = Q_{21}(s)$, respectively. Utilizing the standard Laplace transform property for the derivative,

$$\mathcal{L}\left\{\frac{dP(t)}{dt}\right\} = s\hat{P}(s) - P(0),$$

we apply the Laplace transformation to the system of Chapman–Kolmogorov differential-difference equations. This process yields a system of linear equations in the Laplace domain, which is then expressed in matrix form. The resulting coefficient matrix derived from the Laplace-transformed system is non-singular, i.e., det ≠ 0, ensuring the existence of a unique solution for the state probabilities in the Laplace domain.

$$\mathbf{A}(s)\mathbf{P}(s)=\mathbf{B} \tag{14.1}$$

$$\underbrace{\begin{bmatrix} s+2\lambda+2\alpha & -\mu & -v & 0 & 0 \\ -(2\lambda+2\alpha) & s+2\lambda+\alpha+\mu & 0 & -v & 0 \\ 0 & 0 & s+2\lambda+\alpha+v & -\mu & 0 \\ 0 & -(2\lambda+\alpha) & -(2\lambda+\alpha) & s+2\lambda+\mu+v & 0 \\ 0 & 0 & 0 & -2\lambda & s \end{bmatrix}}_{\mathbf{A}(s)} \cdot \underbrace{\begin{bmatrix} \hat{P}_{22}(s) \\ \hat{P}_{21}(s) \\ \hat{Q}_{21}(s) \\ \hat{P}_{20}(s) \\ \hat{P}_{10}(s) \end{bmatrix}}_{\mathbf{P}(s)} = \underbrace{\begin{bmatrix} 1 \\ 0 \\ 0 \\ 0 \\ 0 \end{bmatrix}}_{\mathbf{B}}$$

Incorporating the given initial conditions and applying Cramer's rule, we obtain the expressions for the desired state probabilities in the Laplace domain. By inverting $\hat{P}_{10}(s)$, $\hat{P}_{20}(s)$, $\hat{P}_{21}(s)$, $\hat{Q}_{21}(s)$, and $\hat{P}_{22}(s)$ to $P_{10}(t)$, $P_{20}(t)$, $P_{21}(t)$, $Q_{21}(t)$, and $P_{22}(t)$ using the inverse Laplace transform, we obtain the state probabilities of the system at time t (see the Appendix for detailed steps). Let Y be the continuous random variable representing the time-to-failure of the system. $P_{00}(t)$ denotes the probability that no machine is operational in the system, indicating system failure at or before time t. Therefore, the reliability function of the system, denoted by $R(t)$, can be expressed as:

$$R(t)=1-P_{10}(t)$$

This function represents the probability that the system remains operational at time t, considering all failure and repair dynamics, including the warm standby units and heterogeneous repairmen. The reliability function is a critical metric for evaluating the system's resilience over time, especially under uncertain and fuzzy conditions.

$$R_Y(t)=1-P_{10}(t), \quad t\ge 0 \tag{14.2}$$

Using the theory of reliability, the failure density $Y(t)$ can be derived as:

$$Y(t) = -\frac{dR_Y(t)}{dt} = -\frac{d(1 - P_{10}(t))}{dt} = \frac{dP_{10}(t)}{dt} \tag{14.3}$$

The Laplace transform of the failure density in equation (14.17) is expressed as $\tilde{Y}(s) = s\tilde{P}_{10}(s) - P_0(0)$. Hence, MTTF (T) of the system can be derived as:

$$\text{MTTF} = T = -\left.\frac{d\tilde{Y}(s)}{ds}\right|_{s=0}$$

$$\text{MTTF} = \frac{\begin{bmatrix} 2\alpha^3 + 16\alpha^2\lambda + 2\alpha^2\mu + 4\alpha^2 v + 36\alpha\lambda^2 + 8\alpha\lambda\mu + 18\alpha\lambda v + 4\alpha\mu v + \\ 2\alpha v^2 + 24\lambda^3 + 8\lambda^2\mu + 16\lambda^2 v + 8\lambda\mu v + 2\lambda v^2 + \mu^2 v + \mu v^2 \end{bmatrix}}{4\lambda\left(\alpha^3 + 5\alpha^2\lambda + \alpha^2 v + 8\alpha\lambda^2 + 3\alpha\lambda v + 4\lambda^3 + 2\lambda^2 v\right)} \tag{14.4}$$

14.5 AVAILABILITY ANALYSIS OF THE SYSTEM

Accurate performance prediction of a storage system necessitates a rigorous analysis of its reliability metrics. This section focuses on determining the steady-state availability, a critical indicator of the system's long-term operational performance. To this end, the Chapman–Kolmogorov differential-difference equations are reformulated into a system of linear equations involving the steady-state probabilities. Let the steady-state probabilities be defined as:

$$P_{ij} = \lim_{t\to\infty} P_{ij}(t), \quad Q_{ij} = \lim_{t\to\infty} Q_{ij}(t), \quad \forall i, j$$

Using these definitions, the balance equations at steady state can be written in matrix form as:

$$\begin{bmatrix} -(2\lambda + 2\alpha) & \mu & v & 0 & 0 \\ 2\lambda + 2\alpha & -(2\lambda + \alpha + \mu) & 0 & v & 0 \\ 0 & 0 & -(2\lambda + \alpha + v) & \mu & 0 \\ 0 & 2\lambda + \alpha & 2\lambda + \alpha & -(2\lambda + \mu + v) & v + \mu \\ 0 & 0 & 0 & 2\lambda & -(\mu + v) \end{bmatrix} \cdot \begin{bmatrix} P_{22} \\ P_{21} \\ Q_{21} \\ P_{20} \\ P_{10} \end{bmatrix} = \begin{bmatrix} 0 \\ 0 \\ 0 \\ 0 \\ 0 \end{bmatrix}$$

To ensure the uniqueness of the steady-state solution, the following normalization condition is applied:

$$P_{22} + P_{21} + Q_{21} + P_{20} + P_{10} = 1 \tag{14.5}$$

The system of linear equations together with the normalization condition (normal) is solved to obtain the steady-state probabilities $P_{22}, P_{21}, Q_{21}, P_{20}, P_{10}$, and P_{00}. These probabilities, as presented in the Appendix, are computed to ensure their sum equals unity, thereby reflecting the steady-state behavior of the system. Finally, the steady-state availability of the storage system is given by: $A(\infty) = 1 - P_{10}$ which quantifies the long-run proportion of time that the system remains operational.

$$A(\infty) = \left[\frac{\begin{array}{l} 2\alpha^3\mu + 2\alpha^3 v + 10\alpha^2\mu\lambda + 10\alpha^2\lambda v + 2\alpha^2\mu^2 + 6\alpha^2\mu v + 4\alpha^2 v^2 \\ +16\alpha\lambda^2\mu + 16\alpha\lambda^2 v + 6\alpha\lambda\mu^2 + 18\alpha\lambda\mu v + 12\alpha\lambda v^2 + 4\alpha\mu^2 v \\ +6\alpha\mu v^2 + 2\alpha v^3 + 8\lambda^3\mu + 8\lambda^3 v + 4\lambda^2\mu^2 + 12\lambda^2\mu v \\ +8\lambda^2 v^2 + 6\lambda\mu^2 v + 8\lambda\mu v^2 + 2\lambda v^3 + \mu^3 v + 2\mu^2 v^2 + \mu v^3 \end{array}}{\begin{array}{l} 4\alpha^3\lambda + 2\alpha^3\mu + 2\alpha^3 v + 20\alpha^2\lambda^2 + 10\alpha^2\mu\lambda + 14\alpha^2\lambda v + 2\alpha^2\mu^2 \\ +6\alpha^2\mu v + 4\alpha^2 v^2 + 32\alpha\lambda^3 + 16\alpha\lambda^2\mu + 28\alpha\lambda^2 v + 6\alpha\lambda\mu^2 \\ +18\alpha\lambda\mu v + 12\alpha\lambda v^2 + 4\alpha\mu^2 v + 6\alpha\mu v^2 + 2\alpha v^3 + 16\lambda^4 \\ +8\lambda^3\mu + 16\lambda^3 v + 4\lambda^2\mu^2 + 12\lambda^2\mu v + 8\lambda^2 v^2 + 6\lambda\mu^2 v \\ +8\lambda\mu v^2 + 2\lambda v^3 + \mu^3 v + 2\mu^2 v^2 + \mu v^3 \end{array}} \right] \tag{14.6}$$

14.6 MEMBERSHIP FUNCTIONS IN FUZZY REDUNDANT REPAIRABLE SYSTEMS

This study focuses on assessing the performance of a repairable system by incorporating linguistic vagueness, estimation approximations, and uncertainty in key system parameters. In this section, we model these parameters using fuzzy specifications, establishing a robust framework for analyzing uncertainty and vagueness. The approximations, vagueness, and uncertainty in the failure rates of operating and standby machines (λ), as well as the service rates of two heterogeneous repairmen, which are exponentially distributed with rate parameters μ and ν ($\nu < \mu$), respectively, are represented using fuzzy numbers $\tilde{\lambda}$, $\tilde{\alpha}$, $\tilde{\mu}$, and $\tilde{\nu}$ respectively, each with its corresponding membership functions: $\kappa_{\tilde{\lambda}}(y)$, $\kappa_{\tilde{\alpha}}(u)$, $\kappa_{\tilde{\mu}}(w)$ and $\kappa_{\tilde{\nu}}(x)$. Therefore, the following fuzzy sets are defined as ordered pairs.

$$\begin{aligned}
\tilde{\lambda} &= \left\{\left(y, \kappa_{\tilde{\lambda}}(y)\right) \mid y \in Y\right\} \\
\tilde{\alpha} &= \left\{\left(u, \kappa_{\tilde{\alpha}}(u)\right) \mid u \in U\right\} \\
\tilde{\mu} &= \left\{\left(w, \kappa_{\tilde{\mu}}(w)\right) \mid w \in V\right\} \\
\tilde{\nu} &= \left\{\left(x, \kappa_{\tilde{\nu}}(x)\right) \mid x \in X\right\}
\end{aligned} \tag{14.7}$$

where Y, U, V, and X are the crisp universe of discourse of the system parameters.

A fuzzy number is a special type of fuzzy set that is normal and convex. It extends the traditional real number line to incorporate vagueness and uncertainty, representing a range of possible values rather than a single precise value. Each potential value is assigned a weight between 0 and 1, forming a fuzzy set. A fuzzy number is thus a specific instance of a convex, normalized fuzzy set on the real line. Calculations involving fuzzy numbers allow for the integration of uncertainty in parameters, properties, geometry, initial conditions, and more. Let $r(y,u,w,x)$ represent the reliability characteristic of the redundant repairable system. Given that $\tilde{\lambda},\tilde{\alpha},\tilde{\mu}$, and $\tilde{\nu}$ are fuzzy numbers, it follows that $r\left(\tilde{\lambda},\tilde{\alpha},\tilde{\mu},\tilde{\nu}\right)$ is also a fuzzy number. Based on Zadeh's extension principle, the membership function for the desired reliability characteristic $r\left(\tilde{\lambda},\tilde{\alpha},\tilde{\mu},\tilde{\nu}\right)$ is derived as follows: $r\left(\tilde{\lambda},\tilde{\alpha},\tilde{\mu},\tilde{\nu}\right) = \left\{\left(r, \kappa_{\tilde{r}}(r)\right) \mid r \in R\right\}$, where $\kappa_{\tilde{r}}(r)$ is the membership function of the fuzzy reliability characteristic. According to Zadeh's extension principle, the membership function $\kappa_{\tilde{r}}(r)$ is given by: $\kappa_{\tilde{r}}(r) = \min\left(\kappa_{\tilde{\lambda}}(\lambda), \kappa_{\tilde{\alpha}}(\alpha), \kappa_{\tilde{\mu}}(\mu), \kappa_{\tilde{\nu}}(\nu)\right)$, where λ, α, μ, and ν are the crisp values corresponding to the fuzzy parameters $\tilde{\lambda}$, $\tilde{\alpha}$, $\tilde{\mu}$, and $\tilde{\nu}$, respectively. This formula uses the minimum operator to combine the individual fuzzy sets, yielding a fuzzy reliability characteristic. Thus, the reliability characteristic $\tilde{r}$ captures the uncertainty in the system parameters $\tilde{\lambda},\tilde{\alpha},\tilde{\mu},\tilde{\nu}$, providing a fuzzy evaluation of the system's reliability under vagueness and uncertainty.

$$\kappa_{r\left(\tilde{\lambda},\tilde{\alpha},\tilde{\mu},\tilde{\nu}\right)}(z) = \sup_{\Omega} \min\left\{\kappa_{\tilde{\lambda}}(y), \kappa_{\tilde{\alpha}}(u), \kappa_{\tilde{\mu}}(w), \kappa_{\tilde{\nu}}(x) \mid z = r(y,u,w,x)\right\} \tag{14.8}$$

For the studied redundant MRP with heterogeneous multiple repair facilities, we derive the reliability characteristics r as MTTF in equation (14.4) and steady-state availability of the system in equation (14.6). Thus, the membership grade function $\kappa_{\tilde{T}}(z)$ for MTTF and $\kappa_{\tilde{A}}(z)$ for steady-state availability of the system, respectively, are

$$\kappa_{\tilde{T}}(z) = \sup_{\Omega} \min\left\{\kappa_{\tilde{\lambda}}(y), \kappa_{\tilde{\alpha}}(u), \kappa_{\tilde{\mu}}(w), \kappa_{\tilde{v}}(x) \mid z = r(y,u,w,x)\right\} \quad (14.9)$$

and

$$\kappa_{\tilde{A}}(z) = \sup_{\Omega} \min\left\{\kappa_{\tilde{\lambda}}(y), \kappa_{\tilde{\alpha}}(u), \kappa_{\tilde{\mu}}(w), \kappa_{\tilde{v}}(x)(s) \mid z = r(y,u,w,x)\right\} \quad (14.10)$$

respectively, where

$$r_{\tilde{T}}(y,u,w,x) = \frac{\begin{bmatrix} 2u^3 + 16u^2y + 2u^2w + 4u^2x + 36uy^2 + 8uyw \\ +18uyx + 4uwx + 2ux^2 + 24y^3 + 8y^2w + 16y^2x \\ +8ywx + 2yx^2 + w^2x + wx^2 \end{bmatrix}}{4y\left(u^3 + 5u^2y + u^2x + 8uy^2 + 3uyx + 4y^3 + 2y^2x\right)}$$

$$r_{\tilde{A}}(y,u,w,x) = \begin{bmatrix} \dfrac{\begin{matrix} 2u^3w + 2u^3x + 10u^2wy + 10u^2yx + 2u^2w^2 + 6u^2wx \\ +4u^2x^2 + 16uy^2w + 16uy^2x + 6uyw^2 + 18uywx \\ +12uyx^2 + 4uw^2x + 6uwx^2 + 2ux^3 + 8y^3w + 8y^3x \\ +4y^2w^2 + 12y^2wx + 8y^2x^2 + 6yw^2x + 8ywx^2 \\ +2yx^3 + w^3x + 2w^2x^2 + wx^3 \end{matrix}}{\begin{matrix} 4u^3y + 2u^3w + 2u^3x + 20u^2y^2 + 10u^2wy + 14u^2yx + 2u^2w^2 \\ +6u^2wx + 4u^2x^2 + 32uy^3 + 16uy^2w + 28uy^2x + 6uyw^2 \\ +18uywx + 12uyx^2 + 4uw^2x + 6uwx^2 + 2ux^3 + 16y^4 \\ +8y^3w + 16y^3x + 4y^2w^2 + 12y^2wx + 8y^2x^2 + 6yw^2x \\ +8ywx^2 + 2yx^3 + w^3x + 2w^2x^2 + wx^3 \end{matrix}} \end{bmatrix}$$

with

$$\Omega = \{y \in Y, u \in U, w \in V, x \in X, \mid y > 0, u > 0, w > 0, x > 0\}$$

The derived membership grade functions are not presented in an explicit form, which makes their practical application challenging. Visualizing the shapes of the memberships and inferring their properties are often complex and not straightforward. To overcome this challenge, we apply parametric NLP based on the extension principle, which will be discussed in detail in the next section.

14.7 PARAMETRIC NONLINEAR PROGRAMS

Parametric NLP is employed to compute the γ-cuts of the fuzzy reliability characteristic $r(\lambda,\alpha,\mu,\nu)$ through Zadeh's extension principle. To express the membership function $\kappa_r(z)$ in a clear and conventional format, we apply Zadeh's method [30], which utilizes the γ-cuts of the fuzzy parameters. The γ-cuts of λ, α, μ, and ν are defined as crisp intervals:

$$\lambda(\gamma)=\left[y_\gamma^L,y_\gamma^U\right]=\left[\min_{y\in U}\left\{y\mid\kappa_{\tilde{\lambda}}(y)\geq\gamma\right\},\ \max_{y\in U}\left\{y\mid\kappa_{\tilde{\lambda}}(y)\geq\gamma\right\}\right] \quad (14.11a)$$

$$\alpha(\gamma)=\left[u_\gamma^L,u_\gamma^U\right]=\left[\min_{u\in U}\left\{u\mid\kappa_{\tilde{\alpha}}(u)\geq\gamma\right\},\ \max_{u\in U}\left\{u\mid\kappa_{\tilde{\alpha}}(U)\geq\gamma\right\}\right] \quad (14.11b)$$

$$\mu(\gamma)=\left[w_\gamma^L,w_\gamma^U\right]=\left[\min_{w\in W}\left\{w\mid\kappa_{\tilde{\mu}}(w)\geq\gamma\right\},\max_{w\in W}\left\{w\mid\kappa_{\tilde{\mu}}(W)\geq\gamma\right\}\right] \quad (14.11c)$$

$$v(\gamma)=\left[x_\gamma^L,x_\gamma^U\right]=\left[\min_{x\in X}\left\{x\mid\kappa_{\tilde{v}}(x)\geq\gamma\right\},\max_{x\in X}\left\{x\mid\kappa_{\tilde{v}}(X)\geq\gamma\right\}\right] \quad (14.11d)$$

The endpoints of these intervals are functions of γ; $y_\gamma^L=\min\kappa_{\tilde{\lambda}}^{-1}(\gamma)$, $y_\gamma^U=\max\kappa_{\tilde{\lambda}}^{-1}(\gamma)$, $w_\gamma^L=\min\kappa_{\tilde{\mu}}^{-1}(\gamma)$, $w_\gamma^U=\max\kappa_{\tilde{\mu}}^{-1}(\gamma)$, $x_\gamma^L=\min\kappa_{\tilde{\beta}}^{-1}(\gamma)$, $x_\gamma^U=\max\kappa_{\tilde{\beta}}^{-1}(\gamma)$, $y_\gamma^L=\min\kappa_{\tilde{c}}^{-1}(\gamma)$, $y_\gamma^U=\max\kappa_{\tilde{c}}^{-1}(\gamma)$ Thus, we have: $y\in\left[y_\gamma^L,y_\gamma^U\right]$, $w\in\left[w_\gamma^L,w_\gamma^U\right]$, $x\in\left[x_\gamma^L,x_\gamma^U\right]$, $y\in\left[y_\gamma^L,y_\gamma^U\right]$, $s\in\left[s_\gamma^L,s_\gamma^U\right]$. This transformation maps the fuzzy redundant repairable system into a family of crisp (conventional) models using the γ-cut approach. In this context, λ denotes the failure rate of the primary (operating) machines, γ represents the failure rate of the warm standby machines ($\gamma<\lambda$), μ and v correspond to the repair rates of the faster and slower repairmen, respectively. To construct the membership function $\kappa_{\tilde{r}}(z)$ for the fuzzy reliability characteristic $\tilde{r}$, we apply Zadeh's extension principle. This requires identifying all values of $z=r(y,u,w,x)$ for which the minimum membership grade among the input fuzzy parameters equals γ, i.e., $\kappa_{\tilde{r}}(z)=\sup\{\gamma\in[0,1]\mid\min\left(\kappa_{\tilde{\lambda}}(y),\kappa_{\tilde{\alpha}}(u),\kappa_{\tilde{\mu}}(w),\kappa_{\tilde{v}}(x)\right)=\gamma,z=r(y,u,w,x)\}$.

Table 14.1 Conditions for $\kappa_{\tilde{r}}(z)=\gamma$ under fuzzy membership cases

Case	$\kappa_{\tilde{\lambda}}(y)$	$\kappa_{\tilde{\mu}}(w)$	$\kappa_{\tilde{v}}(x)$	$\kappa_{\tilde{\alpha}}(u)$
(i)	$=\gamma$	$\geq\gamma$	$\geq\gamma$	$\geq\gamma$
(ii)	$\geq\gamma$	$=\gamma$	$\geq\gamma$	$\geq\gamma$
(iii)	$\geq\gamma$	$\geq\gamma$	$=\gamma$	$\geq\gamma$
(iv)	$\geq\gamma$	$\geq\gamma$	$\geq\gamma$	$=\gamma$

As shown in Table 14.1, the fuzzy membership conditions for different input parameters are categorized into four cases based on which parameter attains the minimum grade.

$$(r)_\gamma^{L_i} = \min_{\Omega_i} r(y,u,w,x), \tag{14.12a}$$

$$(r)_\gamma^{U_i} = \max_{\Omega_i} r(y,u,w,x), \quad i = 1, 2, 3, 4 \tag{14.12b}$$

Due to the nesting property of the γ-cuts (i.e., for $0 < \gamma_2 < \gamma_1 \leq 1$,

$$\left[y_{\gamma_1}^L, y_{\gamma_1}^U\right] \subseteq \left[y_{\gamma_2}^L, y_{\gamma_2}^U\right], \quad \left[w_{\gamma_1}^L, w_{\gamma_1}^U\right] \subseteq \left[w_{\gamma_2}^L, w_{\gamma_2}^U\right], \quad \text{etc.}),$$

we define the overall bounds as:

$$(r)_\gamma^{L} = \min_{\Omega} r(y,w,x,y), \tag{14.13a}$$

$$(r)_\gamma^{U} = \max_{\Omega} r(y,w,x,y), \tag{14.13b}$$

subject to:

$$y \in \left[y_\gamma^L, y_\gamma^U\right], \quad w \in \left[w_\gamma^L, w_\gamma^U\right], \quad x \in \left[x_\gamma^L, x_\gamma^U\right], \quad y \in \left[y_\gamma^L, y_\gamma^U\right], \quad s \in \left[s_\gamma^L, s_\gamma^U\right]$$

Due to the convexity of $\tilde{r}$, we have

$$(r)_{\gamma_1}^L \geq (r)_{\gamma_2}^L \quad \text{and} \quad (r)_{\gamma_1}^U \leq (r)_{\gamma_2}^U \quad \text{for} \quad \gamma_1 > \gamma_2$$

If these bounds are invertible with respect to γ, we can define the left and right shape functions as

$$L(z) = [(r)_\gamma^L]^{-1}, \quad R(z) = [(r)_\gamma^U]^{-1},$$

so that the membership function $\kappa_{\tilde{r}}(z)$ is given by:

$$\kappa_{\tilde{r}}(z) = \begin{cases} L(z), & (r)_0^L \le z \le (r)_1^L, \\ 1, & (r)_1^L \le z \le (r)_1^U, \\ R(z), & (r)_1^U \le z \le (r)_0^U. \end{cases} \tag{14.14}$$

In practice, analytic solutions for $(r)_\gamma^L$ and $(r)_\gamma^U$ are often unavailable; hence, numerical methods are used to determine these bounds across $\gamma \in [0,1]$. The resulting collection

$$\left\{\left[(r)_\gamma^L, (r)_\gamma^U\right] \mid \gamma \in [0,1]\right\}$$

describes the shape of $\kappa_{\tilde{r}}(z)$. A similar approach can be applied to derive the fuzzy membership functions for MTTF and steady-state availability by replacing $r(y,w,x,y,s)$ with the corresponding expressions r_T and r_A. Since users typically require a crisp value for reliability characteristics rather than a fuzzy set, defuzzification is performed using Yager's ranking index method [31]. This method, based on area compensation, converts a convex fuzzy number $\phi[\Lambda]$ with γ-cut $\left((\phi[\Lambda])_\gamma^L, (\phi[\Lambda])_\gamma^U\right)$ into a crisp value:

$$O(\phi[\Lambda]) = \int_0^1 \frac{(\phi[\Lambda])_\gamma^L + (\phi[\Lambda])_\gamma^U}{2} d\gamma \tag{14.15}$$

This ranking technique satisfies the properties of compensation, linearity, and additivity.

14.8 ILLUSTRATIVE EXAMPLE AND NUMERICAL RESULTS

Consider a cloud-based repairable system comprising two identical active virtual machines (VMs) operating in parallel and two warm standby VMs, forming a typical high-availability configuration in mission-critical environments. A centralized orchestrator monitors all VMs and initiates fault detection and recovery. Upon failure of an active VM, a standby is activated, with non-deterministic delays due to resource contention, boot time, or software initialization captured via fuzzy failure rates. The failure rate of active units is denoted by the fuzzy parameter $\tilde{\lambda}$, while standby units have a fuzzy failure rate $\tilde{\alpha}$, with $\tilde{\alpha} < \tilde{\lambda}$. Repairs are handled by two heterogeneous agents: a fast repair agent with fuzzy repair rate $\tilde{\mu}$ and a slow repair agent with fuzzy repair rate $\tilde{v}$. Uncertainty in diagnostics and workloads justifies the use of fuzzy modeling. To evaluate system performance such as availability or MTTF, a fuzzy reliability metric $\tilde{r} = r\left(\tilde{\lambda}, \tilde{\alpha}, \tilde{\mu}, \tilde{v}\right)$ is defined. The extension principle is applied to derive the membership function $\kappa_{\tilde{r}}(z)$ of the output. For each confidence level $\gamma \in [0,1]$, the γ-cuts of fuzzy input parameters are computed. Subsequently, parametric nonlinear optimization is used to obtain the lower and upper bounds of the γ-cut of $\tilde{r}$, considering four boundary scenarios. This framework demonstrates the practical utility of fuzzy set theory in assessing the reliability of complex, redundant cloud-based systems under parameter uncertainty.

14.8.1 Triangular fuzzy-based estimation of MTTF

Let the failure rates of the operating unit and standby units, along with the repair rates of heterogeneous repairmen μ and v, be represented as triangular fuzzy numbers: $\tilde{\lambda} = [4,6,7]$, $\tilde{\alpha} = [5,7,8]$, $\tilde{v} = [6,8,9]$, and $\tilde{\mu} = [8,10,11]$. The corresponding γ-cuts for these fuzzy numbers are obtained as follows: $\tilde{\lambda}_\gamma = [4+2\gamma, 7-\gamma]$, $\tilde{\alpha}_\gamma = [5+2\gamma, 8-\gamma]$, $\tilde{v}_\gamma = [6+2\gamma, 9-\gamma]$, and $\tilde{\mu}_\gamma = [8+2\gamma, 11-\gamma]$, which serve as the basis for evaluating the γ-cuts of the fuzzy MTTF $\tilde{T}$.

$$(T)_\gamma^L = \frac{-29{,}772 + 4{,}682\gamma - 328\gamma^2 + 7\gamma^3}{2(-7+\gamma)\left(-9{,}900 + 3{,}000\gamma - 269\gamma^2 + 6\gamma^3\right)} \tag{14.16a}$$

$$(T)_\gamma^U = \frac{13{,}245 + 9{,}520\gamma + 2{,}431\gamma^2 + 215\gamma^3}{4(2+\gamma)\left(2{,}808 + 3{,}129\gamma + 1{,}106\gamma^2 + 120\gamma^3\right)} \tag{14.17a}$$

Obviously the function $(T)_\gamma^U$ and $(T)_\gamma^L$ is invertible in nature, which yields the membership function:

$$k_{\tilde{T}}(z) = \begin{cases} L(z), & \dfrac{827}{3{,}850} \le z \le \dfrac{25{,}411}{85{,}956} \\ 1, & \dfrac{25{,}411}{85{,}956} \\ R(z), & \dfrac{25{,}411}{85{,}956} \le z \le \dfrac{4{,}415}{7{,}488} \end{cases} \tag{14.18}$$

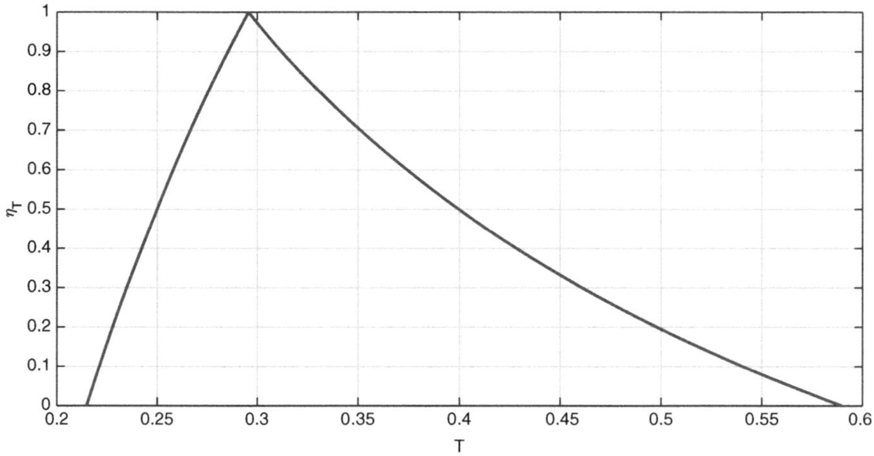

Figure 14.2 The State transition diagram for reliability analysis of the two-unit repairable system with standby.

Table 14.2 α-cuts of the failure rate, the repair rate, and fuzzy MTTF

γ	λ_L	λ_U	α_L	α_U	μ_L	μ_U	v_L	v_U	T_L	T_U
0.0	4.0	7.0	5.0	8.0	6.0	9.0	8.0	11.0	0.21481	0.58961
0.1	4.2	6.9	5.2	7.9	6.2	8.9	8.2	10.9	0.22116	0.54055
0.2	4.4	6.8	5.4	7.8	6.4	8.8	8.4	10.8	0.22783	0.49805
0.3	4.6	6.7	5.6	7.7	6.6	8.7	8.6	10.7	0.23483	0.46098
0.4	4.8	6.6	5.8	7.6	6.8	8.6	8.8	10.6	0.24221	0.42842
0.5	5.0	6.5	6.0	7.5	7.0	8.5	9.0	10.5	0.24996	0.39965
0.6	5.2	6.4	6.2	7.4	7.2	8.4	9.2	10.4	0.25814	0.37409
0.7	5.4	6.3	6.4	7.3	7.4	8.3	9.4	10.3	0.26676	0.35126
0.8	5.6	6.2	6.6	7.2	7.6	8.2	9.6	10.2	0.27585	0.33079
0.9	5.8	6.1	6.8	7.1	7.8	8.1	9.8	10.1	0.28546	0.31233
1.0	6.0	6.0	7.0	7.0	8.0	8.0	10.0	10.0	0.29563	0.29563

The functional form of membership functions $L(z)$ and $R(z)$ is tedious to express. The graphical shape of the membership function is depicted in Figure 14.2. The γ-cut-based bounds of the fuzzy parameters and the corresponding MTTF values computed using triangular fuzzy numbers are shown in Table 14.2.

14.8.2 Triangular fuzzy-based estimation of availability

Let the failure rates of the operating unit and standby units, along with the repair rates of heterogeneous repairmen μ and v, be represented as triangular fuzzy numbers $\tilde{\lambda}=[4,6,7]$, $\tilde{\alpha}=[5,7,8]$, $\tilde{v}=[6,8,9]$, and $\tilde{\mu}=[8,10,11]$. The corresponding γ-cuts for these fuzzy numbers are obtained as follows: $\tilde{\lambda}_\gamma=[4+2\gamma,7-\gamma]$, $\tilde{\alpha}_\gamma=[5+2\gamma,8-\gamma]$, $\tilde{v}_\gamma=[6+2\gamma,9-\gamma]$, and $\tilde{\mu}_\gamma=[8+2\gamma,11-\gamma]$, which serve as the basis for evaluating the γ-cuts of the fuzzy availability $\tilde{A}$ derived as follows:

$$(A)_\gamma^L=-\frac{85\gamma^4+297\gamma^3-5,918\gamma^2-46,115\gamma-94,050}{155\gamma^4+2,395\gamma^3+16,600\gamma^2+64,247\gamma+105,282} \tag{14.19a}$$

$$(A)_\gamma^U=-\frac{4\gamma^4-156\gamma^3+2,377\gamma^2-24,782\gamma-123,144}{2\gamma^4-155\gamma^3+2,506\gamma^2-6,118\gamma+192,444} \tag{14.20a}$$

Obviously, the functions $(A)_\gamma^U$ and $(A)_\gamma^L$ are invertible in nature, which yields the membership function:

$$k_{\tilde{T}}(z)=\begin{cases} L(z), & \frac{1,466}{2,291}\le z\le\frac{48,567}{62,893} \\ 1, & \frac{48,567}{62,893} \\ R(z), & \frac{48,567}{62,893}\le z\le\frac{5,225}{5,849} \end{cases} \tag{14.21}$$

The functional forms of the membership functions $L(z)$ and $R(z)$ are cumbersome to express explicitly. The graphical representation of the membership function is illustrated in Figure 14.3. The γ-cut-based bounds of the fuzzy parameters and the corresponding availability values computed using triangular fuzzy numbers are tabulated in Table 14.3.

14.9 DISCUSSION

This study presents a comprehensive analysis of a 2-out-of-4: F repairable system with warm standby units and heterogeneous repairmen under fuzzy uncertainty. The integration of CTMCs with fuzzy set theory enables a robust evaluation of system performance where precise parameter estimation is not feasible. The use of exponential failure and repair distributions, coupled with a priority-based repair mechanism, allows for realistic modeling of heterogeneous repair scenarios. The application of Zadeh's extension principle and γ-cut method provides a systematic means to derive fuzzy MTTF and availability, while parametric NLP facilitates the computation

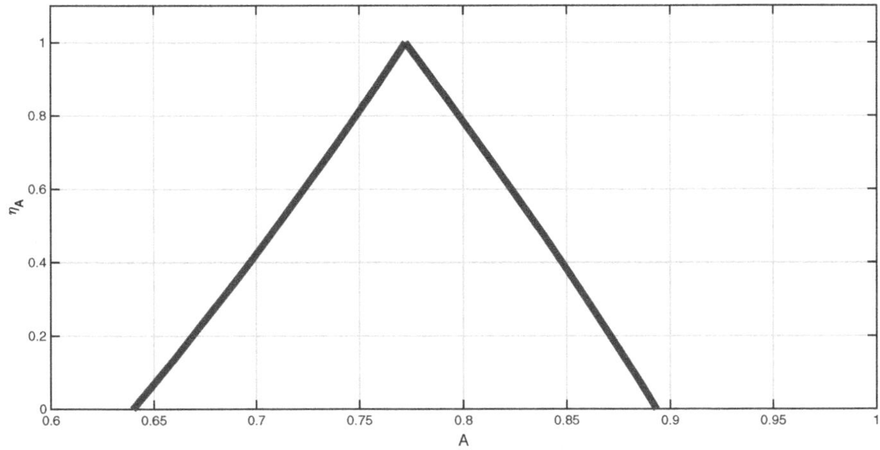

Figure 14.3 The State transition diagram for availability analysis of the two-unit repairable system with standby.

Table 14.3 α-cuts of the failure rate, the repair rate, and fuzzy availability

γ	λ_L	λ_U	α_L	α_U	μ_L	μ_U	v_L	v_U	A_L	A_U
0.0	4.0	7.0	5.0	8.0	6.0	9.0	8.0	11.0	0.63990	0.89332
0.1	4.2	6.9	5.2	7.9	6.2	8.9	8.2	10.9	0.65465	0.88242
0.2	4.4	6.8	5.4	7.8	6.4	8.8	8.4	10.8	0.66907	0.87116
0.3	4.6	6.7	5.6	7.7	6.6	8.7	8.6	10.7	0.68317	0.85959
0.4	4.8	6.6	5.8	7.6	6.8	8.6	8.8	10.6	0.69693	0.84773
0.5	5.0	6.5	6.0	7.5	7.0	8.5	9.0	10.5	0.71034	0.83562
0.6	5.2	6.4	6.2	7.4	7.2	8.4	9.2	10.4	0.72342	0.82328
0.7	5.4	6.3	6.4	7.3	7.4	8.3	9.4	10.3	0.73614	0.81074
0.8	5.6	6.2	6.6	7.2	7.6	8.2	9.6	10.2	0.74852	0.79803
0.9	5.8	6.1	6.8	7.1	7.8	8.1	9.8	10.1	0.76055	0.78518
1.0	6.0	6.0	7.0	7.0	8.0	8.0	10.0	10.0	0.77222	0.77222

of their membership functions. Numerical results confirm that system performance is highly sensitive to the disparity between the repair rates ($\mu > v$) and the failure behavior of standby units ($\alpha < \lambda$). As γ increases, the fuzzy intervals contract, indicating more confident reliability predictions. This reflects the utility of fuzzy methods in bounding risk and uncertainty. Moreover, defuzzification via Yager's index yields crisp reliability metrics that are suitable for decision-making in maintenance planning and redundancy allocation. The findings validate the effectiveness of fuzzy modeling in scenarios where conventional deterministic approaches are inadequate.

14.10 CONCLUSION

This chapter presents a fuzzy-based reliability framework for analyzing a 2-out-of-4: *F* repairable system with warm standby units and heterogeneous repairmen. By modeling system uncertainties using trapezoidal and triangular fuzzy numbers, the approach effectively captures the imprecision inherent in failure and repair rates. The combined use of CTMCs, Zadeh's extension principle, and parametric NLP enables accurate computation of fuzzy MTTF and steady-state availability. The results demonstrate that system performance is significantly influenced by the disparity in repair rates and the failure behavior of standby components. Fuzzy modeling provides bounded reliability metrics, supporting more robust decision-making in uncertain environments. Defuzzified values obtained through Yager's index further enable practical application in real-world systems, such as cloud infrastructures and industrial maintenance settings. Overall, the proposed model offers a powerful tool for evaluating and optimizing fault-tolerant systems under uncertainty and lays the groundwork for future extensions involving imperfect switching, load-sharing mechanisms, and cost-reliability trade-offs.

14.11 FUTURE SCOPE

The present work can be extended in several meaningful directions to enhance its applicability and analytical depth. Future studies may incorporate imperfect switching mechanisms and delayed fault detection to model more realistic operational environments. Additionally, considering time-dependent or non-exponential failure and repair distributions would allow the system to capture aging effects and variable workloads. The integration of load-sharing strategies among components, as well as dynamic repair priority policies based on queue length or component criticality, can provide further insight into optimal maintenance resource allocation. Moreover, cost-based optimization models could be developed by linking fuzzy reliability metrics with economic constraints to support design trade-offs between redundancy, availability, and operational cost. Finally, the framework can be generalized to larger *k*-out-of-*n*: *F* systems, multi-state components, and networked architectures, particularly for cloud data centers, smart grids, and cyber-physical systems where uncertainty and heterogeneity are intrinsic. Incorporating machine learning for adaptive parameter tuning under fuzzy conditions is another promising direction.

APPENDIX

$$\hat{P}_{10}(s) = \frac{4\lambda(\alpha + 2\lambda + v + s)(\alpha^2 + 3\lambda\alpha + 2\lambda^2)}{D_t}$$

$$\hat{P}_{20}(s) = \frac{2(\alpha + 2\lambda + v + s)(\alpha^2 + 3\lambda\alpha + 2\lambda^2)}{D_t}$$

$$\hat{P}_{21}(s) = \frac{2\left[\begin{array}{l} 2\alpha^2\lambda + \alpha^2 v + \alpha^2 s + 6\alpha\lambda^2 + 5\alpha\lambda v + 5\alpha\lambda s \\ +\alpha\mu v + \alpha\mu s + \alpha v^2 + 2\alpha v s + \alpha s^2 \\ +4\lambda^3 + 4\lambda^2 v + 4\lambda^2 s + \lambda\mu v + \lambda\mu s + \lambda v^2 + 2\lambda v s + \lambda s^2 \end{array}\right]}{D_t}$$

$$\hat{P}_{22}(s) = \frac{\left[\begin{array}{l} 2\alpha^2\lambda + \alpha^2 s + 8\alpha\lambda^2 + 2\alpha\lambda\mu + 2\alpha\lambda v + 8\alpha\lambda s \\ +2\alpha\mu v + 2\alpha\mu s + 2\alpha v s + 2\alpha s^2 + 8\lambda^3 + 4\lambda^2\mu + 4\lambda^2 v + 12\lambda^2 s \\ +6\lambda\mu v + 6\lambda\mu s + 6\lambda v s + 6\lambda s^2 + \mu^2 v + \mu^2 s + \mu v^2 \\ +3\mu v s + 2\mu s^2 + v^2 s + 2 v s^2 + s^3 \end{array}\right]}{D_t}$$

$$\hat{Q}_{21}(s) = \frac{2(\alpha^2 + 3\lambda\alpha + 2\lambda^2)\mu}{D_t}$$

where

$$\begin{aligned} D_t = {} & s(4\alpha^3\lambda + 2\alpha^3 s + 20\alpha^2\lambda^2 + 4\alpha^2\lambda v + 20\alpha^2\lambda s + 2\alpha^2\mu s + 4\alpha^2 v s + 5\alpha^2 s^2 \\ & +32\alpha\lambda^3 + 12\alpha\lambda^2 v + 48\alpha\lambda^2 s + 8\alpha\lambda\mu s + 18\alpha\lambda v s + 24\alpha\lambda s^2 \\ & +4\alpha\mu v s + 4\alpha\mu s^2 + 2\alpha v^2 s + 6\alpha v s^2 + 4\alpha s^3 \\ & +16\lambda^4 + 8\lambda^3 v + 32\lambda^3 s + 8\lambda^2\mu s + 16\lambda^2 v s + 24\lambda^2 s^2 \\ & +8\lambda\mu v s + 8\lambda\mu s^2 + 2\lambda v^2 s + 10\lambda v s^2 + 8\lambda s^3 \\ & +\mu^2 v s + \mu^2 s^2 + \mu v^2 s + 3\mu v s^2 + 2\mu s^3 + v^2 s^2 + 2 v s^3 + s^4) \end{aligned}$$

$$P_{10} = \frac{4\lambda(\alpha^3 + 5\lambda\alpha^2 + v\alpha^2 + 8\alpha\lambda^2 + 3\alpha\lambda v + 4\lambda^3 + 2\lambda^2 v)}{D_s}$$

$$P_{20} = \frac{2(2\lambda+\alpha+v)\left(\mu\alpha^2+v\alpha^2+3\alpha\lambda\mu+3\alpha\lambda v+2\lambda^2\mu+2\lambda^2 v\right)}{D_s}$$

$$P_{21} = \frac{2v\begin{pmatrix}\mu\alpha^2+v\alpha^2+3\alpha\lambda\mu+3\alpha\lambda v+\alpha\mu^2+2\alpha\mu v+\alpha v^2\\ +2\lambda^2\mu+2\lambda^2 v+\lambda\mu^2+2\lambda\mu v+\lambda v^2\end{pmatrix}}{D_s}$$

$$P_{22} = \frac{\mu v\begin{pmatrix}2\mu\alpha+2v\alpha+4\mu\lambda+4\lambda v\\ +\mu^2+2\mu v+v^2\end{pmatrix}}{D_s}$$

$$Q_{21} = \frac{2\mu\begin{pmatrix}\mu\alpha^2+v\alpha^2+3\alpha\lambda\mu+3\alpha\lambda v\\ +2\lambda^2\mu+2\lambda^2 v\end{pmatrix}}{D_s}$$

where

$$\begin{aligned}D_s = {} & 4\alpha^3\lambda+2\alpha^3\mu+2\alpha^3 v+20\lambda^2\alpha^2+10\mu\alpha^2\lambda+14\lambda\alpha^2 v+2\mu^2\alpha^2+6\mu\alpha^2 v\\ & +4\alpha^2 v^2+32\alpha\lambda^3+16\alpha\lambda^2\mu+28\alpha\lambda^2 v+6\alpha\lambda\mu^2+18\alpha\lambda\mu v+12\alpha\lambda v^2\\ & +4\alpha\mu^2 v+6\alpha\mu v^2+2\alpha v^3+16\lambda^4+8\lambda^3\mu+16\lambda^3 v+4\lambda^2\mu^2+12\lambda^2\mu v\\ & +8\lambda^2 v^2+6\lambda\mu^2 v+8\lambda\mu v^2+2\lambda v^3+\mu^3 v+2\mu^2 v^2+\mu v^3\end{aligned}$$

REFERENCES

1. A. Kumar and M. Agarwal, "A review of standby systems," *IEEE Transactions on Reliability*, vol. R-29, no. 4, pp. 290–294, 1980.
2. S. Osaki and T. Nakagawa, "Bibliography for reliability and availability of stochastic systems," *IEEE Transactions on Reliability*, vol. 25, no. 4, pp. 284–287, 1976.
3. C. Shekhar, A. Raina, A. Kumar, and J. Iqbal, "A survey on queues in machining system: progress from 2010 to 2017," *Yugoslav Journal of Operations Research*, vol. 27, no. 4, pp. 391–413, 2017.
4. R. D. Yearout, P. Reddy, and D. L. Grosh, "Standby redundancy in reliability – a review," *IEEE Transactions on Reliability*, vol. R-35, no. 3, pp. 285–292, 1986.
5. L. Haque and M. J. Armstrong, "A survey of the machine interference problem," *European Journal of Operational Research*, vol. 179, no. 2, pp. 469–482, 2007.

6. M. Rubin and D. Du, "A review of dynamic fault tree analysis and capacity degradation for complex redundant systems," *Quality and Reliability Engineering International*, vol. 41, no. 3, pp. 1161–1181, 2025.
7. V. Srinivasan, "The effect of standby redundancy in system's failure with repair maintenance," *Operations Research*, vol. 14, no. 6, pp. 1024–1036, 1966.
8. R. E. Barlow and F. Proschan, Mathematical Theory of Reliability. SIAM, 1996.
9. F. Gizatullin and O. Yushkova, "Reliability assessment of capacitive ignition systems with partial redundancy in normal and emergency operating modes," *Russian Engineering Research*, vol. 44, no. 12, pp. 1697–1700, 2024.
10. A. S. Bhandari, S. Rawat, Y. Sharma, M. Ram, and A. Kumar, "Optimizing reliability in total artificial heart design through redundancy strategies and swarm-inspired optimization," *International Journal of Applied and Computational Mathematics*, vol. 9, no. 6, p. 130, 2023.
11. J. Zhong, H. Zhang, and Q. Miao, "Enhancing aircraft reliability with information redundancy: a sensor-modal fusion approach leveraging deep learning," *Reliability Engineering & System Safety*, vol. 261, p. 111068, 2025.
12. Z. Zhang, L. Yang, Y. Xu, R. Zhu, and Y. Cao, "A novel reliability redundancy allocation problem formulation for complex systems," *Reliability Engineering & System Safety*, vol. 239, p. 109471, 2023.
13. J. E. Breneman, C. Sahay, and E. E. Lewis, *Introduction to Reliability Engineering*. John Wiley & Sons, 2022.
14. K.-H. Wang, J.-B. Ke, and J.-C. Ke, "Profit analysis of the M/M/R machine repair problem with balking, reneging, and standby switching failures," *Computers & Operations Research*, vol. 34, no. 3, pp. 835–847, 2007.
15. C. Shekhar, M. Jain, A. Raina, and R. Mishra, "Sensitivity analysis of repairable redundant system with switching failure and geometric reneging," *Decision Science Letters*, vol. 6, no. 4, pp. 337–350, 2017.
16. D. Yue, W. Yue, and H. Qi, "Performance analysis and optimization of a machine repair problem with warm spares and two heterogeneous repairmen," *Optimization and Engineering*, vol. 13, pp. 545–562, 2012.
17. J.-C. Ke, T.-H. Liu, and C.-H. Wu, "An optimum approach of profit analysis on the machine repair system with heterogeneous repairmen," *Applied Mathematics and Computation*, vol. 253, pp. 40–51, 2015.
18. A. Kumar, "Parametric optimization of repairable systems in IoT: addressing detection delays, imperfect coverage, and fuzzy parameters," *Life Cycle Reliability and Safety Engineering*, pp. 1–12, 2025.
19. M. Devanda, C. Shekhar, and S. Kaswan, "Fuzzified imperfect repair redundant machine repair problems," *International Journal of System Assurance Engineering and Management*, vol. 15, no. 4, pp. 1483–1502, 2024.
20. H.-I. Huang, C.-H. Lin, and J.-C. Ke, "Parametric nonlinear programming approach for a repairable system with switching failure and fuzzy parameters," *Applied Mathematics and Computation*, vol. 183, no. 1, pp. 508–517, 2006.
21. J.-C. Ke, H.-I. Huang, and C.-H. Lin, "A redundant repairable system with imperfect coverage and fuzzy parameters," *Applied Mathematical Modelling*, vol. 32, no. 12, pp. 2839–2850, 2008.
22. S. Malik, S. C. Malik, N. Nandal, and A. D. Yadav, "Reliability assessment of a NSP system under constant triangular fuzzy failure rates," *International Journal of Industrial and Systems Engineering*, vol. 49, no. 4, pp. 405–420, 2025.

23. A. D. Yadav, N. Nandal, S. Malik, and S. C. Malik, "Markov approach for reliability-availability-maintainability analysis of a three unit repairable system," *Opsearch*, vol. 60, no. 4, pp. 1731–1756, 2023.
24. M. S. Kadyan, J. Bura, and J. Kumar, "Application of picture fuzzy set theory to reliability and risk analysis of complex industrial systems," *Life Cycle Reliability and Safety Engineering*, vol. 13, no. 3, pp. 293–307, 2024.
25. M. Ram, S. Kharola, and N. Goyal, "Reliability and sensitivity analysis of a maintainable energy system under priority repair," *OPSEARCH*, vol. 62, pp. 1–29, 2024.
26. M. Ram, N. Goyal, and Shivani, "Standby system reliability modeling with cost and sensitivity analysis," *International Journal of Reliability, Quality and Safety Engineering*, vol. 32, p. 2502001, 2024.
27. M. Jain, C. Shekhar, and S. Shukla, "Machine repair problem with an unreliable server and controlled arrival of failed machines," *Opsearch*, vol. 51, no. 3, pp. 416–433, 2014.
28. A. Singh and S. B. Singh, "Dynamic reliability and sensitivity analysis of weighted (k,r)-out-of-n cold standby system with multi-performance multi-state components," *Reliability Engineering & System Safety*, vol. 262, p. 111221, 2025.
29. G. Levitin, L. Xing, and Y. Dai, "1-out-of-n standby systems with storages and activation moment-dependent component operation time limit," *Reliability Engineering & System Safety*, vol. 245, p. 110017, 2024.
30. L. A. Zadeh, "Fuzzy sets as a basis for a theory of possibility," *Fuzzy Sets and Systems*, vol. 1, no. 1, pp. 3–28, 1978.
31. R. R. Yager, "A procedure for ordering fuzzy subsets of the unit interval," *Information Sciences*, vol. 24, no. 2, pp. 143–161, 1981.

Chapter 15

Reliability analysis of an off-grid fire safety system

Mohd Anas, Mangey Ram, and Shristi Kharola

15.1 INTRODUCTION

In this modern world, everyone needs a non-stop operation of the engineering system or equipment. Before buying the equipment, one wants to know about its reliability, such as the possibility of the function, equipment, or system operating successfully under the given interval of time and conditions. Reliability plays a significant role in critical systems, as any disruption in service can lead to a hazardous event. A traditional fire safety system and security subsystem usually depend on the utility grid, which could be unavailable when it is needed during a hazardous event, such as outages during an emergency. This chapter addresses a series-parallel system in which an advanced solar energy system with a fire safety system is commonly introduced to enhance autonomous energy and provide an unstoppable service when the grid is unavailable. The reliability of the fire safety system in modern infrastructure plays a significant role, as its failure during an emergency can result in catastrophic loss of property and life. This chapter highlights the reliability of an autonomous off-grid solar energy fire safety system designed to operate independently from the utility grid. The construction of the system is defined by seven critical components, including photovoltaic (PV) panels, an inverter, a battery, two water pumps, a water controller sensor for pressure or level monitoring, an alarm unit for occupant notification, and a valve for water flow regulation. The complication of the interaction between the solar energy system and the electromechanical loads needs a robust reliability assessment. Andrey Andreyevich Markov was a Russian mathematician who developed the theory of stochastic process, which is a mathematical approach used to know the system transition between various states over time, such as failed, degraded, and operational. It shows the memoryless property, i.e., probability of moving to the next state depends only on the current state. This approach known as the Markov process and is used to evaluate the system reliability and performance, incorporating failure and repair rates for each component to determine the safe functioning of the system when needed. In the past, literature showed that several studies have been done

DOI: 10.1201/9781003787525-15

on the reliability enhancement of complex series-parallel systems. Shivani et al. (2023) investigated the reliability characteristics of many complex series-parallel system based on Markov process. Guilani et al. (2014) assessed the reliability of a non-repairable system based on Markov process and observed that the beta distribution aligns with the distribution of time to failure. Goyal and Ram (2014) assessed the reliability, availability, cost analysis, and mean time to failure (MTTF) of software development life cycle graphically and numerically through the Markov process. Bisht and Singh (2021) examined complex repairable system by the Interval Universal Generating Function (UGF) technique. Also analysed the system components state probabilities, reliability, MTTF, and sensitivity through the Markov process including one numerical example. Juang et al. (2008) optimized the system availability of an industrial series-parallel system through redundancy allocation problem and determined the mean time between failures (MTBF) and mean time to repair (MTTR). Malik et al. (2020) evaluated the reliability of a complex system by converting into an equivalent series-parallel system by logic diagram. Weibull distribution is applied to system components failures, and MTTF and reliability are derived using real-life application. Ram et al. (2013) analysed availability, MTTF, and cost analysis through the supplementary variable and Laplace transform approach, including human error based on Gumbel-Hougaard copula function. Yang and Xie (2000) examined software reliability by NHPP (non-homogeneous Poisson process) models by visually distinguishing between testing reliability and operative reliability. It shows how misuse of these concepts leads to different reliability assessment and decision; their impact on reliability evaluation and optimal software release time is observed. Yalaoui et al. (2005) addressed series-parallel system reliability with diverse technologies operative in parallel. It established theoretical and practical results through a one-stage and various stage techniques, and also proposed an optimal situation, introducing an estimated applied approach and derived through numerical samples. Tian et al. (2008) proposed an optimal model for various state series-parallel systems which jointly examined optimal redundancy at each stage and optimal component state allocation.

15.2 MODEL CONSTRUCTION

In this chapter, a mixed series-parallel system is used, in which subsystems A1, A2, A3, and A4 are connected in series. In subsystem A1, components C1, C2; in subsystem A2, components C3, C4; in subsystem A4, components C6, C7 are connected in parallel except subsystem A3 which consists of a single component C5. Figure 15.1 depicts the proposed system configuration, which is further modelled by Markov process and supplementary approach.

The proposed model is presented in three states: good state, degraded state (partially failed), and fully failed state. It is supposed that a system

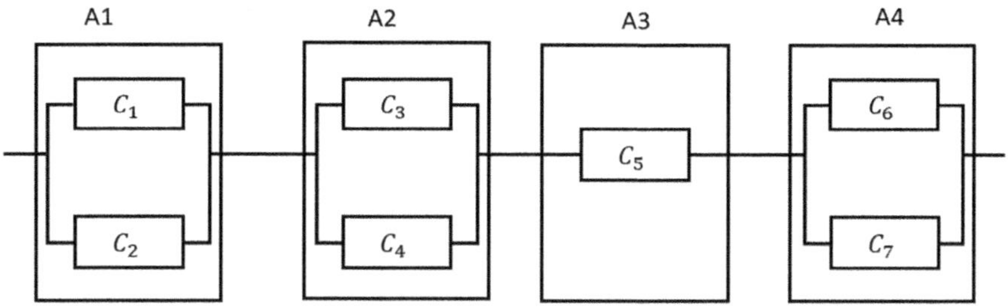

Figure 15.1 System block configuration.

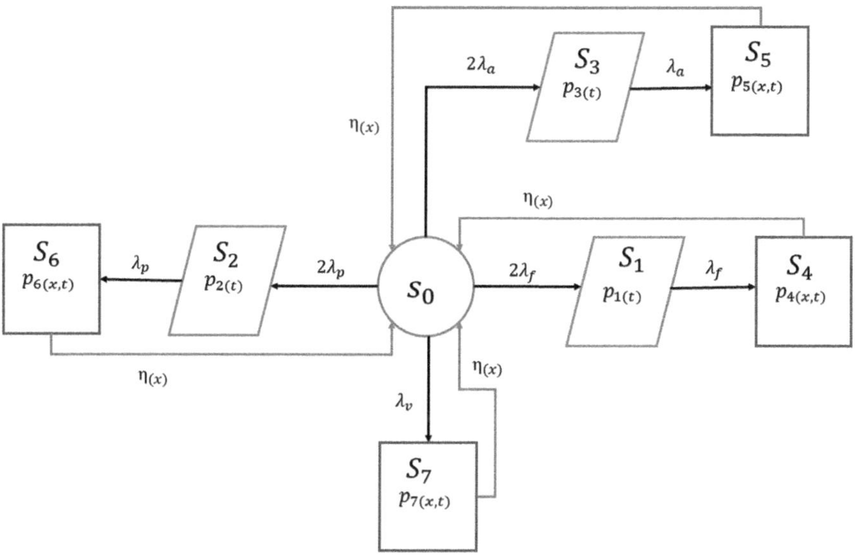

Figure 15.2 State transition diagram.

failure can be partially or fully failed. Whenever the system reaches its partially failed state, it is supposed that the system's internal components may be damaged or not working satisfactory. Fully failed state results from the complete shutdown of the entire system.

The state transition diagram of the system is shown in Figure 15.2.

15.3 ASSUMPTIONS

The proposed model considers the following assumptions:

1. Initially, all the system components are in good condition and perform satisfactorily.

Table 15.1 Nomenclature of the proposed system

Symbols	*Description*
S_i	Transition state for i = 0, 1, 2, 3, 4, 5, 6, 7
t	Time
S	Laplace transform variable
$p_{i(t)}$	The probability of system at state Si = 0, 1, 2, 3
$p_{j(x,t)}$	The probability of system at state Sj = 4, 5, 6, 7
$\lambda_f, \lambda_p, \lambda_a, \lambda_v$	Failure rate of solar, pump, alarm, and valve, respectively
$\eta(x)$	Repair rates of all fully failed states
p_{up}	Probability of the working states
p_{down}	Probability of the fully failed states
C_0	Cost of service

2. After some time, the system faces degraded failure states.
3. It is supposed that all the repair rates are equal.
4. The system failure rates λ_f, λ_p, λ_a, and λ_v are accepted as per their real-life hazard rates from the literature.

15.4 NOMENCLATURE

Table 15.1 shows the notation of the proposed model of transition diagram.

15.5 MATHEMATICAL FORMULATION OF THE MODEL

As per the proposed system transition state diagram, this section presents the set of differential equation, which has been developed through the Markov process.

$$\left(\frac{d}{dt}+2\lambda_f+2\lambda_p+2\lambda_a+\lambda_v\right)P_0(t)=\int_0^\infty\sum_{j=4}^{7}p_{j(x,t)}\eta(x)dx \tag{15.1}$$

$$\left(\frac{d}{dt}+2\lambda_f\right)P_1(t)=2\lambda_f P_0(t) \tag{15.2}$$

$$\left(\frac{d}{dt}+2\lambda_p\right)P_2(t)=2\lambda_p P_0(t) \tag{15.3}$$

$$\left(\frac{d}{dt}+2\lambda_a\right)P_3(t)=2\lambda_a P_0(t) \tag{15.4}$$

$$\left(\frac{d}{dx}+\frac{d}{dt}+\eta(x)\right)P_4(x,\,t)=0 \tag{15.5}$$

$$\left(\frac{d}{dx}+\frac{d}{dt}+\eta(x)\right)P_5(x,\,t)=0 \tag{15.6}$$

$$\left(\frac{d}{dx}+\frac{d}{dt}+\eta(x)\right)P_6(x,\,t)=0 \tag{15.7}$$

$$\left(\frac{d}{dx}+\frac{d}{dt}+\eta(x)\right)P_7(x,\,t)=0 \tag{15.8}$$

Possible boundary conditions:

$$P_4(0,t)=\lambda_f P_1(t) \tag{15.9}$$

$$P_5(0,t)=\lambda_a P_{13}(t) \tag{15.10}$$

$$P_6(0,t)=\lambda_p P_2(t) \tag{15.11}$$

$$P_7(0,t)=\lambda_v P_0(t) \tag{15.12}$$

At initial condition $P_0(t)=1$ and at $t=0$ all other state probabilities are zero.
Taking Laplace transformation in equations (15.1)–(15.12)

$$\left(s+2\lambda_f+2\lambda_p+2\lambda_a+\lambda_v\right)\overline{P}(s)=1+\int_0^{\infty}\sum_{j=4}^{7}\overline{P}_{j(x,s)}\eta(x)\,dx \tag{15.13}$$

$$\left(s+2\lambda_f\right)\overline{P}_1(t)=2\lambda_f\overline{P}_0(s) \tag{15.14}$$

$$\left(s+2\lambda_p\right)\overline{P}_2(t)=2\lambda_f\overline{P}_0(s) \tag{15.15}$$

$$\left(s+2\lambda_a\right)\overline{P}_3(t)=2\lambda_a\overline{P}_0(s) \tag{15.16}$$

$$\left(\frac{d}{dx}+s+\eta(x)\right)\bar{P}_4(x,\ s)=0 \tag{15.17}$$

$$\left(\frac{d}{dx}+s+\eta(x)\right)\bar{P}_5(x,\ s)=0 \tag{15.18}$$

$$\left(\frac{d}{dx}+s+\eta(x)\right)\bar{P}_6(x,\ s)=0 \tag{15.19}$$

$$\left(\frac{d}{dx}+s+\eta(x)\right)\bar{P}_7(x,\ s)=0 \tag{15.20}$$

Boundary conditions:

$$\bar{P}_4(0,s)=\lambda_f\bar{P}_1(s) \tag{15.21}$$

$$\bar{P}_5(0,s)=\lambda_a\bar{P}_3(s) \tag{15.22}$$

$$\bar{P}_6(0,s)=\lambda_p\bar{P}_2(s) \tag{15.23}$$

$$\bar{P}_7(0,s)=\lambda_v\bar{P}_0(s) \tag{15.24}$$

Now solving equations (15.13)–(15.20) based on (15.21)–(15.24) we get the probabilities of transition states.

$$\bar{P}_0(s)=\frac{1}{\left[\left(s+2\lambda_f+2\lambda_p+2\lambda_a+\lambda_v\right)-S\mu(s)\frac{2\lambda_f^{\,2}}{\left(s+\lambda_f\right)}+\frac{2\lambda_a^{\,2}}{\left(s+\lambda_a\right)}+\frac{2\lambda_p^{\,2}}{\left(s+\lambda_p\right)}+\lambda_v\right]} \tag{15.25}$$

$$\bar{P}_1(s)=\frac{2\lambda_f\bar{P}_0(s)}{\left(s+\lambda_f\right)} \tag{15.26}$$

$$\bar{P}_2(s)=\frac{2\lambda_p\bar{P}_0(s)}{\left(s+\lambda_p\right)} \tag{15.27}$$

$$\bar{P}_3(s)=\frac{2\lambda_a\bar{P}_0(s)}{\left(s+\lambda_a\right)} \tag{15.28}$$

$$\bar{P}_4(s)=\frac{2\lambda_f^{\,2}}{(s+\lambda_f)}\bar{P}_0(s)\left[\frac{1-S\mu(s)}{s}\right] \tag{15.29}$$

$$\bar{P}_5(s)=\frac{2\lambda_a^{\,2}}{(s+\lambda_a)}\bar{P}_0(s)\left[\frac{1-S\mu(s)}{s}\right] \tag{15.30}$$

$$\bar{P}_6(s)=\frac{2\lambda_p^{\,2}}{(s+\lambda_p)}\bar{P}_0(s)\left[\frac{1-S\mu(s)}{s}\right] \tag{15.31}$$

$$\bar{P}_7(s)=\lambda_v\bar{P}_0(s)\left[\frac{1-S\mu(s)}{s}\right] \tag{15.32}$$

The upstate probability of the system is given as:

$$\bar{P}_{\text{up}}=\bar{P}_0+\bar{P}_1(s)+\bar{P}_2(s)+\bar{P}_3(s) \tag{15.33}$$

$$\bar{P}_{\text{down}}=\bar{P}_4(s)+\bar{P}_5(s)+\bar{P}_6(s)+\bar{P}_7(s) \tag{15.34}$$

15.6 ASSESSMENT OF RELIABILITY CHARACTERISTICS

Reliability is the probability that a system performs its best operation under a given situation for a given interval of time. To assess the reliability of the proposed system the failure rates for parameters are assumed as $\lambda_f=0.5$, $\lambda_a=0.2$, $\lambda_p=0.08$, $\lambda_v=0.13$ and the repair rate is assumed as $\eta(x)=0$ in equation (15.33).

$$R(t)=e^{(-1.690000000t)}+0.680672269e^{(-1.095000000t)}\sinh(0.5950000000t) \tag{15.35}$$

$$+0.9187577640e^{(-0.88500000000t)}\sinh(0.8050000000t)e^{(-0.94500000000t)}$$

$$\sinh(0.7450000000t)+0.5369127517$$

On varying the time limit 0–10, in equation (15.35), the reliability of the system is depicted in Table 15.2 and Figure 15.3 respectively.

Table 15.2 Reliability analysis

Time (t)	*Reliability*
0	1
1	0.7828
2	0.5666
3	0.4117
4	0.3062
5	0.2343
6	0.1841
7	0.1483
8	0.1219
9	0.1020
10	0.0866

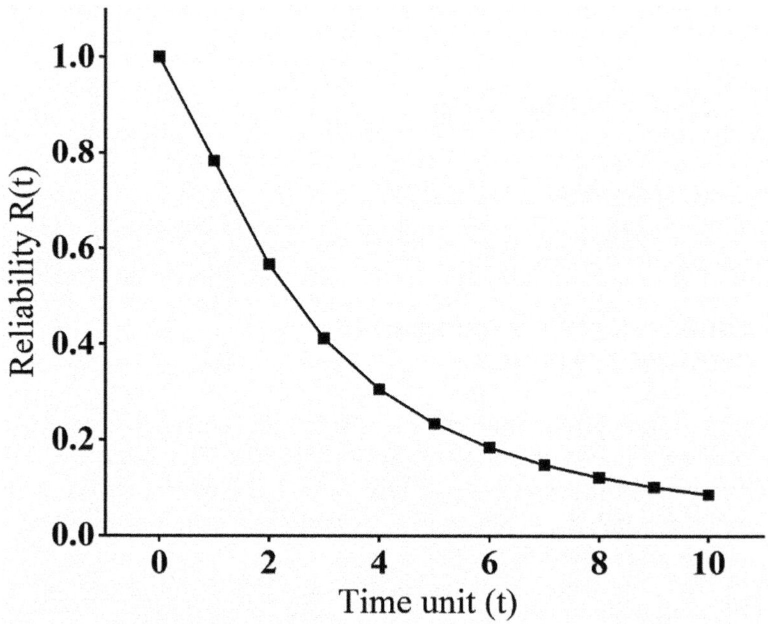

Figure 15.3 Graphical representation of reliability.

15.7 ASSESSMENT OF AVAILABILITY OF THE SYSTEM

Availability is the possibility that a system performs its intended function at any given point of time. To assess the availability of the proposed system the failure rates are supposed as $\lambda_f = 0.5$, $\lambda_a = 0.2$, $\lambda_p = 0.08$, $\lambda_v = 0.13$ and the repair rate as constant $\eta(x) = 1$. Taking the parameters in equation (15.33) and applying the inverse Laplace transformation, equation (15.36) is obtained:

$$A(t) = 0.01126823904 + e^{(-1.9344362688t)} + 0.02634717481e^{(-0.2897545036t)}$$
$$+0.0571908020e^{(-0.1060684154t)} + 0.9277302624e^{(-0.02274081316t)} \tag{15.36}$$

And now varying the time 0–10 in equation (15.36), the availability table and graph of the same is depicted in Table 15.3 and Figure 15.4, respectively.

Table 15.3 Availability analysis

Time (t)	*Availability*
0	1
1	0.9763
2	0.9472
3	0.9191
4	0.8927
5	0.8678
6	0.8442
7	0.8218
8	0.8004
9	0.7799
10	0.7602

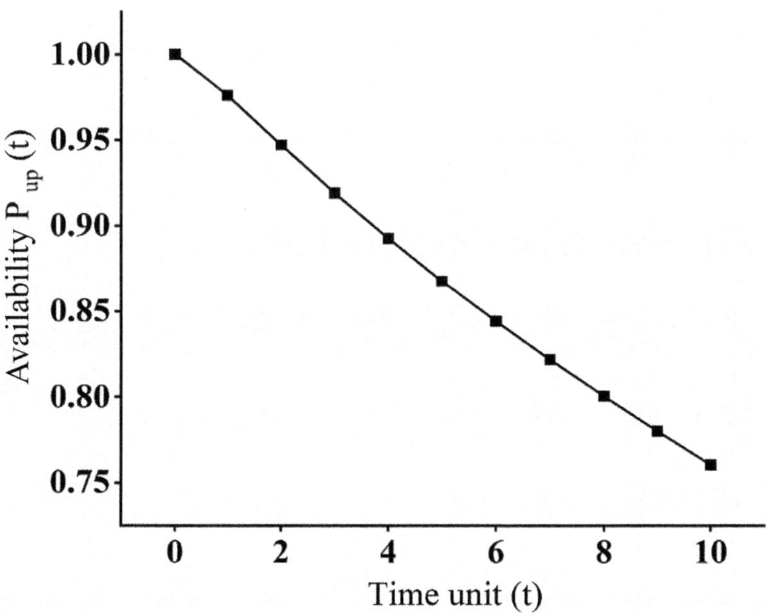

Figure 15.4 Graphical representation of availability.

Table 15.4 MTTF of the system

Fluctuation in failure rate	λ_f	λ_p	λ_a	λ_v
0.1	13.2075	5.1094	4.6979	5.3846
0.2	9.5890	4.4585	4.1420	5.0000
0.3	7.5268	3.9548	3.7037	4.6666
0.4	6.1946	3.5532	3.3492	4.3750
0.5	5.2631	3.2258	3.0567	4.1176
0.6	4.5751	2.9535	2.8112	3.8888
0.7	4.0462	2.7237	2.6022	3.6842
0.8	3.6269	2.5270	2.4221	3.5000
0.9	3.2863	2.3569	2.2653	3.3333

15.8 ASSESSMENT OF MTTF

MTTF is the average time for a non-repairable system before its first breakdown. In equation (15.33), setting the repair rate = 0 and allowing the Laplace variable 's' to approach zero gives the MTTF.

$$\text{MTTF} = \lim_{s \to 0} \overline{P}_{\text{up}} \tag{15.37}$$

or,

$$\text{MTTF} = \frac{7}{\left[2\lambda_f + 2\lambda_p + 2\lambda_a + 2\lambda_v\right]} \tag{15.38}$$

Considering the failure rates as $\lambda_f = 0.5$, $\lambda_p = 0.08$, $\lambda_a = 0.2$, and $\lambda_v = 0.13$ and varying them individually from a scale of 0.1 to 0.9, MTTF is obtained as shown in Table 15.4 and graphically in Figure 15.5.

15.9 ASSESSMENT OF EXPECTED COST

Expected cost is the systematic process of studying, comparing, and estimating the associated costs of the system. If the service facility is assumed to be always accessible, the expected cost accounting for service expenses for the range $[0, t]$ is as follows:

$$E_p(t) = K_1 \int_0^t P_{\text{up}}(t)\,dt - tK_2 \tag{15.39}$$

In equation (15.39), setting $K_1 = 1$ and $K_2 = 0.1, 0.2, 0.3, 0.4, 0.5$ respectively, the expected cost is calculated as given in Table 15.5 and in Figure 15.6.

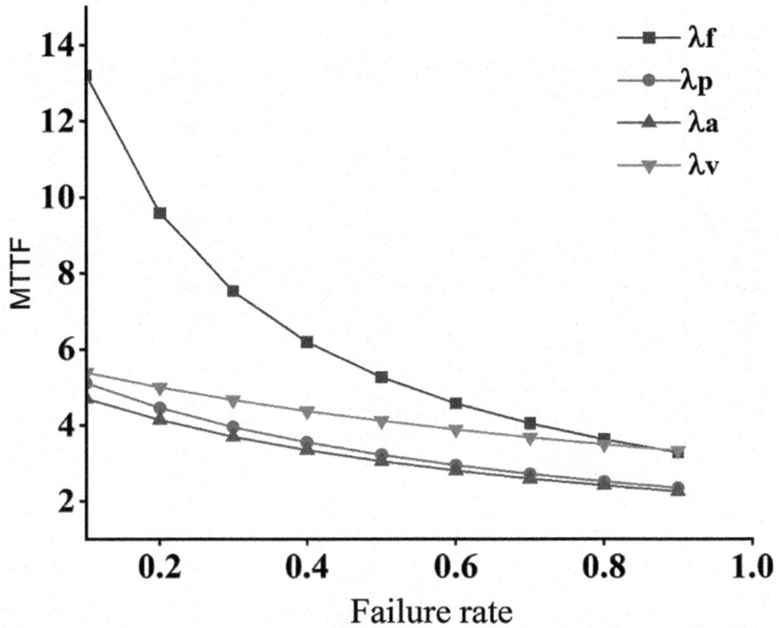

Figure 15.5 Graphical representation of MTTF.

Table 15.5 Expected cost analysis

Time (t)	$K_2 = 0.1$	$K_2 = 0.2$	$K_2 = 0.3$	$K_2 = 0.4$	$K_2 = 0.5$
0	0	0	0	0	0
1	0.8836	0.7836	0.6836	0.5836	0.4836
2	1.7497	1.5497	1.3497	1.1497	0.9497
3	2.6097	2.3097	2.0097	1.7097	1.4097
4	3.4675	3.0675	2.6675	2.2675	1.8675
5	4.3245	3.8245	3.3245	2.8245	2.3245
6	5.1812	4.5812	3.9812	3.3812	2.7812
7	6.0379	5.3379	4.6379	3.9379	3.2379
8	6.8945	6.0945	5.2945	4.4945	3.6945
9	7.7511	6.8511	5.9511	5.0511	4.1511

15.10 RESULTS AND DISCUSSION

The proposed system based on off-grid fire safety system shows a reducing reliability with probability of uninterrupted operation decrease from 1.0 to 0.0866 under the given interval of time (0–10), which highlights an obvious time-dependent vulnerability of the proposed system under selected failure

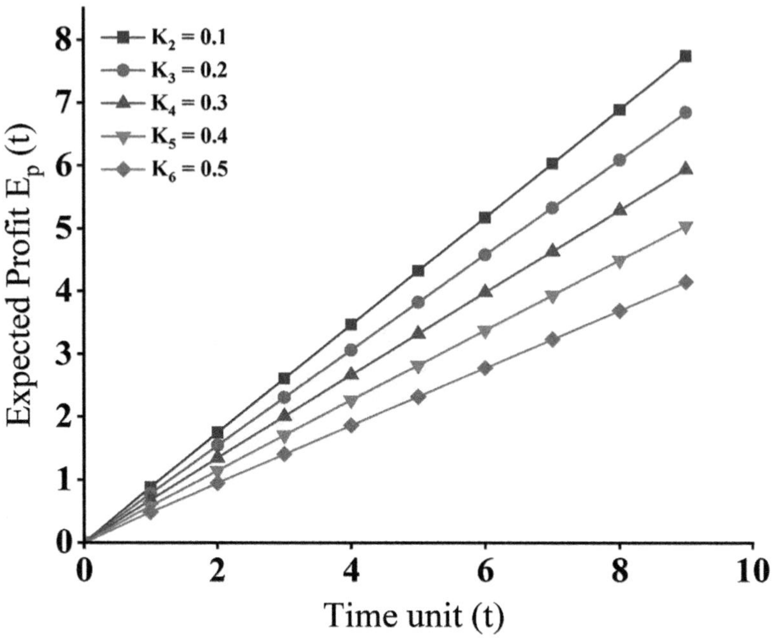

Figure 15.6 Graphical representation of cost analysis.

rate. Moreover, availability remains high whenever repairs are included, provided that with proper maintenance, the system can continue to function over the majority of its life and meet safety regulations in situations where the grid is unreliable. Design efforts should prioritize upgrading PV and related power components, as MTTF assessment displays that enhancing component failure rates significantly shortens system lifespan, with solar subsystem parameters 'λf' having the biggest influence. However, cumulative costs increase over time, high maintenance cost rates K_2 eventually lower long-term operating costs, according to projected cost results, indicating that investing in preventive and corrective maintenance is financially beneficial. In general, these results provide reliance to the feasibility of the solar-powered fire safety system, as long as proactive maintenance is implemented and significant reliability components are upgraded.

15.11 CONCLUSION

In this chapter, authors propose a hybrid system combining a fire safety system with solar energy. The cost, MTTF, reliability, and availability of the proposed system are analysed and presented graphically and numerically. The measured steady-state availability demonstrates that the solar-powered fire safety system remains functional for a considerable duration, even among component failures and maintenance scenarios. Despite irregular solar energy availability, the system ensures adequate availability through energy storage (battery), rendering it appropriate for remote or power-scarce areas. The results indicate that the MTTF of the solar-powered fire safety system diminishes consistently as failure rate fluctuations increase. This suggests that the failure intensities substantially shorten the system's operational lifespan, with λ_f exerting the most significant influence. For all values of K_2, the predicted cost rises with time, but larger values of K_2 consistently result in a lower expected cost, suggesting that greater maintenance and repair expenditures lower the system's long-term operating costs. In order to further improve the sustainability and long-term dependability of the solar-powered fire safety system, further investigations may expand the suggested model by adding battery deterioration and preventive maintenance techniques while assessing the environmental impact and carbon savings.

REFERENCES

Bisht, S., & Singh, S. B. (2021). Reliability evaluation of repairable parallel-series multi-state system implementing interval valued universal generating function. *Journal of Reliability and Statistical Studies*, *14*, 81–120.

Goyal, N., & Ram, M. (2014). Software development life cycle testing analysis: A reliability approach. *Mathematics in Engineering, Science & Aerospace*, 5(3), 313–329.

Guilani, P. P., Sharifi, M., Niaki, S. T. A., & Zaretalab, A. (2014). Reliability evaluation of non-reparable three-state systems using Markov model and its comparison with the UGF and the recursive methods. *Reliability Engineering & System Safety*, 129, 29–35.

Juang, Y. S., Lin, S. S., & Kao, H. P. (2008). A knowledge management system for series-parallel availability optimization and design. *Expert Systems with Applications*, 34(1), 181–193.

Malik, S. C., Chauhan, S. K., & Ahlawat, N. (2020). Reliability analysis of a non series–parallel system of seven components with Weibull failure laws. *International Journal of System Assurance Engineering and Management*, 11(3), 577–582.

Ram, M., Singh, S. B., & Singh, V. V. (2013). Stochastic analysis of a standby system with waiting repair strategy. *IEEE Transactions on Systems, Man, and Cybernetics: Systems*, 43(3), 698–707.

Shivani, S., Ram, M., Goyal, N., & Kumar, A. (2023). Analysis of series–parallel system's sensitivity in context of components failures. *RAIRO-Operations Research*, 57(4), 2131–2149.

Tian, Z., Zuo, M. J., & Huang, H. (2008). Reliability-redundancy allocation for multi-state series-parallel systems. *IEEE Transactions on Reliability*, 57(2), 303–310.

Yalaoui, A., Chu, C., & Chatelet, E. (2005). Reliability allocation problem in a series–parallel system. *Reliability Engineering & System Safety*, 90(1), 55–61.

Yang, B., & Xie, M. (2000). A study of operational and testing reliability in software reliability analysis. *Reliability Engineering & System Safety*, 70(3), 323–329.

Chapter 16

Availability and reliability analysis of an interchangeable three-unit series-parallel system

Venkata Seshachala Sarma Yadavalli, Balakrishnan Thilaka, Sankaralingam Hemanth Kumar, and Michael Ayomoh

16.1 INTRODUCTION

Reliability systems are those consisting of one or more component units that are unreliable and prone to failure and repair. A unit of a reliability system works in tandem with other units (if there are any) of the system. Tremendous progress has been made in the past in modelling and analysis of multi-component reliability systems (see Barlow and Prochan (1965), Gnedenko et al. (1969), Birolini (2017)). The extensive bibliography provided by Osaki and Nakagawa (1976) depicts the early interest and the volume of research work carried out on stochastic models on system reliability using stochastic processes such as Markov chains, Markov processes, renewal processes, and semi-Markov (Markov renewal) processes. Reliability analysis of three-unit repairable systems has been carried out by several researchers by incorporating various inherent features such as differentiable failure times, repair times having general distributions, repairman taking vacation (either single or multiple) when there are no failed units under repair, imperfect switching of repaired units, priority units, random environment (see, for example, Sarma and Parvez (1984), Goel et al. (1989), Song and Deng (1993), Li et al. (1998), Kovalenko (2001), Mwanga et al. (2009), Hu et al. (2010)). Sarma and Parvez (1984) considered a three-unit standby redundant system with arbitrarily distributed discrete lifetime and obtained the availability of the system using the regeneration point technique. Goel et al. (1989) analysed a three-unit redundant system with two types of repairs and inspection, and they obtained several reliability characteristics such as mean time to system failure, availability, busy period, expected number of visits by the repairman by using the regeneration point technique. Song and Deng (1993) studied a three-unit parallel repairable system with one repair facility operating in a random environment by using Markov renewal processes and obtained system reliability function, system availability, and system failure frequency. Li et al. (1998) studied a three-dissimilar unit repairable system with two different repair facilities and obtained explicit expressions for state probabilities and several performance measures such as availability, failure frequency, renewal

 DOI: 10.1201/9781003787525-16

frequency, and reliability of the system. Kovalenko (2001) has obtained stationary probabilities of a three-component system consisting of one master control element and two slave elements with priority servicing by a single repair facility. Mwanga et al. (2009) studied a three-unit system with a preparation time for the repair facility and obtained steady-state availability of the system. Hu et al. (2010) studied a three-unit series-parallel repairable system with a single repairman and single vacation, and they obtained explicit expressions for the steady-state probabilities, steady-state availability, and the mean up-time of the system by using supplementary variable method. Recently, Kadyan et al. (2020) did a stochastic analysis of a repairable system of three units by taking one unit in operation and two identical units in cold standby controlled by a switch. They obtained some important reliability measures including mean sojourn times, transition probabilities, mean time to system failure, availability, expected number of repairs for both kinds of units separately, expected number of visits by the server, and busy period analysis of the server due to repair. Hu et al. (2020) investigated the steady-state availability of a repairable series-parallel system with redundant dependency by Markov theory and matrix analysis method. Choudhary et al. (2024) obtained numerous reliability aspects of a designed system which is made up of three subsystems that are interconnected in a series-parallel manner.

It is observed from the foregoing literature survey that transient analysis of three-unit series-parallel repairable system needs to be given a fair attention. The present chapter fills the gap by presenting a time-dependent solution for the problem of finding the reliability characteristics of a three-unit series-parallel reliability system with a single repairman.

The chapter is organised as follows: Section 16.2 describes the stochastic model of the three-unit series-parallel system with repair. A system of interconnected integral equations is derived in Section 16.3 using renewal-theoretic method. Time-dependent state probabilities are explicitly obtained in Section 16.4. In Section 16.5, availability function of the system is found. Reliability function of the system is derived in Section 16.6. The mean down-time of the system is found in Section 16.7. A numerical example is discussed in Section 16.8. Section 16.9 provides a conclusion of the chapter.

16.2 MODEL DESCRIPTION

Consider a system consisting of three unreliable units and a single repairman. One of the units is used to function in series and the other two are to work in parallel. If a unit fails while working, it is immediately taken to the repairman and the repair starts immediately under first-come-first-serve policy. In other words, there are three subsystems in the system, namely series subsystem, parallel subsystem, and repair facility. While a unit functions in the series subsystem, its lifetime is exponentially distributed with mean $\frac{1}{\mu_1}$. On the other hand, while a unit is functioning in the parallel subsystem, its lifetime is exponentially distributed with mean $\frac{1}{\mu_2}$. When the functioning unit in the series subsystem fails, the failed unit is immediately taken for repair and the repair of it commences immediately, and one of the functioning units (if any) in the parallel subsystem is switched immediately to function in series subsystem. The repair time for any failed unit is exponentially distributed with mean $\frac{1}{\gamma}$. When the series subsystem is functioning, if either of the units in the parallel subsystem fails, then the system becomes a two-unit series system, and the failed unit (say W) is immediately taken for repair and the repair of it commences immediately. If during the repair time of the unit W, neither the unit in the series subsystem nor the unit in the parallel subsystem fails, then the unit W is switched immediately after its repair completion to function in the parallel subsystem and the system now becomes a three-unit series-parallel system. If during the repair time of W, the unit in the series subsystem fails, then it is immediately queued in the repair facility and the system enters into down state. During downtime, the only unit in parallel subsystem remains cold (does not fail). If during the repair time of the failed parallel unit W, the unit in the series subsystem does not fail and the unit in the parallel subsystem fails, then it is immediately queued before the repairman and the system enters into down state. During every downtime, the only unit in the series subsystem remains cold. When one unit is in the parallel subsystem and two units are in the repair facility, the unit undergoing repair is switched to function as the unit in the series subsystem so that the system gets released from the down state and begins to function as a two-unit series system. It is impossible to have all units together in the repair facility at any time.

16.2.1 Notation

$X_S(t)$: Number of units in the series subsystem at time t.

$X_P(t)$: Number of units in the parallel subsystem at time t.

$X_R(t)$: Number of units in the repair facility at time t.

$Z(t): (X_S(t), X_P(t), X_R(t))$

μ_1 : Failure rate of a unit while working in series subsystem.

μ_2 : Failure rate of a unit while working in parallel subsystem.

γ : Repair rate of a unit in the repair facility.

$\phi(t) \copyright \psi(t) : \int_0^t \phi(u)\psi(t-u)\,du =$ Convolution of $\phi(t)$ and $\psi(t)$.

$v^*(s) : \int_0^\infty e^{-st} v(t)\,dt =$ Laplace transform of $v(t)$.

16.3 GOVERNING EQUATIONS

By definition, we have

$$X_S(t) = \begin{cases} 1 & \text{if one unit is in the series subsystem at time } t, \\ 0 & \text{if no unit is in the series subsystem at time } t. \end{cases}$$

$$X_P(t) = \begin{cases} 2 & \text{if two units are in the parallel subsystem at time } t, \\ 1 & \text{if one unit is in the parallel subsystem at time } t, \\ 0 & \text{if no unit is in the parallel subsystem at time } t. \end{cases}$$

$$X_R(t) = \begin{cases} 2 & \text{if two units are in the repair facility at time } t, \\ 1 & \text{if one unit is in the repair facility at time } t, \\ 0 & \text{if no unit is in the repair facility at time } t. \end{cases}$$

The three-dimensional stochastic process $\{Z(t) \mid t \geq 0\}$ is a Markov process with state space given by $\Omega = \{(1,2,0),(1,1,1),(1,0,2),(0,1,2)\}$. The state transition diagram is as shown in Figure 16.1.

We assume that the system starts from the state $(1,2,0)$ at time $t = 0$. We define the state probability function for the system by

$$P(i,j,k,t) = Pr\left[X_S(t) = i, X_P(t) = j, X_R(t) = k\right], (i,j,k) \in \Omega.$$

Using probability laws, we obtain the following system of integral equations:

$$P(1,2,0,t) = e^{-(\mu_1+2\mu_2)t} + \gamma \int_0^t P(1,1,1,u) e^{-(\mu_1+2\mu_2)(t-u)}\, du, \tag{16.1}$$

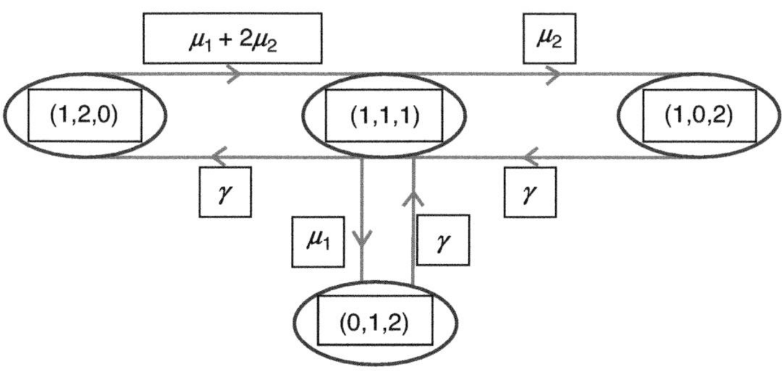

Figure 16.1 State transition diagram.

$$P(1,1,1,t)=\int_0^t\left[(\mu_1+2\mu_2)P(1,2,0,u)+\gamma\left\{P(0,1,2,u)+P(1,0,2,u)\right\}\right] \tag{16.2}$$

$$e^{-(\mu_1+\mu_2+\gamma)(t-u)}du,$$

$$P(0,1,2,t)=\mu_1\int_0^t P(1,1,1,u)e^{-\gamma(t-u)}\,du, \tag{16.3}$$

$$P(1,0,2,t)=\mu_2\int_0^t P(1,1,1,u)e^{-\gamma(t-u)}\,du \tag{16.4}$$

16.4 STATE PROBABILITIES

Taking Laplace transform of (16.1)–(16.4), we have

$$(s+\mu_1+2\mu_2)P^*(1,2,0,s)=1+\gamma P^*(1,1,1,s), \tag{16.5}$$

$$(s+\mu_1+\mu_2+\gamma)P^*(1,1,1,s)=(\mu_1+2\mu_2)P^*(1,2,0,s)$$
$$+\gamma\left[P^*(0,1,2,s)+P^*(1,0,2,s)\right], \tag{16.6}$$

$$(s+\gamma)P^*(0,1,2,s)=\mu_1P^*(1,1,1,s), \tag{16.7}$$

$$(s+\gamma)P^*(1,0,2,s)=\mu_2P^*(1,1,1,s) \tag{16.8}$$

Using (16.5), (16.7), and (16.8), we have

$$P^*(1,2,0,s)=\frac{1}{(s+\mu_1+2\mu_2)}+\frac{\gamma P^*(1,1,1,s)}{(s+\mu_1+2\mu_2)}, \tag{16.9}$$

$$P^*(0,1,2,s)=\frac{\mu_1P^*(1,1,1,s)}{(s+\gamma)}, \tag{16.10}$$

$$P^*(1,0,2,s)=\frac{\mu_2P^*(1,1,1,s)}{(s+\gamma)} \tag{16.11}$$

By total probability axiom, we have

$$P^*(1,2,0,s)+P^*(1,1,1,s)+P^*(1,0,2,s)+P^*(0,1,2,s)=\frac{1}{s} \tag{16.12}$$

Substituting (16.9), (16.10), and (16.11) into (16.12) and solving for $P^*(1,1,1,s)$, we have

$$P^*(1,1,1,s)=\frac{(\mu_1+2\mu_2)(s+\gamma)}{s\left[s^2+(2\mu_1+3\mu_2+2\gamma)s+(\mu_1+2\mu_2)(\mu_1+\mu_2+\gamma)+\gamma^2\right]} \tag{16.13}$$

The zeros of $s^2+(2\mu_1+3\mu_2+2\gamma)s+(\mu_1+2\mu_2)(\mu_1+\mu_2+\gamma)+\gamma^2$ are given by

$$\alpha=\frac{-(2\mu_1+3\mu_2+2\gamma)+\sqrt{\mu_2^2+4\gamma(\mu_1+\mu_2)}}{2},$$

$$\beta=\frac{-(2\mu_1+3\mu_2+2\gamma)-\sqrt{\mu_2^2+4\gamma(\mu_1+\mu_2)}}{2}$$

They satisfy the following properties:

(i) α and β are real and distinct,

(ii) $\alpha\beta=(\mu_1+2\mu_2)(\mu_1+\mu_2+\gamma)+\gamma^2>0$,

(iii) $0>\alpha>\beta$,

(iv) $|\alpha|=-\alpha, |\beta|=-\beta$

Using the zeros α and β in (16.13), we get

$$P^*(1,1,1,s)=\frac{(\mu_1+2\mu_2)(s+\gamma)}{s(s-\alpha)(s-\beta)} \tag{16.14}$$

Splitting (16.14) into partial fractions, we get

$$P^*(1,1,1,s)=\frac{\gamma(\mu_1+2\mu_2)}{\alpha\beta s}+\frac{(\mu_1+2\mu_2)}{(\alpha-\beta)}\left[\frac{(\alpha+\gamma)}{\alpha(s-\alpha)}-\frac{(\beta+\gamma)}{\beta(s-\beta)}\right] \tag{16.15}$$

Taking inverse Laplace transform of both sides of (16.15), we get

$$P(1,1,1,t)=\frac{\gamma(\mu_1+2\mu_2)}{\alpha\beta}+\frac{(\mu_1+2\mu_2)}{(\alpha-\beta)}\left[\frac{(\alpha+\gamma)}{\alpha}e^{-|\alpha|t}-\frac{(\beta+\gamma)}{\beta}e^{-|\beta|t}\right] \tag{16.16}$$

Substituting (16.14) into (16.9), we get

$$P^*(1,2,0,s)=\frac{s(s-\alpha)(s-\beta)+\gamma(\mu_1+2\mu_2)(s+\gamma)}{s(s-\alpha)(s-\beta)(s+\mu_1+2\mu_2)} \tag{16.17}$$

Splitting into partial fractions, (16.17) yields

$$\begin{aligned}P^*(1,2,0,s)=&\frac{\gamma^2}{s(\alpha\beta)}+\frac{\gamma(\mu_1+2\mu_2)(\alpha+\gamma)}{\alpha(\alpha+\mu_1+2\mu_2)(s-\alpha)(\alpha-\beta)}\\&+\frac{\gamma(\mu_1+2\mu_2)(\beta+\gamma)}{\beta(\beta+\mu_1+2\mu_2)(\beta-\alpha)(s-\beta)}\\&+\frac{(\mu_1+2\mu_2+\alpha)(\mu_1+2\mu_2+\beta)+\gamma(\mu_1+2\mu_2-\gamma)}{(s+\mu_1+2\mu_2)(\mu_1+2\mu_2+\alpha)(\mu_1+2\mu_2+\beta)}\end{aligned} \tag{16.18}$$

Taking the inverse Laplace transform of both sides of (16.18), we have

$$\begin{aligned}P(1,2,0,t)=&\frac{\gamma^2}{(\alpha\beta)}+\frac{\gamma(\mu_1+2\mu_2)}{(\alpha-\beta)}\left\{\frac{(\alpha+\gamma)e^{-|\alpha|t}}{\alpha(\alpha+\mu_1+2\mu_2)}-\frac{(\beta+\gamma)e^{-|\beta|t}}{\beta(\beta+\mu_1+2\mu_2)}\right\}\\&+\frac{(\mu_1+2\mu_2+\alpha)(\mu_1+2\mu_2+\beta)+\gamma(\mu_1+2\mu_2-\gamma)}{(\mu_1+2\mu_2+\alpha)(\mu_1+2\mu_2+\beta)}e^{-(\mu_1+2\mu_2)t}\end{aligned} \tag{16.19}$$

Substituting (16.14) into (16.10), we get

$$P^*(0,1,2,s)=\frac{\mu_1(\mu_1+2\mu_2)}{s(s-\alpha)(s-\beta)} \tag{16.20}$$

Splitting into partial fractions, (16.20) gives

$$P^*(0,1,2,s)=\mu_1(\mu_1+2\mu_2)\left[\frac{1}{\alpha\beta}\frac{1}{s}+\frac{1}{(\alpha-\beta)}\left\{\frac{1}{\alpha(s-\alpha)}-\frac{1}{\beta(s-\beta)}\right\}\right] \tag{16.21}$$

Taking inverse Laplace transform of both sides of (16.21), we have

$$P(0,1,2,t) = \mu_1(\mu_1 + 2\mu_2)\left[\frac{1}{\alpha\beta} + \frac{1}{(\alpha-\beta)}\left\{\frac{e^{-|\alpha|t}}{\alpha} - \frac{e^{-|\alpha|t}}{\beta}\right\}\right] \tag{16.22}$$

Similarly, (16.11) gives

$$P(1,0,2,t) = \mu_2(\mu_1 + 2\mu_2)\left[\frac{1}{\alpha\beta} + \frac{1}{(\alpha-\beta)}\left\{\frac{e^{-|\alpha|t}}{\alpha} - \frac{e^{-|\alpha|t}}{\beta}\right\}\right] \tag{16.23}$$

The time-dependent state probabilities are given by (16.16), (16.19), (16.22), and (16.23).

Let the steady-state probability that the system is in the state (i,j,k) be denoted by

$$\pi(i,j,k),(i,j,k) \in \Omega$$

By using the final-value theorem of Laplace transform, we have

$$\pi(i,j,k) = \lim_{t\to\infty} P(i,j,k,t) = \lim_{s\to 0} s\left[P^*(i,j,k,s)\right],(i,j,k) \in \Omega$$

Multiplying both sides of (16.13), (16.9), (16.10), and (16.11) by s and taking limit as $s \to 0$, we have

$$\pi(1,1,1) = \frac{\gamma(\mu_1 + 2\mu_2)}{(\mu_1 + 2\mu_2)(\mu_1 + \mu_2 + \gamma) + \gamma^2}, \tag{16.24}$$

$$\pi(1,2,0) = \frac{\gamma^2}{(\mu_1 + 2\mu_2)(\mu_1 + \mu_2 + \gamma) + \gamma^2}, \tag{16.25}$$

$$\pi(0,1,2) = \frac{\mu_1(\mu_1 + 2\mu_2)}{(\mu_1 + 2\mu_2)(\mu_1 + \mu_2 + \gamma) + \gamma^2}, \tag{16.26}$$

$$\pi(1,0,2) = \frac{\mu_2(\mu_1 + 2\mu_2)}{(\mu_1 + 2\mu_2)(\mu_1 + \mu_2 + \gamma) + \gamma^2} \tag{16.27}$$

16.5 AVAILABILITY ANALYSIS

We use the notation $e_{12} = (1,2,0)$, $e_{11} = (1,1,1)$, $e_{01} = (0,1,2)$, $e_{10} = (1,0,2)$. In order that the system be available (that is, functioning) at an arbitrary time t, it must be either in the state e_{12} or e_{11} at the instant t. Let $\rho(t)$ be the probability that the system is available at time t. Then, $\rho(t)$ is called the availability of the system. For obtaining an expression for $\rho(t)$, we require the following probability functions:

$$
\begin{aligned}
P_{e12,e12}(t) &= \Pr\left[Z(t) = e_{12} \mid Z(0) = e_{12}\right], \\
P_{e12,e11}(t) &= \Pr\left[Z(t) = e_{11} \mid Z(0) = e_{12}\right], \\
P_{e12,e01}(t) &= \Pr\left[Z(t) = e_{01} \mid Z(0) = e_{12}\right], \\
P_{e12,e10}(t) &= \Pr\left[Z(t) = e_{10} \mid Z(0) = e_{12}\right];
\end{aligned}
$$

$$
\begin{aligned}
P_{e11,e12}(t) &= \Pr\left[Z(t) = e_{12} \mid Z(0) = e_{11}\right], \\
P_{e11,e11}(t) &= \Pr\left[Z(t) = e_{11} \mid Z(0) = e_{11}\right], \\
P_{e11,e01}(t) &= \Pr\left[Z(t) = e_{01} \mid Z(0) = e_{11}\right], \\
P_{e11,e10}(t) &= \Pr\left[Z(t) = e_{10} \mid Z(0) = e_{11}\right];
\end{aligned}
$$

$$
\begin{aligned}
P_{e01,e12}(t) &= \Pr\left[Z(t) = e_{12} \mid Z(0) = e_{01}\right], \\
P_{e01,e11}(t) &= \Pr\left[Z(t) = e_{11} \mid Z(0) = e_{01}\right], \\
P_{e01,e01}(t) &= \Pr\left[Z(t) = e_{01} \mid Z(0) = e_{01}\right], \\
P_{e01,e10}(t) &= \Pr\left[Z(t) = e_{10} \mid Z(0) = e_{01}\right];
\end{aligned}
$$

$$
\begin{aligned}
P_{e10,e12}(t) &= \Pr\left[Z(t) = e_{12} \mid Z(0) = e_{10}\right], \\
P_{e10,e11}(t) &= \Pr\left[Z(t) = e_{11} \mid Z(0) = e_{10}\right], \\
P_{e10,e01}(t) &= \Pr\left[Z(t) = e_{01} \mid Z(0) = e_{10}\right], \\
P_{e10,e10}(t) &= \Pr\left[Z(t) = e_{10} \mid Z(0) = e_{10}\right].
\end{aligned}
$$

Using regenerative principles, we have

$$P_{e12,e12}(t) = e^{-(\mu_1+2\mu_2)t} + \mu_1 \int_0^t e^{-(\mu_1+2\mu_2)u} P_{e11,e12}(t-u)\,du + 2\mu_2 \int_0^t e^{-(\mu_1+2\mu_2)u} P_{e11,e12}(t-u)\,du, \tag{16.28}$$

$$P_{e12,e11}(t) = \mu_1 \int_0^t e^{-(\mu_1+2\mu_2)u} P_{e11,e11}(t-u)\,du + 2\mu_2 \int_0^t e^{-(\mu_1+2\mu_2)u} P_{e11,e11}(t-u)\,du, \tag{16.29}$$

$$P_{e12,e01}(t) = \mu_1 \int_0^t e^{-(\mu_1+2\mu_2)u} P_{e11,e01}(t-u)\,du + 2\mu_2 \int_0^t e^{-(\mu_1+2\mu_2)u} P_{e11,e01}(t-u)\,du, \tag{16.30}$$

$$P_{e12,e10}(t) = \mu_1 \int_0^t e^{-(\mu_1+2\mu_2)u} P_{e11,e10}(t-u)\,du + 2\mu_2 \int_0^t e^{-(\mu_1+2\mu_2)u} P_{e11,e10}(t-u)\,du, \tag{16.31}$$

$$P_{e11,e12}(t) = \mu_1 \int_0^t e^{-(\mu_1+\mu_2+\gamma)u} P_{e01,e12}(t-u)\,du + \mu_2 \int_0^t e^{-(\mu_1+\mu_2+\gamma)u} P_{e10,e12}(t-u)\,du + \gamma \int_0^t e^{-(\mu_1+\mu_2+\gamma)u} P_{e12,e12}(t-u)\,du, \tag{16.32}$$

$$P_{e11,e11}(t) = e^{-(\mu_1+\mu_2+\gamma)t} + \mu_1 \int_0^t e^{-(\mu_1+\mu_2+\gamma)u} P_{e01,e11}(t-u)\,du \\ +\mu_2 \int_0^t e^{-(\mu_1+\mu_2+\gamma)u} P_{e10,e11}(t-u)\,du \\ +\gamma \int_0^t e^{-(\mu_1+\mu_2+\gamma)u} P_{e12,e11}(t-u)\,du, \tag{16.33}$$

$$P_{e11,e01}(t) = \mu_1 \int_0^t e^{-(\mu_1+\mu_2+\gamma)u} P_{e01,e01}(t-u)\,du \\ +\mu_2 \int_0^t e^{-(\mu_1+\mu_2+\gamma)u} P_{e10,e01}(t-u)\,du \\ +\gamma \int_0^t e^{-(\mu_1+\mu_2+\gamma)u} P_{e12,e01}(t-u)\,du, \tag{16.34}$$

$$P_{e11,e10}(t) = \mu_1 \int_0^t e^{-(\mu_1+\mu_2+\gamma)u} P_{e01,e10}(t-u)\,du \\ +\mu_2 \int_0^t e^{-(\mu_1+\mu_2+\gamma)u} P_{e10,e10}(t-u)\,du \\ +\gamma \int_0^t e^{-(\mu_1+\mu_2+\gamma)u} P_{e12,e01}(t-u)\,du, \tag{16.35}$$

$$P_{e01,e12}(t) = \gamma \int_0^t e^{-\gamma u} P_{e11,e12}(t-u)\,du, \tag{16.36}$$

$$P_{e01,e11}(t) = \gamma \int_0^t e^{-\gamma u} P_{e11,e11}(t-u)\,du, \tag{16.37}$$

$$P_{e01,e01}(t) = e^{-\gamma t} + \gamma \int_0^t e^{-\gamma u} P_{e11,e01}(t-u)\,du, \tag{16.38}$$

$$P_{e01,e10}(t) = \gamma \int_0^t e^{-\gamma u} P_{e11,e10}(t-u)\,du, \tag{16.39}$$

$$P_{e10,e12}(t) = \gamma \int_0^t e^{-\gamma u} P_{e11,e12}(t-u)\,du, \tag{16.40}$$

$$P_{e10,e11}(t) = \gamma \int_0^t e^{-\gamma u} P_{e11,e11}(t-u)\,du, \tag{16.41}$$

$$P_{e10,e01}(t) = \gamma \int_0^t e^{-\gamma u} P_{e11,e01}(t-u)\,du, \tag{16.42}$$

$$P_{e10,e10}(t) = e^{-\gamma t} + \gamma \int_0^t e^{-\gamma u} P_{e11,e10}(t-u)\,du \tag{16.43}$$

Taking Laplace transform of both sides of (16.28)–(16.43), we have

$$(s+\mu_1+2\mu_2)P^*_{e12,e12}(s) = 1+(\mu_1+2\mu_2)P^*_{e11,e12}(s), \tag{16.44}$$

$$(s+\mu_1+2\mu_2)P^*_{e12,e11}(s) = (\mu_1+2\mu_2)P^*_{e11,e11}(s), \tag{16.45}$$

$$(s+\mu_1+2\mu_2)P^*_{e12,e01}(s) = (\mu_1+2\mu_2)P^*_{e11,e01}(s), \tag{16.46}$$

$$(s+\mu_1+2\mu_2)P^*_{e12,e10}(s) = (\mu_1+2\mu_2)P^*_{e11,e10}(s), \tag{16.47}$$

$$(s+\mu_1+\mu_2+\gamma)P^*_{e11,e12}(s) = \mu_1 P^*_{e01,e12}(s) + \mu_2 P^*_{e10,e12}(s) + \gamma P^*_{e12,e12}(s), \tag{16.48}$$

$$(s+\mu_1+\mu_2+\gamma)P^*_{e11,e11}(s) = 1+\mu_1 P^*_{e01,e11}(s) + \mu_2 P^*_{e10,e11}(s) + \gamma P^*_{e12,e11}(s), \tag{16.49}$$

$$(s+\mu_1+\mu_2+\gamma)P^*_{e11,e01}(s) = \mu_1 P^*_{e01,e01}(s) + \mu_2 P^*_{e10,e01}(s) + \gamma P^*_{e12,e01}(s), \tag{16.50}$$

$$(s+\mu_1+\mu_2+\gamma)P^*_{e11,e10}(s) = \mu_1 P^*_{e01,e10}(s) + \mu_2 P^*_{e10,e10}(s) + \gamma P^*_{e12,e10}(s), \tag{16.51}$$

$$(s+\gamma)P^*_{e01,e12}(s)=\gamma P^*_{e11,e12}(s), \tag{16.52}$$

$$(s+\gamma)P^*_{e01,e11}(s)=\gamma P^*_{e11,e11}(s), \tag{16.53}$$

$$(s+\gamma)P^*_{e01,e01}(s)=1+\gamma P^*_{e11,e01}(s), \tag{16.54}$$

$$(s+\gamma)P^*_{e01,e10}(s)=\gamma P^*_{e11,e10}(s), \tag{16.55}$$

$$(s+\gamma)P^*_{e10,e12}(s)=\gamma P^*_{e11,e12}(s), \tag{16.56}$$

$$(s+\gamma)P^*_{e10,e11}(s)=\gamma P^*_{e11,e11}(s), \tag{16.57}$$

$$(s+\gamma)P^*_{e10,e01}(s)=\gamma P^*_{e11,e01}(s), \tag{16.58}$$

$$(s+\gamma)P^*_{e10,e10}(s)=1+\gamma P^*_{e11,e10}(s) \tag{16.59}$$

By (16.44), (16.52), and (16.56), we get

$$P^*_{e12,e12}(s)=\frac{1}{(s+\mu_1+2\mu_2)}+\frac{(\mu_1+2\mu_2)}{(s+\mu_1+2\mu_2)}P^*_{e11,e12}(s), \tag{16.60}$$

$$P^*_{e01,e12}(s)=\frac{\gamma}{(s+\gamma)}P^*_{e11,e12}(s), \tag{16.61}$$

$$P^*_{e10,e12}(s)=\frac{\gamma}{(s+\gamma)}P^*_{e11,e12}(s) \tag{16.62}$$

Substituting (16.60), (16.61), and (16.62) into (16.48) and solving for $P^*_{e11,e12}(s)$, we get

$$P^*_{e11,e12}(s)=\frac{\gamma(s+\gamma)}{s\left[s^2+(2\mu_1+3\mu_2+2\gamma)s+(\mu_1+2\mu_2)(\mu_1+\mu_2+\gamma)+\gamma^2\right]} \tag{16.63}$$

Splitting into partial fractions, (16.63) yields

$$P^*_{e11,e12}(s)=\frac{\gamma(s+\gamma)}{s(s-\alpha)(s-\beta)}=\left(\frac{\gamma^2}{\alpha\beta}\right)\frac{1}{s}+\frac{\gamma}{\alpha-\beta}\left\{\frac{(\alpha+\gamma)}{\alpha(s-\alpha)}-\frac{(\beta+\gamma)}{\beta(s-\beta)}\right\} \tag{16.64}$$

Taking inverse Laplace transform of both sides of (16.64), we get

$$P_{e11,e12}(t)=\left(\frac{\gamma^2}{\alpha\beta}\right)+\frac{\gamma}{\alpha-\beta}\left\{\frac{(\alpha+\gamma)}{\alpha}e^{-|\alpha|t}-\frac{(\beta+\gamma)}{\beta}e^{-|\beta|t}\right\} \tag{16.65}$$

Taking inverse Laplace transform of both sides of (16.60), (16.61), and (16.62), we get

$$P_{e12,e12}(t)=e^{-(\mu_1+2\mu_2)t}+(\mu_1+2\mu_2)e^{-(\mu_1+2\mu_2)t}©P_{e11,e12}(t), \tag{16.66}$$

$$P_{e01,e12}(t)=\gamma e^{-\gamma t}©P_{e11,e12}(t), \tag{16.67}$$

$$P_{e10,e12}(t)=\gamma e^{-\gamma t}©P_{e11,e12}(t) \tag{16.68}$$

By (16.45), (16.53), and (16.57), we get

$$P^*_{e12,e11}(s)=\frac{(\mu_1+2\mu_2)}{(s+\mu_1+2\mu_2)}P^*_{e11,e11}(s), \tag{16.69}$$

$$P^*_{e01,e11}(s)=\frac{\gamma}{(s+\gamma)}P^*_{e11,e11}(s), \tag{16.70}$$

$$P^*_{e10,e11}(s)=\frac{\gamma}{(s+\gamma)}P^*_{e11,e11}(s) \tag{16.71}$$

Substituting (16.69), (16.70), and (16.71) into (16.49) and solving for $P^*_{e11,e11}(s)$, we get

$$P^*_{e11,e11}(s)=\frac{(s+\gamma)(s+\mu_1+2\mu_2)}{s(s-\alpha)(s-\beta)} \tag{16.72}$$

$$=\frac{\gamma(\mu_1+2\mu_2)}{\alpha\beta}\frac{1}{s}$$

$$+\frac{1}{(\alpha-\beta)}\left\{\frac{(\alpha+\gamma)(\alpha+\mu_1+2\mu_2)}{\alpha(s-\alpha)}-\frac{(\beta+\gamma)(\beta+\mu_1+2\mu_2)}{\beta(s-\beta)}\right\}$$

Taking inverse Laplace transform of both sides of (16.72), we get

$$P_{e11,e11}(t) = \frac{\gamma(\mu_1 + 2\mu_2)}{\alpha\beta} + \frac{(\alpha+\gamma)(\alpha+\mu_1+2\mu_2)}{\alpha(\alpha-\beta)} e^{-|\alpha|t} - \frac{(\beta+\gamma)(\beta+\mu_1+2\mu_2)}{\beta(\alpha-\beta)} e^{-|\beta|t} \tag{16.73}$$

Taking inverse Laplace transform of both sides of (16.69), (16.70), and (16.71), we get

$$P_{e12,e11}(t) = (\mu_1 + 2\mu_2) e^{-(\mu_1+2\mu_2)t} \copyright P_{e11,e11}(t), \tag{16.74}$$

$$P_{e01,e11}(t) = \gamma e^{-\gamma t} \copyright P_{e11,e11}(t), \tag{16.75}$$

$$P_{e10,e11}(t) = \gamma e^{-\gamma t} \copyright P_{e11,e11}(t) \tag{16.76}$$

By (16.46), (16.54), and (16.58), we get

$$P^*_{e12,e01}(s) = \frac{(\mu_1 + 2\mu_2)}{(s+\mu_1+2\mu_2)} P^*_{e11,e01}(s), \tag{16.77}$$

$$P^*_{e01,e01}(s) = \frac{1}{(s+\gamma)} + \frac{\gamma}{(s+\gamma)} P^*_{e11,e01}(s), \tag{16.78}$$

$$P^*_{e10,e01}(s) = \frac{\gamma}{(s+\gamma)} P^*_{e11,e01}(s) \tag{16.79}$$

Substituting (16.77), (16.78), and (16.79) into (16.50) and solving for $P^*_{e11,e01}(s)$, we get

$$P^*_{e11,e01}(s) = \frac{\mu_1(s+\mu_1+2\mu_2)}{s(s-\alpha)(s-\beta)} = \frac{\mu_1(\mu_1+2\mu_2)}{\alpha\beta s} + \frac{\mu_1(\alpha+\mu_1+2\mu_2)}{\alpha(\alpha-\beta)(s-\alpha)} - \frac{\mu_1(\beta+\mu_1+2\mu_2)}{\beta(\alpha-\beta)(s-\beta)} \tag{16.80}$$

Taking inverse Laplace transform of both sides of (16.80), we get

$$P_{e11,e01}(t) = \frac{\mu_1(\mu_1 + 2\mu_2)}{\alpha\beta} + \frac{\mu_1(\alpha + \mu_1 + 2\mu_2)}{\alpha(\alpha - \beta)} e^{-|\alpha|t} - \frac{\mu_1(\beta + \mu_1 + 2\mu_2)}{\beta(\alpha - \beta)} e^{-|\beta|t} \tag{16.81}$$

Taking inverse Laplace transform of both sides of (16.77), (16.78), and (16.79), we get

$$P_{e12,e01}(t) = (\mu_1 + 2\mu_2) e^{-(\mu_1 + 2\mu_2)t} © P_{e11,e01}(t), \tag{16.82}$$

$$P_{e01,e01}(t) = e^{-\gamma t} + \gamma e^{-\gamma t} © P_{e11,e01}(t), \tag{16.83}$$

$$P_{e10,e01}(t) = \gamma e^{-\gamma t} © P_{e11,e01}(t) \tag{16.84}$$

By (16.47), (16.55), and (16.59), we get

$$P^*_{e12,e10}(s) = \frac{(\mu_1 + 2\mu_2)}{(s + \mu_1 + 2\mu_2)} P^*_{e11,e10}(s), \tag{16.85}$$

$$P^*_{e01,e10}(s) = \frac{\gamma}{(s + \gamma)} P^*_{e11,e10}(s), \tag{16.86}$$

$$P^*_{e10,e10}(s) = \frac{1}{(s + \gamma)} + \frac{\gamma}{(s + \gamma)} P^*_{e11,e10}(s) \tag{16.87}$$

Substituting (16.85), (16.86), and (16.87) into (16.51) and solving for $P^*_{e11,e10}(s)$, we get

$$P^*_{e11,e10}(s) = \frac{\mu_2(s + \mu_1 + 2\mu_2)}{s(s - \alpha)(s - \beta)} = \frac{\mu_2(\mu_1 + 2\mu_2)}{\alpha\beta s} + \frac{\mu_2(\alpha + \mu_1 + 2\mu_2)}{\alpha(\alpha - \beta)(s - \alpha)} - \frac{\mu_2(\beta + \mu_1 + 2\mu_2)}{\beta(\alpha - \beta)(s - \beta)} \tag{16.88}$$

Taking inverse Laplace transform of both sides of (16.88), we get

$$P_{e11,e10}(t)=\frac{\mu_2(\mu_1+2\mu_2)}{\alpha\beta}+\frac{\mu_2(\alpha+\mu_1+2\mu_2)}{\alpha(\alpha-\beta)}e^{-|\alpha|t}-\frac{\mu_2(\beta+\mu_1+2\mu_2)}{\beta(\alpha-\beta)}e^{-|\beta|t} \tag{16.89}$$

Taking inverse Laplace transform of both sides of (16.85), (16.86), and (16.87), we get

$$P_{e12,e10}(t)=(\mu_1+2\mu_2)e^{-(\mu_1+2\mu_2)t}©P_{e11,e10}(t), \tag{16.90}$$

$$P_{e01,e10}(t)=\gamma e^{-\gamma t}©P_{e11,e10}(t), \tag{16.91}$$

$$P_{e10,e10}(t)=e^{-\gamma t}+\gamma e^{-\gamma t}©P_{e11,e10}(t) \tag{16.92}$$

Now, the availability function $\rho(t)$ is given by

$$\rho(t)=P_{e12,e12}(t)+P_{e12,e11}(t) \tag{16.93}$$

From (16.65) and (16.66), we get

$$P_{e12,e12}(t)=\frac{\gamma^2}{\alpha\beta}+\left[1+\frac{\gamma(\mu_1+2\mu_2-\gamma)}{(\alpha+\mu_1+2\mu_2)(\beta+\mu_1+2\mu_2)}\right]e^{-(\mu_1+2\mu_2)t}+\frac{\gamma(\mu_1+2\mu_2)}{\alpha-\beta}\left[\frac{(\alpha+\gamma)}{\alpha(\alpha+\mu_1+2\mu_2)}e^{-|\alpha|t}-\frac{(\beta+\gamma)}{\beta(\beta+\mu_1+2\mu_2)}e^{-|\beta|t}\right] \tag{16.94}$$

From (16.73) and (16.74), we get

$$P_{e12,e11}(t)=\frac{\gamma(\mu_1+2\mu_2)}{\alpha\beta}+\frac{(\mu_1+2\mu_2)}{(\alpha-\beta)}\left[\frac{(\alpha+\gamma)}{\alpha}e^{-|\alpha|t}-\frac{(\beta+\gamma)}{\beta}e^{-|\beta|t}\right] \tag{16.95}$$

The steady-state availability $\rho(\infty)$ of the system is given by

$$\rho(\infty)=\frac{\gamma(\mu_1+2\mu_2+\gamma)}{\gamma^2+(\mu_1+2\mu_2)(\mu_1+\mu_2+\gamma)} \tag{16.96}$$

16.6 RELIABILITY OF THE SYSTEM

The reliability function $R_0(t)$ of the system is the probability that the system enters into down state only after time t given that the system started in the state e_{12} at time $t = 0$. To obtain an expression for $R_0(t)$, we need to obtain the auxiliary functions $P_{e12,e01}(t), P_{e12,e10}(t), P_{e11,e01}(t)$, and $P_{e11,e10}(t)$ which are defined by

$$P_{e12,e01}(t) = Pr\{Z(t) = e_{01}, Z(u) \notin \{e_{01}, e_{10}\} \forall u \in (0,t) \mid Z(0) = e_{12}\},$$

$$P_{e12,e10}(t) = Pr\{Z(t) = e_{10}, Z(u) \notin \{e_{01}, e_{10}\} \forall u \in (0,t) \mid Z(0) = e_{12}\}.$$

$$P_{e11,e01}(t) = Pr\{Z(t) = e_{01}, Z(u) \notin \{e_{01}, e_{10}\} \forall u \in (0,t) \mid Z(0) = e_{11}\},$$

$$P_{e11,e10}(t) = Pr\{Z(t) = e_{10}, Z(u) \notin \{e_{01}, e_{10}\} \forall u \in (0,t) \mid Z(0) = e_{11}\}.$$

These auxiliary functions have the following interpretations:

i. $P_{e12,e01}(t)dt$ represents the conditional probability that the system has not visited any of the down states e_{01} and e_{10} throughout the interval $(0,t)$ and it enters into the down state e_{01} between t and $t+dt$ given that the system is switched from e_{12} at $t = 0$.
ii. $P_{e12,e10}(t)dt$ represents the conditional probability that the system has not visited any of the down states e_{01} and e_{10} throughout the interval $(0,t)$ and it enters into the down state e_{10} between t and $t+dt$ given that the system is switched from e_{12} at $t = 0$.
iii. $P_{e11,e01}(t)dt$ represents the conditional probability that the system has not visited any of the down states e_{01} and e_{10} throughout the interval $(0,t)$ and it enters into the down state e_{01} between t and $t+dt$ given that the system is switched from e_{11} at $t = 0$.
iv. $P_{e11,e10}(t)dt$ represents the conditional probability that the system has not visited any of the down states e_{01} and e_{10} throughout the interval $(0,t)$ and it enters into the down state e_{10} between t and $t+dt$ given that the system is switched from e_{11} at $t = 0$. Using renewal-theoretic arguments, we obtain

$$P_{e12,e01}(t) = (\mu_1 + 2\mu_2)\int_0^t e^{-(\mu_1+2\mu_2)u} P_{e11,e01}(t-u)\,du, \tag{16.97}$$

$$P_{e12,e10}(t) = (\mu_1 + 2\mu_2)\int_0^t e^{-(\mu_1+2\mu_2)u} P_{e11,e10}(t-u)\,du, \tag{16.98}$$

$$P_{e11,e01}(t) = e^{-(\mu_1+\mu_2+\gamma)t}\mu_1 + \gamma\int_0^t e^{-(\mu_1+\mu_2+\gamma)u}P_{e12,e01}(t-u)du, \tag{16.99}$$

$$P_{e11,e10}(t) = e^{-(\mu_1+\mu_2+\gamma)t}\mu_2 + \gamma\int_0^t e^{-(\mu_1+\mu_2+\gamma)u}P_{e12,e10}(t-u)du \tag{16.100}$$

Taking Laplace transform of both sides of (16.97)–(16.100), we have

$$(s+\mu_1+2\mu_2)P^*_{e12,e01}(s) = (\mu_1+2\mu_2)P^*_{e11,e01}(s), \tag{16.101}$$

$$(s+\mu_1+2\mu_2)P^*_{e12,e10}(s) = (\mu_1+2\mu_2)P^*_{e11,e10}(s), \tag{16.102}$$

$$(s+\mu_1+\mu_2+\gamma)P^*_{e11,e01}(s) = \mu_1 + \gamma P^*_{e12,e01}(s), \tag{16.103}$$

$$(s+\mu_1+\mu_2+\gamma)P^*_{e11,e10}(s) = \mu_2 + \gamma P^*_{e12,e10}(s) \tag{16.104}$$

From (16.101), we get

$$P^*_{e12,e01}(s) = \frac{(\mu_1+2\mu_2)}{(s+\mu_1+2\mu_2)}P^*_{e11,e01}(s) \tag{16.105}$$

Substituting (16.105) into (16.103), we get

$$P^*_{e11,e01}(s) = \frac{\mu_1(s+\mu_1+2\mu_2)}{s^2+(2\mu_1+3\mu_2+\gamma)s+(\mu_1+\mu_2)(\mu_1+2\mu_2)} \tag{16.106}$$

The zeros of $s^2+(2\mu_1+3\mu_2+\gamma)s+(\mu_1+\mu_2)(\mu_1+2\mu_2)$ are given by

$$\delta_1 = \frac{-(2\mu_1+3\mu_2+\gamma)+\sqrt{\mu_2^2+\gamma^2+2\gamma(2\mu_1+3\mu_2)}}{2},$$

$$\delta_2 = \frac{-(2\mu_1+3\mu_2+\gamma)-\sqrt{\mu_2^2+\gamma^2+2\gamma(2\mu_1+3\mu_2)}}{2}$$

They satisfy the following properties:

(i) δ_1 and δ_2 are real and distinct,

(ii) $\delta_1\delta_2 = (\mu_1+\mu_2)(\mu_1+2\mu_2) > 0$,

(iii) $0 > \delta_1 > \delta_2$,

(iv) $|\delta_1| = -\delta_1, |\delta_2| = -\delta_2$.

Now, (16.106) becomes

$$P^*_{e11,e01}(s) = \frac{\mu_1(s+\mu_1+2\mu_2)}{(s-\delta_1)(s-\delta_2)}$$
$$= \frac{\mu_1}{(\delta_1-\delta_2)}\left[\frac{(\delta_1+\mu_1+2\mu_2)}{(s-\delta_1)} - \frac{(\delta_2+\mu_1+2\mu_2)}{(s-\delta_2)}\right] \tag{16.107}$$

Taking inverse transform of both sides of (16.107), we get

$$P_{e11,e01}(t) = \frac{\mu_1}{(\delta_1-\delta_2)}\left[(\delta_1+\mu_1+2\mu_2)e^{-|\delta_1|t} - (\delta_2+\mu_1+2\mu_2)e^{-|\delta_2|t}\right] \tag{16.108}$$

Taking inverse Laplace transform of both sides of (16.105), we get

$$P_{e12,e01}(t) = (\mu_1+2\mu_2)e^{-(\mu_1+2\mu_2)t} © P_{e11,e01}(t) \tag{16.109}$$

From (16.102), we get

$$P^*_{e12,e10}(s) = \frac{(\mu_1+2\mu_2)}{(s+\mu_1+2\mu_2)} P^*_{e11,e10}(s) \tag{16.110}$$

Substituting (16.110) into (16.104), we get

$$P^*_{e11,e10}(s) = \frac{\mu_2(s+\mu_1+2\mu_2)}{s^2+(2\mu_1+3\mu_2+\gamma)s+(\mu_1+\mu_2)(\mu_1+2\mu_2)} \tag{16.111}$$
$$= \frac{\mu_2(s+\mu_1+2\mu_2)}{(s-\delta_1)(s-\delta_2)}$$
$$= \frac{\mu_2}{(\delta_1-\delta_2)}\left[\frac{(\delta_1+\mu_1+2\mu_2)}{(s-\delta_1)} - \frac{(\delta_2+\mu_1+2\mu_2)}{(s-\delta_2)}\right]$$

Taking inverse transform of both sides of (16.111), we get

$$P_{e11,e10}(t)=\frac{\mu_2}{(\delta_1-\delta_2)}\left[(\delta_1+\mu_1+2\mu_2)e^{-|\delta_1|t}-(\delta_2+\mu_1+2\mu_2)e^{-|\delta_2|t}\right] \tag{16.112}$$

Taking inverse transform of both sides of (16.110), we get

$$P_{e12,e10}(t)=(\mu_1+2\mu_2)e^{-(\mu_1+2\mu_2)t}©P_{e11,e10}(t) \tag{16.113}$$

The reliability function $R_0(t)$ of the system is the probability that the system enters into down state only after time t given that the system started in the state e_{12} at time $t=0$. Then, we have

$$R_0(t)=\int_t^{\infty}\left\{P_{e12,e01}(u)+P_{e12,e10}(u)\right\}du \tag{16.114}$$

Taking Laplace transform of both sides of (16.114), we get

$$R_0^*(s)=\frac{1-P^*_{e12,e01}(s)-P^*_{e12,e10}(s)}{s} \tag{16.115}$$

Substituting (16.105) and (16.110) and simplifying, we get

$$R_0^*(s)=\frac{1}{\delta_1-\delta_2}\left[\frac{(\delta_1+2\mu_1+3\mu_2+\gamma)}{s-\delta_1}-\frac{(\delta_2+2\mu_1+3\mu_2+\gamma)}{s-\delta_2}\right] \tag{16.116}$$

Taking inverse Laplace transform of both sides of (16.116), we get the reliability function

$$R_0(t)=\frac{1}{\delta_1-\delta_2}\begin{bmatrix}(\delta_1+2\mu_1+3\mu_2+\gamma)e^{-|\delta_1|t}\\-(\delta_2+2\mu_1+3\mu_2+\gamma)e^{-|\delta_2|t}\end{bmatrix} \tag{16.117}$$

The mean lifetime of the system is given by

$$T_0=\lim_{s\to 0}R_0^*(s)=\frac{(2\mu_1+3\mu_2+\gamma)}{(\mu_1+\mu_2)(\mu_1+2\mu_2)} \tag{16.118}$$

16.7 MEAN DOWN TIME OF THE SYSTEM

Let the system be started in the state (1,2,0) at time $t = 0$. We define a two-valued stochastic process $\{K(t) \mid t \geq 0\}$, where

$$K(t) = \begin{cases} 0, & \text{if } Z(t) \in \{(1,2,0),(1,1,1)\}, \\ 1, & \text{otherwise} \end{cases}$$

If $D(t)$ represents the down time of the system in the interval $[0,t]$, then $D(t)$ is given by the stochastic integral

$$D(t) = \int_0^t K(u)\,du$$

Consequently, the mean down time of the system in $[0,t]$ is given by

$$E[D(t)] = E\left[\int_0^t K(u)\,du\right] = \int_0^t E[K(u)]\,du \tag{16.119}$$

$$= \int_0^t Pr[K(u)=1]\,du = \int_0^t [P(0,1,2,u)+P(1,0,2,u)]\,du$$

Using (16.40) and (16.41) in (16.119), we get

$$E[D(t)] = (\mu_1+\mu_2)\int_0^t e^{-\gamma u} \copyright P(1,1,1,u)\,du \tag{16.120}$$

Taking Laplace transform of both sides of (16.120) and substituting for $P^*(1,1,1,s)$, we get

$$L[E\{D(t)\}] = \frac{(\mu_1+\mu_2)P^*(1,1,1,s)}{s(s+\gamma)} = \frac{(\mu_1+\mu_2)(\mu_1+2\mu_2)}{s^2(s-\alpha)(s-\beta)} \tag{16.121}$$

Splitting into partial fractions, (16.121) gives

$$L[E\{D(t)\}] = \frac{(\mu_1+\mu_2)(\mu_1+2\mu_2)}{\alpha\beta}\begin{bmatrix} \dfrac{1}{s^2} + \dfrac{(\alpha+\beta)}{\alpha\beta}\dfrac{1}{s} + \dfrac{1}{(\alpha-\beta)\alpha\beta} \\ \left\{\dfrac{\beta^2}{(s-\alpha)} - \dfrac{\alpha^2}{(s-\beta)}\right\} \end{bmatrix} \tag{16.122}$$

Taking inverse Laplace transform of both sides of (16.122), we get

$$E\left[D(t)\right]=\frac{(\mu_1+\mu_2)(\mu_1+2\mu_2)}{\alpha\beta}\left[\begin{array}{l} t+\dfrac{(\alpha+\beta)}{\alpha\beta} \\ +\dfrac{1}{(\alpha-\beta)\alpha\beta}\left(\beta^2e^{-|\alpha|t}-\alpha^2e^{-|\beta|t}\right)\end{array}\right] \tag{16.123}$$

16.8 NUMERICAL ILLUSTRATION

Let us fix $\mu_1 = 0.4, \mu_2 = 0.5$, and $\gamma = 1.6$. The above choice is made with the intention of keeping the mean time to repair smaller than the mean time to failure of the units. We have computed the state probabilities for a range of time points until a steady-state condition is reached. These values are tabulated in Tables 16.1–16.3. It is observed from the tables that the steady-state condition is reached not before ten units of time. This behaviour is in agreement with the stationary values

Table 16.1 The probabilities $P(i,j,k,t)$

t	P(1,1,1,t)	P(1,2,0,t)	P(0,1,2,t)	P(1,0,2,t)
0.1	0.1159616	0.8787762	0.0023388	0.0029234
0.2	0.1946308	0.7876720	0.0078654	0.0098318
0.3	0.2481355	0.7181705	0.0149751	0.0187189
0.4	0.2846401	0.6643562	0.0226683	0.0283354
0.5	0.3096435	0.6220884	0.0303414	0.0379268
0.6	0.3268516	0.5884426	0.0376470	0.0470588
0.7	0.3387641	0.5613319	0.0444018	0.0555022
0.8	0.3470689	0.5392488	0.0505255	0.0631568
0.9	0.3529070	0.5210906	0.0560010	0.0700013
1.0	0.3570512	0.5060391	0.0608487	0.0760609
1.1	0.3600260	0.4934780	0.0651094	0.0813867
1.2	0.3621880	0.4829363	0.0688337	0.0860421
1.3	0.3637808	0.4740487	0.0720758	0.0900947
1.4	0.3649714	0.4665275	0.0748894	0.0936117
1.5	0.3658747	0.4601436	0.0773252	0.0966565
1.6	0.3665703	0.4547118	0.0794302	0.0992877
1.7	0.3671139	0.4500813	0.0812466	0.1015582
1.8	0.3675445	0.4461279	0.0828122	0.1035153
1.9	0.3678900	0.4427485	0.0841607	0.1052008
2.0	0.3681704	0.4398569	0.0853212	0.1066515
2.1	0.3684002	0.4373810	0.0863195	0.1078994
2.2	0.3685901	0.4352597	0.0871779	0.1089724

(Continued)

Table 16.1 (Continued) The probabilities $P(i, j, k, t)$

t	P(1,1,1,t)	P(1,2,0,t)	P(0,1,2,t)	P(1,0,2,t)
2.3	0.3687482	0.4334413	0.0879158	0.1098947
2.4	0.3688807	0.4318821	0.0885499	0.1106874
2.5	0.3689922	0.4305447	0.0890947	0.1113684
2.6	0.3690864	0.4293973	0.0895628	0.1119535
2.7	0.3691663	0.4284127	0.0899649	0.1124561
2.8	0.3692342	0.4275678	0.0903102	0.1128878
2.9	0.3692921	0.4268426	0.0906068	0.1132585
3.0	0.3693415	0.4262201	0.0908615	0.1135769
3.1	0.3693837	0.4256857	0.0910803	0.1138503
3.2	0.3694198	0.4252270	0.0912681	0.1140851
3.3	0.3694507	0.4248332	0.0914294	0.1142867
3.4	0.3694771	0.4244951	0.0915679	0.1144599
3.5	0.3694998	0.4242048	0.0916868	0.1146085
3.6	0.3695193	0.4239556	0.0917890	0.1147362
3.7	0.3695359	0.4237416	0.0918766	0.1148458
3.8	0.3695502	0.4235579	0.0919519	0.1149399
3.9	0.3695625	0.4234002	0.0920166	0.1150207
4.0	0.3695731	0.4232647	0.0920721	0.1150901

Table 16.2 The probabilities $P(i, j, k, t)$

t	P(1,1,1,t)	P(1,2,0,t)	P(0,1,2,t)	P(1,0,2,t)
4.1	0.3695821	0.4231484	0.0921198	0.1151497
4.2	0.3695899	0.4230486	0.0921607	0.1152009
4.3	0.3695965	0.4229629	0.0921958	0.1152448
4.4	0.3696022	0.4228893	0.0922260	0.1152825
4.5	0.3696072	0.4228261	0.0922519	0.1153149
4.6	0.3696114	0.4227718	0.0922741	0.1153427
4.7	0.3696150	0.4227252	0.0922932	0.1153666
4.8	0.3696181	0.4226852	0.0923096	0.1153871
4.9	0.3696208	0.4226509	0.0923237	0.1154047
5.0	0.3696231	0.4226214	0.0923358	0.1154198
5.1	0.3696250	0.4225960	0.0923462	0.1154327
5.2	0.3696267	0.4225743	0.0923551	0.1154439
5.3	0.3696282	0.4225556	0.0923628	0.1154535
5.4	0.3696294	0.4225396	0.0923693	0.1154617
5.5	0.3696305	0.4225258	0.0923750	0.1154687
5.6	0.3696314	0.4225140	0.0923798	0.1154748
5.7	0.3696322	0.4225039	0.0923840	0.1154800
5.8	0.3696329	0.4224952	0.0923875	0.1154844
5.9	0.3696334	0.4224877	0.0923906	0.1154883
6.0	0.3696339	0.4224813	0.0923932	0.1154916

(Continued)

Table 16.2 (Continued) The probabilities $P(i,j,k,t)$

t	P(1,1,1,t)	P(1,2,0,t)	P(0,1,2,t)	P(1,0,2,t)
6.1	0.3696344	0.4224757	0.0923955	0.1154944
6.2	0.3696347	0.4224710	0.0923975	0.1154968
6.3	0.3696350	0.4224669	0.0923991	0.1154989
6.4	0.3696353	0.4224634	0.0924005	0.1155007
6.5	0.3696356	0.4224604	0.0924018	0.1155022
6.6	0.3696357	0.4224579	0.0924028	0.1155035
6.7	0.3696359	0.4224557	0.0924037	0.1155047
6.8	0.3696361	0.4224538	0.0924045	0.1155056
6.9	0.3696362	0.4224521	0.0924052	0.1155065
7.0	0.3696363	0.4224507	0.0924058	0.1155072
7.1	0.3696364	0.4224495	0.0924063	0.1155078
7.2	0.3696365	0.4224485	0.0924067	0.1155083
7.3	0.3696365	0.4224476	0.0924070	0.1155088
7.4	0.3696366	0.4224469	0.0924073	0.1155092
7.5	0.3696367	0.4224462	0.0924076	0.1155095
7.6	0.3696367	0.4224456	0.0924078	0.1155098
7.7	0.3696367	0.4224452	0.0924080	0.1155101
7.8	0.3696368	0.4224448	0.0924082	0.1155103
7.9	0.3696368	0.4224444	0.0924084	0.1155104
8.0	0.3696368	0.4224441	0.0924085	0.1155106

Table 16.3 The probabilities $P(i,j,k,t)$

t	P(1,1,1,t)	P(1,2,0,t)	P(0,1,2,t)	P(1,0,2,t)
8.1	0.3696368	0.4224438	0.0924086	0.1155107
8.2	0.3696369	0.4224436	0.0924087	0.1155109
8.3	0.3696369	0.4224434	0.0924088	0.1155110
8.4	0.3696369	0.4224432	0.0924088	0.1155110
8.5	0.3696369	0.4224431	0.0924089	0.1155111
8.6	0.3696369	0.4224430	0.0924089	0.1155112
8.7	0.3696369	0.4224429	0.0924090	0.1155112
8.8	0.3696369	0.4224428	0.0924090	0.1155113
8.9	0.3696369	0.4224427	0.0924090	0.1155113
9.0	0.3696369	0.4224426	0.0924091	0.1155113
9.1	0.3696369	0.4224426	0.0924091	0.1155114
9.2	0.3696369	0.4224425	0.0924091	0.1155114
9.3	0.3696369	0.4224425	0.0924091	0.1155114
9.4	0.3696369	0.4224425	0.0924092	0.1155114
9.5	0.3696369	0.4224424	0.0924092	0.1155115
9.6	0.3696370	0.4224424	0.0924092	0.1155115
9.7	0.3696370	0.4224424	0.0924092	0.1155115
9.8	0.3696370	0.4224424	0.0924092	0.1155115
9.9	0.3696370	0.4224423	0.0924092	0.1155115
10.0	0.3696370	0.4224423	0.0924092	0.1155115

$$\pi(1,1,1) = 0.3696370,$$

$$\pi(1,2,0) = 0.4224422,$$

$$\pi(0,1,2) = 0.0924092$$

$$\pi(1,0,2) = 0.1155116$$

as calculated with (16.24)–(16.27)

Using the table values, the graphs for $P(1,1,1,t), P(1,2,0,t), P(0,1,2,t)$, and $P(1,0,2,t)$ as functions of t are depicted in Figures 16.2–16.5. We find from the figures that the three probabilities $P(1,1,1,t), P(0,1,2,t)$, and $P(1,0,2,t)$ are increasing, and the other probability $P(1,2,0,t)$ is decreasing with respect to time t.

For the prescribed values $\mu_1 = 0.4, \mu_2 = 0.5$, and $\gamma = 1.6$, we have computed the values of the availability function $\rho(t)$ for various time points t. These values are tabulated in Table 16.4. We observe from the table that the availability function becomes stationary after time $t = 10.0$. The stationary value as read from Table 16.4 is 0.7920793 which agrees with the value obtained by using (16.66). Using the data from Table 16.4, we have presented the graph of $\rho(t)$ in Figure 16.6 which exhibits the stationary behaviour.

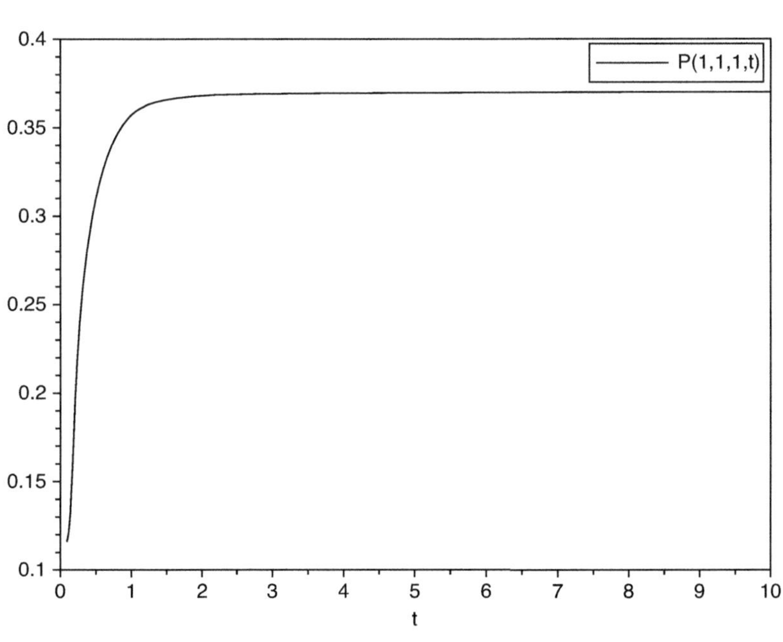

Figure 16.2 Variation of $P(1,1,1,t)$.

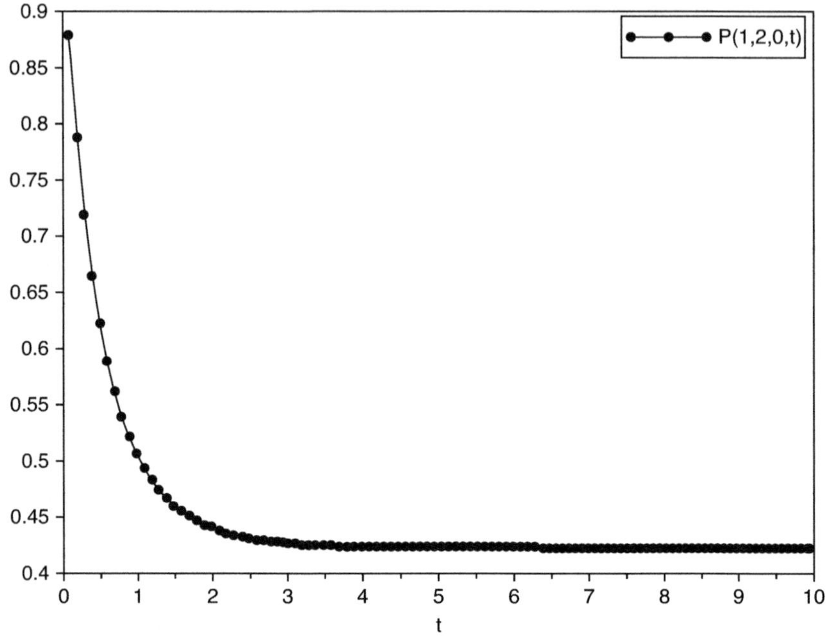

Figure 16.3 Variation of $P(1,2,0,t)$.

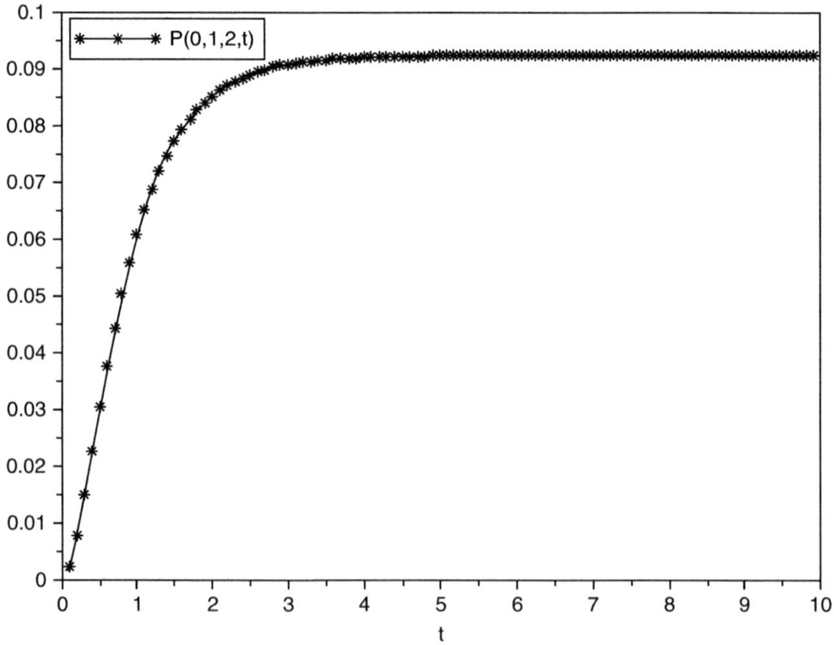

Figure 16.4 Variation of $P(0,1,2,t)$.

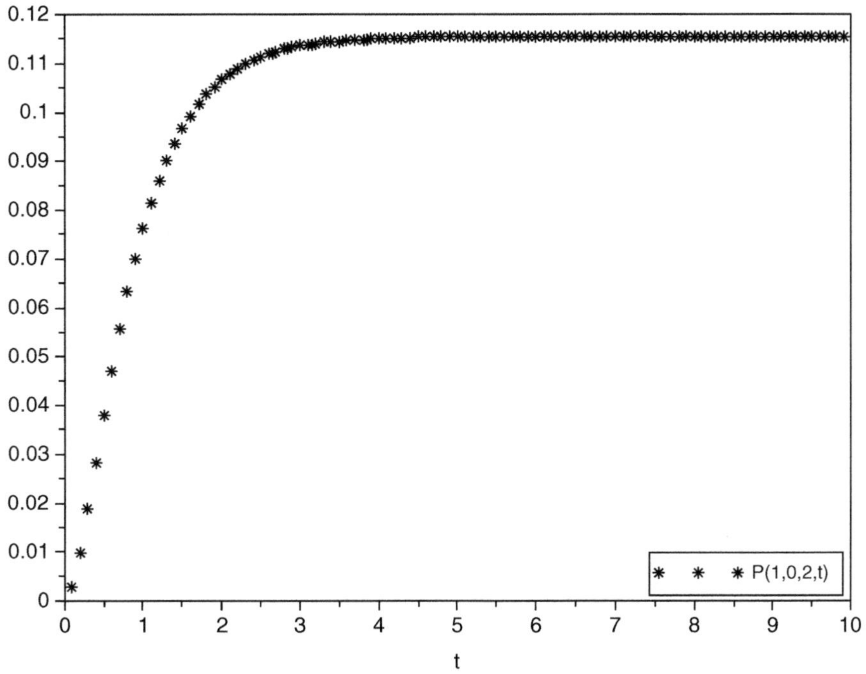

Figure 16.5 Variation of *P*(1,0,2,*t*).

Next, for the same prescribed values $\mu_1 = 0.4, \mu_2 = 0.5$, and $\gamma = 1.6$, we have computed the values of the reliability function $R_0(t)$ for various values of time t. The values $R_0(t)$ for $t = 0.1, 0.2, \ldots, 10.0$ are tabulated in Table 16.5. Using Table 16.5, we have plotted $R_0(t)$ as a function of t in Figure 16.7 which exhibits a decreasing behaviour in the time interval $(0,10)$. The values $R_0(t)$ for $t = 10, 11, \ldots, 50$ are tabulated in Table 16.6. Using Table 16.6, we have presented the graph of $R_0(t)$ in Figure 16.8 which exhibits the characteristic property of repairable systems that $\lim_{t\to\infty} R_0(t) = 0$.

An important measure of performance of a repairable system is the proportion of downtime in an arbitrary time interval. It is quite interesting to study numerically the proportion of time the system is in the down state in any time interval. Assuming the same values $\mu_1 = 0.4, \mu_2 = 0.5$, and $\gamma = 1.6$, we have computed the proportion $E[D(t)]/t$ by varying the time t, and the ratios are listed in Table 16.7. We observe that the proportion $E[D(t)]/t$ is increasing with respect to time. This is in agreement with the observation that the reliability of the system is decreasing with respect to time. The decreasing nature of $E[D(t)]/t$ is highlighted in Figure 16.9. The impact of repair rate on proportion of downtime can be studied by fixing a time point

Table 16.4 Availability $\rho(t)v$ of the system

t	(t)	t	(t)	t	(t)	t	(t)
0.1	0.9947378	2.1	0.8057812	4.1	0.7927306	6.1	0.7921101
0.2	0.9823027	2.2	0.8038498	4.2	0.7926385	6.2	0.7921057
0.3	0.9663060	2.3	0.8021895	4.3	0.7925594	6.3	0.7921020
0.4	0.9489963	2.4	0.8007628	4.4	0.7924915	6.4	0.7920988
0.5	0.9317318	2.5	0.7995368	4.5	0.7924332	6.5	0.7920960
0.6	0.9152942	2.6	0.7984837	4.6	0.7923832	6.6	0.7920936
0.7	0.9000960	2.7	0.7975790	4.7	0.7923402	6.7	0.7920916
0.8	0.8863177	2.8	0.7968020	4.8	0.7923033	6.8	0.7920898
0.9	0.8739976	2.9	0.7961347	4.9	0.7922716	6.9	0.7920883
1.0	0.8630903	3.0	0.7955616	5.0	0.7922444	7.0	0.7920870
1.1	0.8535039	3.1	0.7950694	5.1	0.7922211	7.1	0.7920859
1.2	0.8451243	3.2	0.7946468	5.2	0.7922010	7.2	0.7920850
1.3	0.8378295	3.3	0.7942839	5.3	0.7921838	7.3	0.7920842
1.4	0.8314989	3.4	0.7939722	5.4	0.7921690	7.4	0.7920835
1.5	0.8260183	3.5	0.7937046	5.5	0.7921563	7.5	0.7920829
1.6	0.8212821	3.6	0.7934749	5.6	0.7921454	7.6	0.7920823
1.7	0.8171952	3.7	0.7932776	5.7	0.7921361	7.7	0.7920819
1.8	0.8136725	3.8	0.7931082	5.8	0.7921280	7.8	0.7920815
1.9	0.8106385	3.9	0.7929627	5.9	0.7921211	7.9	0.7920812
2.0	0.8080273	4.0	0.7928378	6.0	0.7921152	8.0	0.7920809

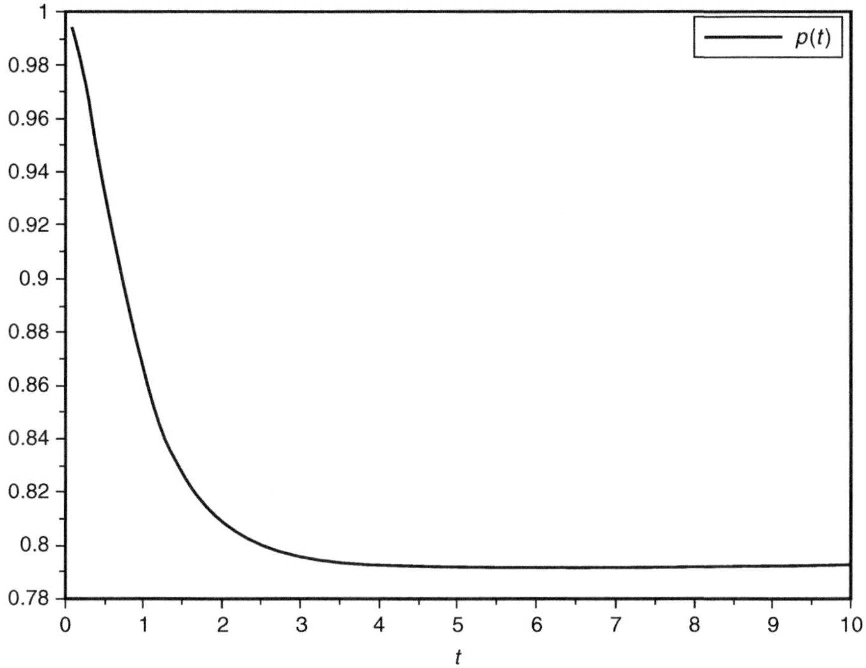

Figure 16.6 Variation of availability function $\rho(t)$.

arbitrarily and plotting $E\left[D(t)/t\right]$ against γ. Taking $\mu_1 = 0.4, \mu_2 = 0.5$, and $t = 10$, we have computed the value of $E\left[D(10)/10\right]$ for $\gamma = 0.2, 0.4, \ldots, 10.0$ and obtained Table 16.8. Using Table 16.8, we have plotted $E\left[D(10)/10\right]$ against γ in Figure 16.10.

16.9 CONCLUSION

In this chapter, we proposed a three-dimensional stochastic model of an interchangeable three-unit series-parallel repairable system and obtained explicit expressions for the transient state probabilities. We deduced steady-state probabilities also. We delved into availability and reliability of the repairable system and explored the mean downtime of the system. As a future study, we can include vacation for the repairman whenever he enters into idle state. However, giving vacation to the repairman would affect the availability and reliability of the system.

Table 16.5 Reliability $R_0(t)$ of the system for $t = 0.1, 0.2, \ldots, 10.0$

t	R_0(t)	t	R_0(t)	t	R_0(t)	t	R_0(t)	t	R_0(t)
0.1	0.9944507	2.1	0.5267940	4.1	0.2587796	6.1	0.1271059	8.1	0.0624312
0.2	0.9803287	2.2	0.5084137	4.2	0.2497422	6.2	0.1226670	8.2	0.0602509
0.3	0.9605497	2.3	0.4906703	4.3	0.2410204	6.3	0.1183830	8.3	0.0581467
0.4	0.9371467	2.4	0.4735428	4.4	0.2326031	6.4	0.1142487	8.4	0.0561160
0.5	0.9115341	2.5	0.4570110	4.5	0.2244798	6.5	0.1102587	8.5	0.0541562
0.6	0.8846923	2.6	0.4410548	4.6	0.2166402	6.6	0.1064081	8.6	0.0522649
0.7	0.8572980	2.7	0.4256546	4.7	0.2090744	6.7	0.1026920	8.7	0.0504397
0.8	0.8298153	2.8	0.4107913	4.8	0.2017728	6.8	0.0991056	8.8	0.0486781
0.9	0.8025590	2.9	0.3964465	4.9	0.1947262	6.9	0.0956445	8.9	0.0469781
1.0	0.7757402	3.0	0.3826023	5.0	0.1879257	7.0	0.0923043	9.0	0.0453375
1.1	0.7494971	3.1	0.3692412	5.1	0.1813627	7.1	0.0890807	9.1	0.0437542
1.2	0.7239173	3.2	0.3563466	5.2	0.1750289	7.2	0.0859697	9.2	0.0422261
1.3	0.6990532	3.3	0.3439021	5.3	0.1689163	7.3	0.0829673	9.3	0.0407514
1.4	0.6749329	3.4	0.3318921	5.4	0.1630172	7.4	0.0800698	9.4	0.0393282
1.5	0.6515674	3.5	0.3203014	5.5	0.1573241	7.5	0.0772735	9.5	0.0379548
1.6	0.6289566	3.6	0.3091155	5.6	0.1518298	7.6	0.0745749	9.6	0.0366293
1.7	0.6070925	3.7	0.2983203	5.7	0.1465274	7.7	0.0719705	9.7	0.0353500
1.8	0.5859617	3.8	0.2879020	5.8	0.1414101	7.8	0.0694570	9.8	0.0341155
1.9	0.5655477	3.9	0.2778475	5.9	0.1364716	7.9	0.0670313	9.9	0.0329241
2.0	0.5458318	4.0	0.2681441	6.0	0.1317055	8.0	0.0646904	10.0	0.0317742

Table 16.6 Reliability $R_0(t)$ of the system for $t = 10, 11, \ldots, 50$

t	*R_0(t)*	*t*	*R_0(t)*	*t*	*R_0(t)*	*t*	*R_0(t)*	*t*	*R_0(t)*
1.0	0.7757402	11.0	0.0222686	21.0	0.0006366	31.0	0.0000182	41.0	0.0000005
2.0	0.5458318	12.0	0.0156067	22.0	0.0004462	32.0	0.0000128	42.0	0.0000004
3.0	0.3826023	13.0	0.0109378	23.0	0.0003127	33.0	0.0000089	43.0	0.0000003
4.0	0.2681441	14.0	0.0076656	24.0	0.0002191	34.0	0.0000063	44.0	0.0000002
5.0	0.1879257	15.0	0.0053724	25.0	0.0001536	35.0	0.0000044	45.0	0.0000001
6.0	0.1317055	16.0	0.0037651	26.0	0.0001076	36.0	0.0000031	46.0	0.0000000
7.0	0.0923043	17.0	0.0026388	27.0	0.0000754	37.0	0.0000022	47.0	0.0000000
8.0	0.0646904	18.0	0.0018493	28.0	0.0000529	38.0	0.0000015	48.0	0.0000000
9.0	0.0453375	19.0	0.0012961	29.0	0.0000371	39.0	0.0000011	49.0	0.0000000
10.0	0.0317742	20.0	0.0009084	30.0	0.0000260	40.0	0.0000007	50.0	0.0000000

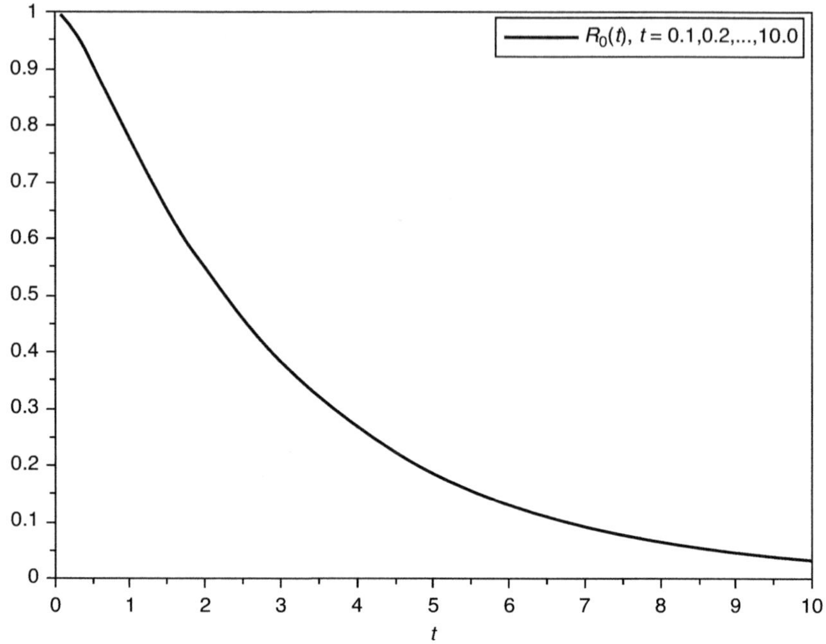

Figure 16.7 Variation of reliability function $R_0(t)$, $0 < t \leq 10$.

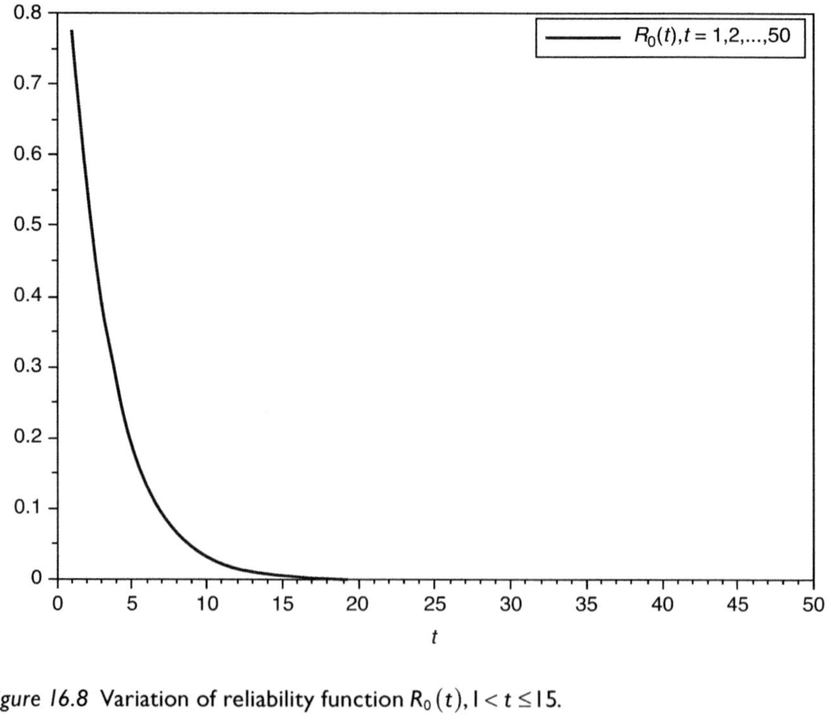

Figure 16.8 Variation of reliability function $R_0(t), 1 < t \leq 15$.

Table 16.7 Proportion of downtime $E\left[D(t)/t\right]$ of the system for $1 \le t \le 100$

t	$E[D(t)/t]$	t	$E[D(t)/t]$	t	$E[D(t)/t]$	t	$E[D(t)/t]$	t	$E[D(t)/t]$
1.0	0.0667832	21.0	0.1989347	41.0	0.2033182	61.0	0.2048272	81.0	0.2055911
2.0	0.1188080	22.0	0.1993432	42.0	0.2034278	62.0	0.2048771	82.0	0.2056195
3.0	0.1457801	23.0	0.1997161	43.0	0.2035323	63.0	0.2049254	83.0	0.2056472
4.0	0.1608685	24.0	0.2000580	44.0	0.2036320	64.0	0.2049722	84.0	0.2056743
5.0	0.1702011	25.0	0.2003725	45.0	0.2037273	65.0	0.2050176	85.0	0.2057007
6.0	0.1764736	26.0	0.2006628	46.0	0.2038185	66.0	0.2050616	86.0	0.2057265
7.0	0.1809634	27.0	0.2009316	47.0	0.2039057	67.0	0.2051043	87.0	0.2057517
8.0	0.1843326	28.0	0.2011813	48.0	0.2039894	68.0	0.2051457	88.0	0.2057764
9.0	0.1869534	29.0	0.2014137	49.0	0.2040696	69.0	0.2051859	89.0	0.2058005
10.0	0.1890501	30.0	0.2016306	50.0	0.2041467	70.0	0.2052250	90.0	0.2058240
11.0	0.1907656	31.0	0.2018335	51.0	0.2042207	71.0	0.2052629	91.0	0.2058471
12.0	0.1921952	32.0	0.2020237	52.0	0.2042918	72.0	0.2052999	92.0	0.2058696
13.0	0.1934049	33.0	0.2022024	53.0	0.2043603	73.0	0.2053358	93.0	0.2058917
14.0	0.1944417	34.0	0.2023706	54.0	0.2044262	74.0	0.2053707	94.0	0.2059133
15.0	0.1953403	35.0	0.2025292	55.0	0.2044898	75.0	0.2054047	95.0	0.2059344
16.0	0.1961266	36.0	0.2026789	56.0	0.2045510	76.0	0.2054378	96.0	0.2059551
17.0	0.1968204	37.0	0.2028206	57.0	0.2046101	77.0	0.2054701	97.0	0.2059754
18.0	0.1974371	38.0	0.2029548	58.0	0.2046672	78.0	0.2055015	98.0	0.2059952
19.0	0.1979888	39.0	0.2030822	59.0	0.2047224	79.0	0.2055321	99.0	0.2060147
20.0	0.1984854	40.0	0.2032031	60.0	0.2047757	80.0	0.2055620	100.0	0.2060337

Table 16.8 Proportion of downtime $E\left[D(10)/10\right]$ of the system for $\gamma = 0.2, 0.4, \ldots, 10.0$

	E[D(10)/10]		*E[D(10)/10]*		*E[D(10)/10]*		*E[D(10)/10]*		*E[D(10)/10]*
0.2	0.6612249	2.2	0.1272374	4.2	0.0486519	6.2	0.0252525	8.2	0.0153856
0.4	0.5367457	2.4	0.1130843	4.4	0.0450999	6.4	0.0238926	8.4	0.0147309
0.6	0.4393269	2.6	0.1011110	4.6	0.0419193	6.6	0.0226392	8.6	0.0141169
0.8	0.3633408	2.8	0.0909023	4.8	0.0390607	6.8	0.0214813	8.8	0.0135405
1.0	0.3038166	3.0	0.0821353	5.0	0.0364824	7.0	0.0204097	9.0	0.0129985
1.2	0.2568025	3.2	0.0745558	5.2	0.0341491	7.2	0.0194159	9.2	0.0124883
1.4	0.2192946	3.4	0.0679624	5.4	0.0320311	7.4	0.0184926	9.4	0.0120075
1.6	0.1890501	3.6	0.0621937	5.6	0.0301028	7.6	0.0176335	9.6	0.0115539
1.8	0.1644021	3.8	0.0571197	5.8	0.0283424	7.8	0.0168325	9.8	0.0111254
2.0	0.1441084	4.0	0.0526347	6.0	0.0267310	8.0	0.0160848	10.0	0.0107203

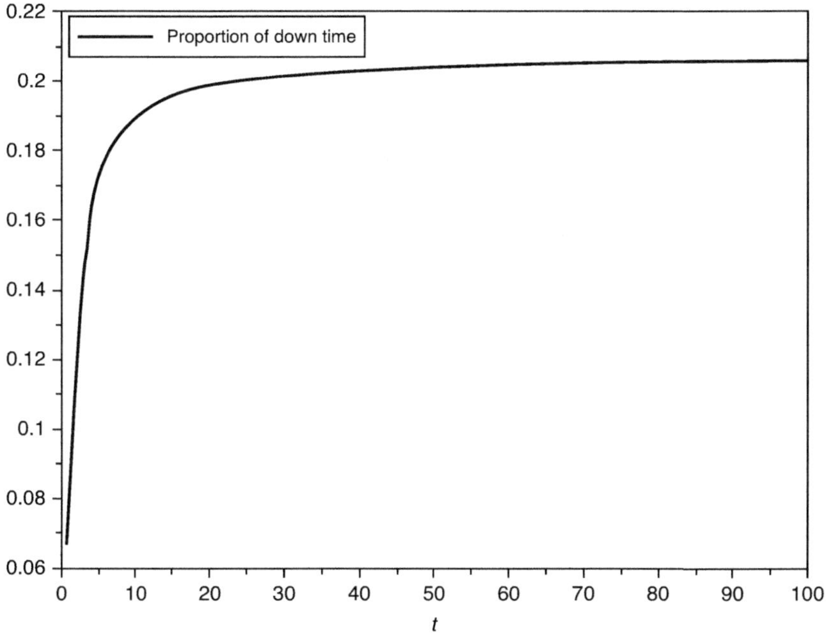

Figure 16.9 Variation of $E(t)/t$, $0 < t \leq 100$.

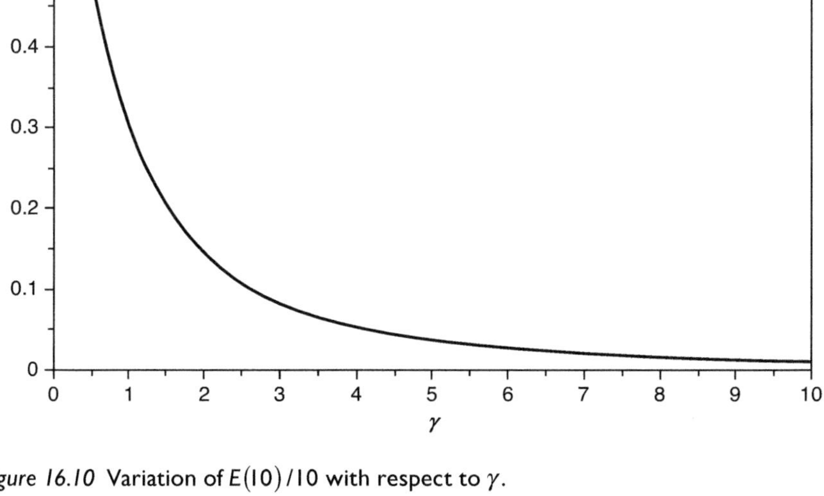

Figure 16.10 Variation of $E(10)/10$ with respect to γ.

REFERENCES

Barlow, R.E. and Proschan, F., (1965), *Mathematical Theory of Reliability*, New York: Wiley.

Birolini, A., (2017), *Reliability Engineering: Theory and Practice*. 8th Edition, Berlin: Springer-Verlag.

Choudhary, S., Ram, M., Goyal, N. and Saini, S., (2024). Reliability and cost optimization of series–parallel system with metaheuristic algorithm, *International Journal of System Assurance Engineering and Management*, 15, pp. 1456–1466.

Gnedenko, B.V., Belyayev, Y.K. and Solovyev, A.D., (1969) *Mathematical Methods of Reliability Theory*. Russian original and English translation edited by R.E. Barlow. New York: Academic Press.

Goel, L.R., Gupta, R. and Agnihotri, R.K., (1989), Analysis of a three-unit redundant system with two types of repair and inspection, *Microelectronics and Reliability*, 29(5), pp. 769–773.

Hu, L., Yue, D. and Li, J., (2010), Probabilistic analysis of a series-parallel repairable system with three units and vacation, *Applied Mathematical Modelling*, 34(10), pp. 2711–2721.

Hu, L., Yue, D. and Ma, Z. (2020), Availability analysis of a repairable series-parallel system with redundant dependency, *Journal of Systems Science and Complexity*, 33, pp. 446–460.

Kadyan, S., Malik, S.C. and Gitanjali, (2020), Stochastic analysis of a three-unit non-identical repairable system with simultaneous working of cold standby units, *Journal of Reliability and Statistical Studies*, 13(2–4), pp. 385–400.

Kovalenko, A.I., (2001) Analysis of the reliability of a three-component system with renewal, *Journal of Mathematical Science*, 103, pp. 273–277.

Li, W., Alfa, A.S. and Zhao, Y.Q., (1998), Stochastic analysis of a repairable system with three units and two repair facilities, *Microelectronics and Reliability*, 38(4), pp. 585–595.

Mwanga, Y.A., Joubert, J.W. and Yadavalli, V.S.S., (2009), A complex three-unit parallel system with preparation time for repair, *South African Journal of Industrial Engineering*, 20(1), pp. 59–68.

Osaki, S. and Nakagawa, T., (1976), Bibliography for reliability and availability of stochastic systems, *IEEE Transactions on Reliability*, R-25(4), pp. 284–287.

Sarma, Y.V.S. and Parvez, M., (1984), A complex three-unit system, *Microelectronics Reliability*, 24(4), pp. 661–662.

Song, S.B. and Deng, Y.L., (1993), Reliability analysis of a three-unit system in a changing environment, *Microelectronics and Reliability*, 33(5), pp. 637–640.

Index

availability analysis 39, 57, 58, 62, 66, 71, 72, 82, 189, 191, 212, 239, 271, 281, 295, 308, 336

Bayesian availability 58, 62, 66, 70
broadcast performance 204
broadcast reliability 204, 208

cloud security 74
cold-standby 58, 72, 158, 160, 162, 163, 165, 166, 171, 175, 178, 180, 186–188, 216, 219, 220, 239, 264, 286, 301, 336
communication and computing networks 109
conditional probability 62, 317

delayed S-shaped SGRMS 7
design space 143, 145, 147
deviation 31

evolutionary search strategy 145
expected profit analysis 234
exponential distribution 14, 15, 58, 72, 238, 261, 264, 267
exponential total testing effort 14

failure mode 127, 129, 131, 134, 135, 139, 144, 183, 188, 216, 256
FCP model 12
FDP model 12–14
feeding system of sugar industry 218
finite element modeling 144
fuzzy systems 131

genetic algorithm 137, 139, 145, 159
governing equations 303
Gumbel–Hougaard copula 159, 216, 222

hazard rate 4, 59, 160, 265, 290

integral risk 28, 30, 34
inverse Laplace transform 40, 82, 83, 270, 294, 306, 307, 313–316, 319, 320, 322

Laplace transform 221
lifecycle cost modeling 142
limit state functions 145
linear total testing effort 15
loads and uncertainties 147
Lz-transform of multi-state component 163

Markov model 160, 193, 212, 263
Markov process 77, 79, 194, 196, 215–217, 220, 223, 287, 288, 290, 303
mathematical modelling 215–217, 220, 221, 238
mean down time 321
minimal cut 95, 159, 216
Monte Carlo simulation 137, 141, 159
MTBF 40, 45, 47, 48, 50–53, 59, 61, 62, 288
MTTF 85, 166, 179, 180, 231, 296
MTTR 59, 61, 288
multistage interconnection network 2, 191, 213, 214
Mvi-based approach 102

network performance 207
network reliability 98, 101, 108, 139, 191
node labelling algorithm 116
non-homogeneous Poisson process 1

optimization architecture 143

Pareto 139, 141, 143, 145–147, 150–155
peeling machine 41, 42, 47–49, 52–55
posterior distribution 63, 64, 66
prior distribution 62, 63
properties of Lz-transform 194

repairable system 57, 58, 66, 71, 85, 161, 190, 193, 214, 216, 222, 238, 261, 262, 264–266, 268, 272, 273, 275, 277, 280, 282, 285, 288, 296, 300, 301, 327, 329
risk assessment 25, 33, 141, 144
robustness 140, 142, 145, 154, 187

sequential policy 165
series-parallel system 160
software release policies 17
standby redundancy 160, 216, 261, 262, 284, 285
state-space enumeration 99
steady state 9, 40, 43, 53, 271
stochastic network 115, 116, 118
structural modeling of uncertainty 144
supplementary variable technique 40, 159, 220
supply chain 94, 95, 110, 115, 127, 129

terminal performance 199
terminal reliability 95, 97, 113, 203
transportation and logistics networks 110

universal generating function 158, 181, 191, 288

Weibull distribution 3, 4, 7, 288

For Product Safety Concerns and Information please contact our EU representative GPSR@taylorandfrancis.com Taylor & Francis Verlag GmbH, Kaufingerstraße 24, 80331 München, Germany

Batch number: 10725800

Printed by Printforce, the Netherlands